Essential Topics in Vibration Engineering and Structural Dynamics

Essential Topics in Vibration Engineering and Structural Dynamics

Edited by **Rene Sava**

Clanrye International

New Jersey

Published by Clanrye International,
55 Van Reypen Street,
Jersey City, NJ 07306, USA
www.clanryeinternational.com

Essential Topics in Vibration Engineering and Structural Dynamics
Edited by Rene Sava

International Standard Book Number: 978-1-63240-227-1 (Hardback)

Printed in the United States of America.

Contents

Preface

This book was inspired by the evolution of our times; to answer the curiosity of inquisitive minds. Many developments have occurred across the globe in the recent past which has transformed the progress in the field.

All the essential topics in vibration engineering and structural dynamics are explained in this book. Significant features of vibration engineering and structural dynamics have been presented, in addition to their impact on study and application in engineering. This book focuses on various issues which pertain to modelling, rotor dynamics, vibration control, assessment and recognition, modal analysis, dynamic arrangements, finite element study, numerical techniques and other practical engineering functions and academic improvements in this vast discipline. The book is meant for those indulged in research, academics, professional practice and teaching who are keen on understanding the current challenges and problems, looking for innovative solutions in vibration engineering and structural dynamics.

This book was developed from a mere concept to drafts to chapters and finally compiled together as a complete text to benefit the readers across all nations. To ensure the quality of the content we instilled two significant steps in our procedure. The first was to appoint an editorial team that would verify the data and statistics provided in the book and also select the most appropriate and valuable contributions from the plentiful contributions we received from authors worldwide. The next step was to appoint an expert of the topic as the Editor-in-Chief, who would head the project and finally make the necessary amendments and modifications to make the text reader-friendly. I was then commissioned to examine all the material to present the topics in the most comprehensible and productive format.

I would like to take this opportunity to thank all the contributing authors who were supportive enough to contribute their time and knowledge to this project. I also wish to convey my regards to my family who have been extremely supportive during the entire project.

Editor

Chapter 1

Rotordynamic Stabilization of Rotors on Electrodynamic Bearings

J. G. Detoni, F. Impinna, N. Amati and A. Tonoli

Additional information is available at the end of the chapter

1. Introduction

In the last decades the advance in the semiconductors technology for power electronics has dictated a growing interest for high rotational speed machines. The use of high rotational speeds allows increasing the power density of the machine, but introduces some critical aspects from the mechanical point of view. One of the most critical issues to be dealt with is the difficulty in operating common mechanical bearings in this condition. For this reason alternatives for classical ball and roller bearings must be found. In this context, active magnetic bearings represent an advantageous alternative because they are capable of supporting the rotating shaft in absence of contact. Nevertheless, the high cost associated with this kind of system reduces their applicability.

A promising system for supporting high rotational speed machines in absence of contact and with relatively low costs, widening the range of applications, is the electrodynamic suspension of rotors [1], [2], [3], [4], [5]. Systems capable of realizing this concept are commonly referred to as electrodynamic bearings (EDB). They exploit repulsive forces due to eddy currents arising between conductors in motion relative to a magnetic field. The supporting forces are generated in a completely passive process, thus representing an increase in the overall reliability of the suspension with respect to active magnetic bearings. Nevertheless, electrodynamic bearings have drawbacks. The eddy current forces that provide levitation produce an energy dissipation that may cause negative damping resulting in rotordynamic instability.

Because the rotor may present an unstable behavior, it is necessary to study the dynamic response of the suspension in order to guarantee stable operation in the working range of speed. This can be achieved by introducing nonrotating damping in the system, but the choice of the damping elements is not obvious, requiring an accurate modeling phase. The present paper presents the development of a dynamic model of the entire suspension that is used to study the mechanical properties of the supports that allow guaranteeing rotordynamic stability. A simple optimization procedure is used in order to identify the characteristics of an elastic support placed in between the electrodynamic bearing's stator and the casing of the machine. The use of anisotropic supports to improve the stabilization characteristics is also investigated, and optimal conditions are identified.

2. Dynamic model of EDBs

To describe the dynamics of the eddy currents inside the coils and also the dynamic effects of electrodynamic bearings on rotors supported by them, we make the assumption that the rotor rotates at constant angular speed Ω ($\theta = \Omega t$). Assuming constant or slowly varying rotational speed is commonly done in rotordynamics[6] and does not reduce the validity of the model.

The systems under analysis are shown in Figure 1a and Figure 1b. The first presents a schematic representation of a heteropolar EDB with the magnetic field generated using a two pole pair Halbach array. The second is a scheme of a homopolar EDB having a radial flux configuration. Different configurations are possible and can be studied using the same models presented in this paper.

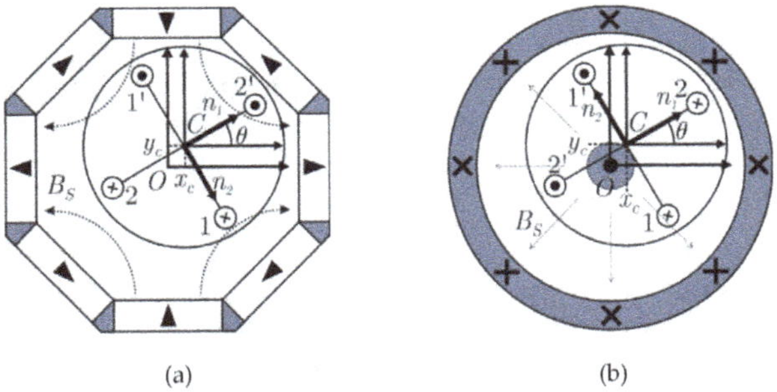

Figure 1. Scheme of possible configurations of electrodynamic bearings.(a) Heteropolar configuration; (b) homopolar configuration.

To write an equation that describes the behavior of the current in the electric circuit of the coils the electric circuit where the current flows must be defined. Figure 2a presents the electric circuit where the terms R_c and L_c are the resistance and inductance of the coil. For some applications it may be interesting to connect inductive loaded circuit in series with the coil [1], [2]. For this reason the terms of a generic passive shunt R_{add} and L_{add} are introduced in the model. The mutual inductance between the coils has been neglected. The orthogonality between the coils justifies this assumption.

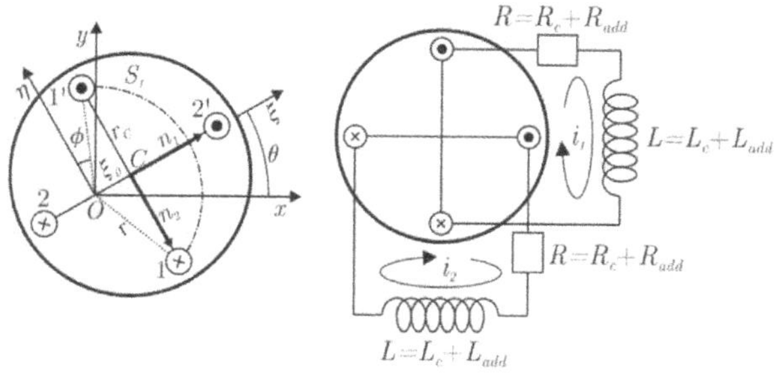

(a) (b)

Figure 2. Schemes of the EDB's rotor describing the main model's variables. (a) Geometrical variables; (b) Electric circuit of the rotor's coils taking into account of a generic passive shunt.

2.1. Eddy currents and bearing's forces

Developing the modeling of the electrical equations in terms of complex quantities allows a strong simplification of the system's equations. Hence we define the main geometrical and electrical variables described in Figure 2a and Figure 2b and necessary for the modeling. The Lagrangian coordinate representing the displacement of the geometric center of the rotor C relative to the axis of the magnetic field O in the Cartesian reference frame is given by q_c in the form:

$$q_c = x + jy \tag{1}$$

The electric current inside the coils in complex coordinates is written as:

$$i = i_1 + j i_2 \tag{2}$$

Considering the above defined variables, the state equation describing the dynamics of the eddy currents inside the coils can be written as:

$$\frac{di}{dt} = \frac{\dot{\lambda}}{L} - \frac{R}{L} i \tag{3}$$

where λ is the magnetic flux generated by the permanent magnets and linking the rotor's coils. The expression describing the flux linkage can be expressed in complex coordinates as:

$$\lambda = q_c \Lambda_0 e^{j(p-1)\Omega t} \tag{4}$$

In this expression the term Λ_0 is a coefficient that gives the variation of magnetic flux linkage due to a lateral displacement of the rotor and has units of Wb/m whereas p represents the number of pole pairs of the magnetic field and is given by an even number ($p=0, 2, 4, \ldots$). Notice that a magnetic field where $p=0$ is a homopolar magnetic field while $p\neq 0$ gives place to heteropolar magnetic fields.

Combining Eq. (3) and Eq. (4) allows writing the system of equations coupling the rotor's motion and the induced current as:

$$\begin{aligned} \frac{di}{dt} &= \frac{\Lambda_0}{L}(\dot{q}_c + \mathrm{j}(p-1)q_c\Omega)e^{\mathrm{j}(p-1)\Omega t} - \frac{R}{L}i \\ F_q &= i\Lambda_0 e^{-\mathrm{j}(p-1)\Omega t} \end{aligned} \tag{5}$$

In the equation the current i is the state variable and the bearing's reaction force F_q is the output equation. Although linear, this equation has periodically time-varying coefficients. Performing a change of variables, substituting the state variable i with the output variable F_q it is possible to obtain a set of equations having constant coefficients as:

$$\begin{aligned} \dot{F}_q &= \frac{\Lambda_0^2}{L}(\dot{q}_c + \mathrm{j}(p-1)q_c\Omega) - F_q\left(\frac{R}{L} + \mathrm{j}(p-1)\Omega\right) \\ i &= \frac{F_q}{\Lambda_0}e^{\mathrm{j}(p-1)\Omega t} \end{aligned} \tag{6}$$

This set of equations allows calculating the reaction forces generated by EDBs of both homopolar and heteropolar configurations, and can be used to study the dynamics of rotors supported by EDBs.

3. Jeffcott rotor on EDBs

Due to the nature of the phenomena, studying the dynamics of a rotor on magnetic bearings requires one to consider that the center of the rotor is moving relative to the stator. In the specific case of the electrodynamic bearing, this means that the center of the conductor (point C) is moving relative to the magnetic field (point O). Equation (6) takes this into account. The new state variable F_q can be used to find the coupling with the dynamic equation of the rotor mass m . In this way it is possible to study the rotordynamic implications of supporting rotors with different types of electrodynamic bearings.

The simplest model that can be used to study the dynamic behavior of a rotor is the Jeffcott rotor model. It consists of a point mass attached to a massless shaft. This model represents an oversimplification as it neglects many aspects present in real world rotors, but, nevertheless it allows gaining insight into important phenomena especially in the case of rotors supported by EDBs.

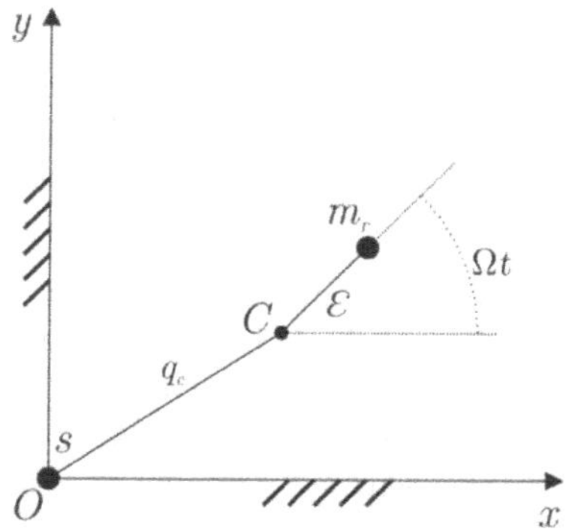

Figure 3. Schematic model of a Jeffcott rotor on an electrodynamic bearing showing the fundamental variables used to describe the dynamics of the whole suspension.

In this section we will study the stability of the Jeffcott rotor model supported exclusively by EDBs. The stability of a linear system is determined by its eigenvalues. Briefly, a system is stable if the real part of all the eigenvalues is negative [7]. This means that the system will exhibit a bounded output for respective bounded inputs. In the rotordynamics context this means that the rotor will respond to any disturbance forces with orbits of bounded radius.

Graphical representations are used to demonstrate the concepts, and the values of the parameters used to obtain the graphs are given in Table 1. A simplified model of a Jeffcott rotor is shown in Figure 3.

Parameter	Symbol	Value	Unit
Rotor's mass	m_r	2.025	kg
Flux linkage constant	Λ_0	10	Wb/m
Bearing's resistance	R	0.286	ohm
Bearing's inductance	L	0.33	mH

Table 1. Parameters describing the dynamics of a Jeffcott rotor on EDBs.

3.1. Undamped Jeffcott rotor

The equation of motion of the Jeffcott rotor supported by EDBs is

$$m_r \ddot{q}_c + F_q = F_{\text{ext}} \tag{7}$$

where F_q is the force introduced in the system by the electrodynamic bearing and F_{ext} is a generic disturbance force acting on the rotor's mass. The external force can be due to gravity or rotor's unbalance for example. Since the equation is linear, the response of the system is given by the superposition of the solution for each force. The response to a constant force, e. g. rotor weight, causes a constant displacement between the rotation axis and the symmetry axis of the magnetic field. The response to unbalance, on the other hand,is given by a whirling of the rotor and depends on the number of pole pairs [9]. Note that the force introduced by the bearing is seen as a reaction by the rotor mass.

The EDB of Eq. (6) and the rotor of Eq. (7) are interacting subsystems. The rotor responds to forces and moments with velocities and displacements. The bearing responds to the rotor's outputs with forces. As a consequence, to study the dynamic behavior of the rotor running on EDBs, Eq. (6) and Eq. (7) must be solved together. Given the linear time invariant form of the equations a state-space model can be used for this purpose. The state space model has the form:

$$\begin{Bmatrix} \ddot{q}_c \\ \dot{q}_c \\ \dot{F}_q \end{Bmatrix} = \mathbf{A} \begin{Bmatrix} \dot{q}_c \\ q_c \\ F_q \end{Bmatrix} + \mathbf{B}\{F_{\text{ext}}\} \tag{8}$$

The dynamic matrix A is

$$\mathbf{A} = \begin{bmatrix} 0 & 0 & -\frac{1}{m_r} \\ 1 & 0 & 0 \\ \frac{\Lambda_0^2}{L} & \mathrm{j}\frac{\Lambda_0^2}{L}(p-1)\Omega & -(\frac{R}{L}+\mathrm{j}(p-1)\Omega) \end{bmatrix} \tag{9}$$

And the input gain matrix B is equal to

$$\mathrm{B} = \begin{bmatrix} \frac{1}{m_r} \\ 0 \\ 0 \end{bmatrix} \tag{10}$$

The state-space modeling allows studying the rotordynamic stability, frequency response, unbalance response, and enables developing other tools to study the dynamics of the suspension in a fast and easy way. The analysis of different systems can be performed as simple parametric studies.

To study the rotor's stability we calculate the eigenvalues of the dynamic matrix A of the suspension's model (rotor supported by EDB) and analyze the evolution of the system's poles in a root loci plot. Figure 5a shows the root loci plot obtained by calculating the eigenvalues of Eq. (9) for increasing values of rotating speed Ω . Note that the figure shows the evolution of the poles for the homopolar case and for the heteropolar with $p=2$.

It can be seen how the system presents a root that is in the right half plane for any value of rotating speed different from zero. This is true for both homopolar and heteropolar cases, representing that the Jeffcott rotor supported by EDBs is unstable for any value of rotating speed if the system is not modified. The reason for this unstable behavior has been identified to be the presence of rotating damping in the system. The eddy currents induced in the conducting part of the EDB dissipate energy associated to the motion of the rotating part. Rotating damping forces are known to destabilize the free whirling motion of the rotors for speeds above the first critical. In particular, if the rotating damping is of viscous type, the instability threshold of the undamped system (no external non rotating damping) is equal to the first critical speed [6].

Intuitively one can think that the instability arises from the fact that the system is always operating in supercritical regime because the electrodynamic supports are unable to give radial stiffness at zero rotating speed. Actually this statement is only partially valid since the behavior of the EDB cannot be correctly represented by a rotating viscous damper. The frequency dependence of the bearing's forces must be taken into account, modifying the overall behavior. In the next sections the suspension model will be used to study the dynamic response and analyze different stabilization techniques proposed previously in the literature [1], [2], [3].

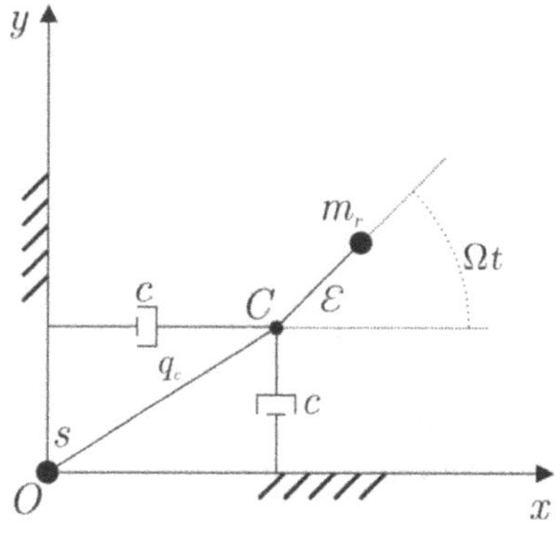

Figure 4. Scheme of the Jeffcott rotor model with electromagnetic damping associated to rotor's translational velocity.

3.2. Damped Jeffcott rotor

The most straightforward way to introduce non-rotating damping in the system is to do it by means of an electromagnetic damper, associating non rotating damping to the rotor's translational velocity $\dot{q}_c$. This stabilization technique is shown in Figure 4, and has been proposed almost from the beginning of the interest in electrodynamic suspension of rotors.

From the modeling point of view it consists simply in introducing a viscous damping element associated to the translational speed.

Notice that in this case the viscous damper is used as an approximated representation of the behavior of the electromagnetic damper.

As a result of this operation non-rotating damping is introduced in the model of Eq. (7) and the new equation of motion of the rotor's mass is:

$$m_r\ddot{q}_c + c\dot{q}_c + F_q = F_{\text{ext}} \tag{11}$$

The dynamic matrix of the state-space model is also updated

$$\mathbf{A} = \begin{bmatrix} -\dfrac{c}{m_r} & 0 & -\dfrac{1}{m_r} \\ 1 & 0 & 0 \\ \dfrac{\Lambda_0^2}{L} & \mathrm{j}\dfrac{\Lambda_0^2}{L}(p-1)\Omega & -(\dfrac{R}{L} + \mathrm{j}(p-1)\Omega) \end{bmatrix} \tag{12}$$

Figure 5b shows the influence of the non-rotating damping on the system's poles. It is readily seen that the presence of damping allows stabilizing the dynamic behavior above a certain value of rotating speed Ω_S . This value represents a stability threshold, being the system unstable for spin speeds below it and stable for speeds above it. Another conclusion that arises from this diagram is that the bearing's rotating damping contribution reduces for higher values of spin speed, when the stabilizing stiffness contribution becomes dominant.

Notice that if the rotor's spin speed is not constant it is necessary to introduce a further equation to express the dependence between angular displacement and driving torque. Since this additional degree of freedom is related to the rotation about the rotor's axis, the rotor's polar moment of inertia cannot be neglected. However for the present study it is acceptable to neglect this behavior and develop the study considering only constant rotational speed, thus elimination this further degree of freedom.

4. EDB's stator on elastic supports

An alternative to the previous solution that allows introducing non-rotating damping in an effective way is to introduce a stabilizing element between the stator of the EDB and a rigid base. This element can be devised in different ways, for example, using viscoelastic elements, spring elements associated to passive eddy current dampers or even using active dampers [8]. In general, the introduction of stiffness and damping contemporarily is needed.

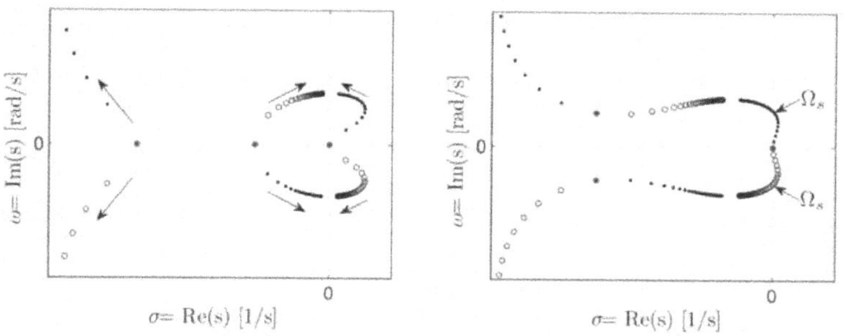

Figure 5. (a) Root loci of the undamped Jeffcott rotor on EDBs; (b) root loci of the damped Jeffcott rotor on EDBs. Point markers (·) represent the roots for homopolar bearings; circular markers (∘) represent the roots for heteropolar bearings with two pole pairs ($p=2$).

Within the EDB's context this system has been analyzed by Tonoli *et al.* [3]. It was shown that the stability boundaries of a Jeffcott rotor on this type of support allow operating at reduced rotational speeds with respect to the most common electromagnetic damping system proposed in literature [1], [2]. Furthermore, this stabilization technique avoids increasing the rotor's mass and complexity because all the additional subsystems are placed on the stator part. In addition, the possibility of introducing non-rotating damping between two stationary parts allows using classical damping technologies, such as, viscoelastic materials or squeeze film dampers. However, the choice of appropriate values of stiffness and damping of the stabilizing element is not obvious, requiring the solution of an optimization problem.

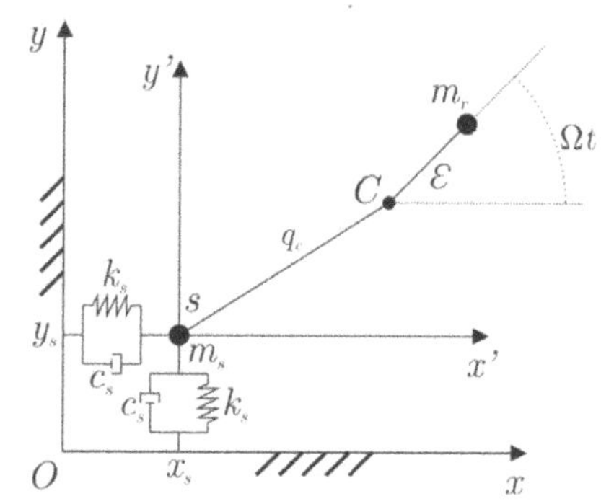

Figure 6. Schematic representation of a Jeffcott rotor supported by EDBs having elastic connections between EDB's stator and casing of the machine.

From the modeling point of view this case is interpreted as shown inFigure 6. In the figure the xy frame is inertial as if attached to an infinitely stiff base while $x'y'$ is a translating reference frame attached to the center of the stator. The axes of the two systems remain parallel to each other. It must be noticed that in this case there are three interacting subsystems, namely, the rotor, the EDB, and the stator. With respect to the figure the associated degrees of freedom are:

- The displacement of the rotor geometric centerC in the inertial frame of reference.

$$q = x + \mathrm{j}y \tag{13}$$

- The displacement of the stator mass m_s represented by the point S in the inertial frame.

$$q_s = x_s + \mathrm{j}y_s \tag{14}$$

- The relative displacement between the displacement between point C and S.

$$q_c = q - q_s \tag{15}$$

The equations of the rotor mass and EDB are given by Eq. (7) and Eq. (6) respectively. The displacements and speeds considered in the EDB's equations are the relative ones (q_c and $\dot{q}_c$). The stator's mass dynamics is described by the following equation:

$$m_s\ddot{q}_s + c_s\dot{q}_s + k_s q_s - F_q = 0 \tag{16}$$

The presence of the negative sign on the bearing's force F_q means that the stator mass sees this force as an external force while the rotor mass perceives it as a reaction force.

The state space model of this system can be written as:

$$\begin{Bmatrix} \ddot{q} \\ \ddot{q}_s \\ \dot{q} \\ \dot{q}_s \\ \dot{F}_q \end{Bmatrix} = \mathbf{A} \begin{Bmatrix} \dot{q} \\ \dot{q}_s \\ q \\ q_s \\ F_q \end{Bmatrix} + \mathbf{B}\{F_{\text{ext}}\} \tag{17}$$

The dynamic matrix A of this state-space model is:

$$\begin{bmatrix} 0 & 0 & 0 & 0 & -\frac{1}{m_r} \\ 0 & -\frac{c_s}{m_s} & 0 & -\frac{k_s}{m_s} & \frac{1}{m_s} \\ 1 & 0 & 0 & 0 & 0 \\ 0 & 1 & 0 & 0 & 0 \\ \frac{\Lambda_0^2}{L} & -\frac{\Lambda_0^2}{L} & \mathrm{j}\frac{\Lambda_0^2}{L}(p-1)\Omega & -\mathrm{j}\frac{\Lambda_0^2}{L}(p-1)\Omega & -(\frac{R}{L} + \mathrm{j}(p-1)\Omega) \end{bmatrix} \tag{18}$$

And the input gain matrix B is equal to

$$\mathbf{B} = \begin{bmatrix} \frac{1}{m_r} \\ 0 \\ 0 \\ 0 \\ 0 \end{bmatrix} \tag{19}$$

The root loci of this system considering the same bearing's characteristics of the previous case are shown in Figure 7a. The values of stiffness k_s and damping c_s are 240 kN/m and 510 N s/m. The choice of these values must be done performing an optimization to minimize the stabilization threshold speed. To have an objective view of the problem one can resort to a plot showing the value of the stabilization threshold speed in terms of different values of k_s and c_s . Figure 7b shows a contour plot of the stabilization threshold speed. The presence of a minimum is clear in the figure leaving to the designer the task of optimizing the system's properties to minimize the stabilization threshold speed guaranteeing that the region surrounding the minimum lies within a region of physically feasible property values.

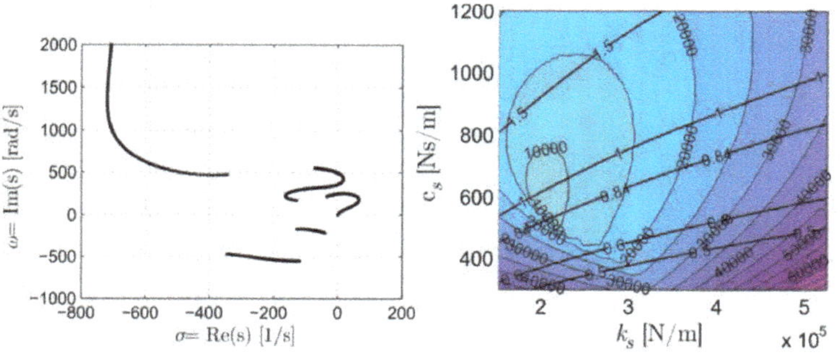

Figure 7. (a) Root loci of the Jeffcott rotor on EDBs having elastic elements in between the stator and the casing of the machine. (b) Contour map of the stabilization threshold speed for different sets of values of damping c_s and stiffness k_s of the elastic element connecting EDB's stator and casing. The lines crossing the graph from lower left to upper right denote the value of loss factor $\eta = c_s / \sqrt{k_s m_s}$ associated to that configuration.

4.1. Anisotropy of heteropolar bearings

In the preceding sections both rotor and stator were assumed to be axial symmetric. Considering the difficulty in insuring stability of the whirling motion of the rotor, a stabilizing technique for transverse whirl modes introducing anisotropy into the bearing stiffness can be considered [2]. This can be achieved in different ways, but one simple strategy is the use of an anisotropic Halbach array of magnets, where the gradient of the flux density in one direction is different from the other, thus modifying the value of the parameter Λ_0 in one direction relative to the other. Another possible strategy is to use rotating magnets and fixed conductors, and to have a different set of properties of the electrical circuit in each direction. For the present study only the first strategy is considered. To study this type of physical problem the state-space model given in terms of complex coordinates in Eq. (17) must be split into its representation in terms of real coordinates as:

$$\begin{Bmatrix} \ddot{x}_c \\ \ddot{y}_c \\ \dot{x}_c \\ \dot{y}_c \\ \dot{F}_x \\ \dot{F}_y \end{Bmatrix} = \mathbf{A} \begin{Bmatrix} \dot{x}_c \\ \dot{y}_c \\ x_c \\ y_c \\ F_x \\ F_y \end{Bmatrix} + \mathbf{B}\{F_{\text{ext}}\} \tag{20}$$

where the dynamic matrix assumes the form:

$$\begin{bmatrix} -\dfrac{c}{m_r} & 0 & 0 & 0 & -\dfrac{1}{m_r} & 0 \\ 0 & -\dfrac{c}{m_r} & 0 & 0 & 0 & -\dfrac{1}{m_r} \\ 1 & 0 & 0 & 0 & 0 & 0 \\ 0 & 1 & 0 & 0 & 0 & 0 \\ \dfrac{\Lambda_{0x}^2}{L_x} & 0 & 0 & (p-1)\dfrac{\Lambda_{0x}^2}{L_x}\Omega & -\dfrac{R_x}{L_x} & -(p-1)\Omega \\ 0 & \dfrac{\Lambda_{0y}^2}{L_y} & -(p-1)\dfrac{\Lambda_{0x}^2}{L_x}\Omega & 0 & (p-1)\Omega & -\dfrac{R_y}{L_y} \end{bmatrix} \tag{21}$$

Calculating the eigenvalues of the dynamic matrix for different values of spin speeds it is possible to fine the stabilization threshold speed. If different values of the ration between the properties in x direction and those in y direction, and finding the stabilization threshold speed in every case it is possible to study how the anisotropy of these properties influence the stabilization speed. Figure 8 shows the graphs obtained performing this operation for different values of non-rotating damping between rotor and stator. It can be noticed that the anisotropy has a strong influence on the stabilization speed. In fact, one of the worst cases is precisely when the properties of the bearing are isotropic; this is evidenced by the peak in the stabilization threshold speed. The non-rotating damping has the effect of reducing the stabilization speed in the entire speed range. One important aspect is the evidence that for larger values of anisotropy combined to large values of bearing's stiffness, the stabilization threshold reduces strongly towards zero also for low values of non-rotating damping.

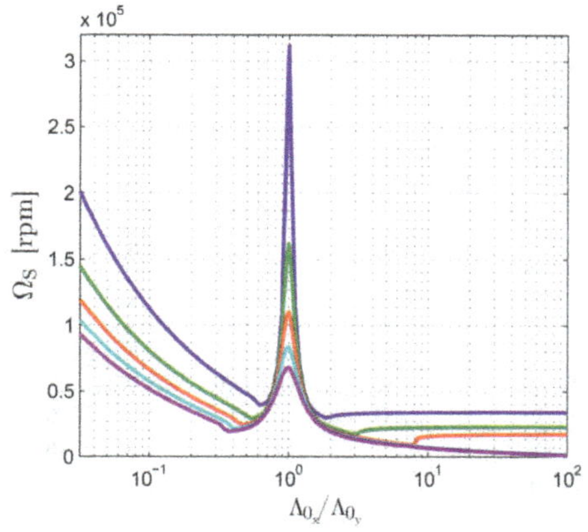

Figure 8. Stabilization speed of the rotor on heteropolar EDB with anisotropic properties of the bearing.

4.2. Anisotropy of stator-casing connections

The homopolar concept was first devised to eliminate unnecessary eddy-current losses generated by AC electrodynamic bearings [5]. The concept itself presupposes axial symmetry of both rotor and stator; hence the introduction of anisotropy of the bearing is not possible. On the other hand, considering the configuration presented in Sec. 4, it is possible to imagine a system where the stiffness and damping of the connection between EDB's stator and casing are different in each direction.

Similarly to the previous case, this system is more conveniently represented in real coordinates. The representation in complex coordinates is possible as well but creates difficulties for the state-space modeling.

In the first paragraph the homopolar concept was cited to motivate this section, however, as a consequence of the unified modeling, the effect of anisotropy can be appreciated in both homopolar and heteropolar configurations. The state-space model can be written as:

$$\begin{Bmatrix} \ddot{x}_c \\ \ddot{y}_c \\ \ddot{x} \\ \ddot{y} \\ \dot{x}_c \\ \dot{y}_c \\ \dot{x} \\ \dot{y} \\ \dot{F}_x \\ \dot{F}_y \end{Bmatrix} = \mathbf{A} \begin{Bmatrix} \dot{x}_c \\ \dot{y}_c \\ \dot{x} \\ \dot{y} \\ x_c \\ y_c \\ x \\ y \\ F_x \\ F_y \end{Bmatrix} + \mathbf{B}\{F_{\text{ext}}\} \tag{22}$$

where the dynamic matrix is:

$$\mathbf{A}=\begin{bmatrix} 0 & 0 & 0 & 0 & 0 & 0 & 0 & 0 & -\frac{1}{m_r} & 0 \\ 0 & 0 & 0 & 0 & 0 & 0 & 0 & 0 & 0 & -\frac{1}{m_r} \\ 0 & 0 & -\frac{c_x}{m_s} & 0 & 0 & 0 & -\frac{k_x}{m_s} & 0 & \frac{1}{m_s} & 0 \\ 0 & 0 & 0 & -\frac{c_y}{m_s} & 0 & 0 & 0 & -\frac{k_y}{m_s} & 0 & \frac{1}{m_s} \\ 1 & 0 & 0 & 0 & 0 & 0 & 0 & 0 & 0 & 0 \\ 0 & 1 & 0 & 0 & 0 & 0 & 0 & 0 & 0 & 0 \\ 0 & 0 & 1 & 0 & 0 & 0 & 0 & 0 & 0 & 0 \\ 0 & 0 & 0 & 1 & 0 & 0 & 0 & 0 & 0 & 0 \\ \frac{\Lambda_0^2}{L} & 0 & -\frac{\Lambda_0^2}{L} & 0 & 0 & \frac{\Lambda_0^2}{L}\Omega & 0 & -\frac{\Lambda_0^2}{L}\Omega & -\frac{R}{L} & -\Omega \\ 0 & \frac{\Lambda_0^2}{L} & 0 & -\frac{\Lambda_0^2}{L} & -\frac{\Lambda_0^2}{L}\Omega & 0 & \frac{\Lambda_0^2}{L}\Omega & 0 & \Omega & -\frac{R}{L} \end{bmatrix} \quad (23)$$

From the stability point of view the inputs of the linear system are irrelevant and the input gain matrix doesn't have to be defined.

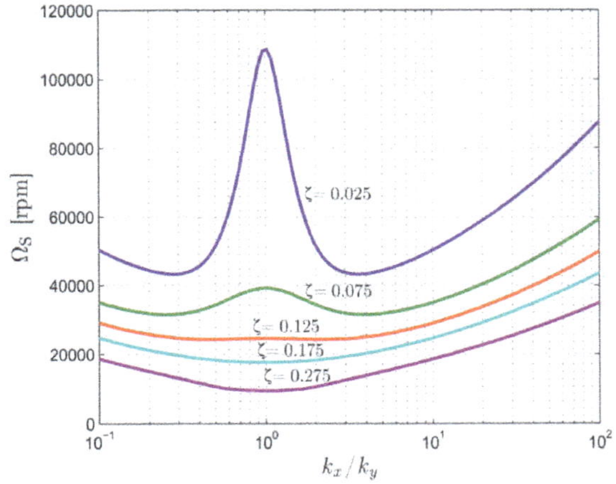

Figure 9. Stabilization speed of the rotor on homopolar EDB with anisotropic connections between bearing stator and casing of the machine.

To study the possibility of taking advantage of anisotropy of the connections to reduce the stabilization threshold speed, the stabilization speed is calculated for different values of the anisotropy ratio ($\alpha = k_x / k_y$). Considering constant values of rotor and stator masses, m and m_s respectively, and the system scheme of Figure 6, the effect of anisotropy is studied in terms of the damping ratio ζ . Figure 9 shows how anisotropy and damping ratios affect the stabilization speed. For systems having low damping ratios the anisotropy can have a beneficial role, reducing the stabilization threshold speed.Increasing the damping ratio eliminates the positive effects due to anisotropy in the elastic connections, but effectively reduces the stabilization threshold.

The anisotropy in this case has a different effect with respect to that illustrated in Figure 8. The increase in the anisotropy ratio with a respective increase of stiffness of one direction results to increase the stabilization threshold speed. It is obvious that this diagram is case dependent, but in general it is expected that the anisotropy has a positive contribution only when the value of damping is low [6]. Furthermore, within physically feasible margins it is always more advantageous to increase the value of damping than to use effects of anisotropy because the stabilization threshold is more sensitive to the first than to the latter.

5. Conclusions

The present paper presents the development of a dynamic model of the radial suspension using electrodynamic bearings that is adopted to study the mechanical properties of the supports that allow guaranteeing rotordynamic stability. A simple procedure is used to identify the characteristics of the bearing, in case of heteropolar bearings, and of the elastic support that allow obtaining the best performance in terms of minimization of stabilization speed.

The effect of anisotropy of the supports in the stabilization threshold speed is also investigated. It is noticed that the anisotropy of the EDB's properties in case of heteropolar configurations can be advantageous independently of the amount of nonrotating damping that can be introduced. The anisotropy allows obtaining stabilization speeds that are lower than the isotropic case. In fact the isotropic bearing represents a critical case, with extremely high stabilization speeds with respect to an anisotropic configuration.

In case of homopolar EDB configurations it is not possible to devise an anisotropic bearing because of the intrinsically axisymmetric distribution of the magnetic field. Hence the anisotropy of the elastic elements connecting the EDB's stator to the casing of the machine has been analyzed under the same hypothesis assumed in the case of heteropolar configurations. It has been observed that anisotropic characteristics of the supports can be advantageous only at low damping levels. For higher values of damping of the connection element the advantages of anisotropy vanish, and the isotropic configuration becomes optimal. Furthermore, it has been observed that is more advantageous to increase damping instead of resorting to anisotropic configurations in the case of anisotropy of the elastic connections because the stabilization threshold speed is more sensitive to the first than to the latter.

Author details

J. G. Detoni*, F. Impinna, N. Amati and A. Tonoli

*Address all correspondence to: joaquimd@gmail.com

Department of Mechanical and Aerospace Engineering, Politecnico di Torino, Turin, Italy

References

[1] Post, R. F., & Ryutov, D. D. (1998). Ambient-temperature passive magnetic bearings: Theory and design equations. Massachusetts, USA. *Proceedings of the 6th International Symposium on Magnetic Bearings*, Cambrige.

[2] Filatov, A., & Maslen, E. H. (2001). Passive magnetic bearing for flywheel energy storage systems. *IEEE Transactions on Magnetics*, 37(6).

[3] Tonoli, A., Amati, N., Impinna, F., & Detoni, J. G. (2011). A solution for the stabilization of electrodynamic bearings: modeling and experimental validation. *ASME Journal of Vibration and Acoustics*, 133.

[4] Lembke, T. A. (2005). Design and analysis of a novel low loss homopolar electrodynamic bearing. *PhD thesis*, Royal Institute of Technology, Stockholm, Sweden.

[5] Kluyskens, V., & Dehez, B. (2009). Parameterized electromechanical model for magnetic bearings with induced currents. *Journal of System Design and Dynamics*, 3(4).

[6] Genta, G. (2005). *Dynamics of rotating systems*, Springer, New York.

[7] Dorf, R. C., & Bishop, R. H. (2010). *Modern Control Systems*, Prentice Hall.

[8] Tonoli, A., Amati, N., Bonfitto, A., Silvagni, M., Staples, B., & Karpenko, E. (2010). Design of Electromagnetic Dampers for Aero-Engine Applications. *Journal of Engineering for Gas Turbines and Power*, 132(11).

[9] Detoni, J. G., Impinna, F., Tonoli, A., & Amati, N. (2012). Unified Modelling of Passive Homopolar and Heteropolar Electrodynamic Bearings. *Journal of Sound and Vibration*, 331(19), 4219-4232.

Chapter 2

Rotors on Active Magnetic Bearings: Modeling and Control Techniques

Andrea Tonoli, Angelo Bonfitto, Mario Silvagni and
Lester D. Suarez

Additional information is available at the end of the chapter

1. Introduction

In the last decades the deeper and more detailed understanding of rotating machinery dynamic behavior facilitated the study and the design of several devices aiming at friction reduction, vibration damping and control, rotational speed increase and mechanical design optimization. Among these devices a promising technology is represented by magnetic actuators used as bearings which found a great spread in rotordynamics and in high precision applications. A first classification of magnetic bearings according to the physical working principle allows to pick out two main families: a) Active Magnetic Bearings [1], [2], making use of an electronic control unit to regulate the current flowing in the coils of the actuators. They need external source of energy. b) Passive Magnetic Bearings [3], [4], [5]: they do not need any electronic equipment. The control of the mechanical structure is achieved without the introduction of any external energy source. They exploit the reluctance force or the Lorentz force due to the generation of eddy currents developed in a conductor in a relative motion in a magnetic field.Active Magnetic Bearings require sensors an electronic equipment but, although more expensive respect to classical ball bearings, they offer several technological advantages:

- The absence of all fatigue and tribology issues due to contact: it allows the use of such bearings in vacuum systems, in clean and sterile rooms, or for the transport of aggressive or very pure media, and at high temperatures;
- No lubrication needed;
- No contamination by the dust created by friction between the rotor and the stator;
- Low bearing losses: at high operating speeds are 5 to 20 times less than in conventional ball or journal bearings, result in lower operating costs;

- Viscous friction can be avoided if the rotor is confined in high vacuum;
- Low vibration level;
- Dynamics adaptable to the desired application by tuning of the control loop;
- Precise positioning of the rotor due to the control loop: this is mainly determined by the quality of the measurement signal within the control loop. Conventional inductive sensors, for example, have a measurement resolution of about 1 ÷ 1000μm of a millimeter;
- Achievable fast positioning and/or high rotational speed of the rotor;
- The small sensitivity to the operating conditions;
- The predictability of the behavior.
- Further statements about the technology of realization can be done:
- The gap between rotor and bearing amounts typically to a few tenths of a millimeter, but for specific applications it can be as large as 20 mm. In that case, of course, the bearing becomes much larger;
- The rotor can be allowed to rotate at high speeds. The high circumferential speed in the bearing, only limited by the strength of material of the rotor, offers the possibilities of designing new machines with higher power density and of realizing novel constructions. Actually, about 350 m/s are achievable, for example by using amorphous metals which can sustain high stresses and at the same time have very good soft-magnetic properties, or by binding the rotor laminations with carbon fibers. Design advantages result from the absence of lubrication seals and from the possibility of having a higher shaft diameter at the bearing site. This makes the shaft stiffer and less sensitive to vibrations;
- The specific load capacity of the bearing depends on the type of ferromagnetic material and the design of the bearing electromagnet. It will be about 20 N/cm2 and can be as high as 40 N/cm2. The reference area is the cross sectional area of the bearing. Thus the maximum bearing load is mainly a function of the bearing size;
- The bearing and the rotor can be integrated on the same shaft by realizing bearingless configurations which allow to reduce the size of the system and to perform a cost saving solution.
- Retainer bearings are additional ball or journal bearings, which in normal operation are not in contact with the rotor. In case of overload or malfunction of the AMB they have to operate for a very short time: they keep the spinning rotor from touching the housing until the rotor comes to rest or until the AMB regains control of the rotor. The design of such retainer bearings depends on the specific application and despite a variety of good solutions still needs special attention;
- The unbalance compensation and the force-free rotation are control features where the vibrations due to residual unbalance are measured and identified by the AMB. The signal is used to either generate counteracting and compensating bearing forces or to shift the rotor axis in such a way that the rotor is rotating force-free;

- Diagnostics are readily performed, as the states of the rotor are measured for the operation of the AMB anyway, and this information can be used to check operating conditions and performance. Even active diagnostics are feasible, by using the AMB as actuators for generating well defined test signals simultaneously with their bearing function;
- The lower maintenance costs and higher life time of an AMB have been demonstrated under severe conditions. Essentially, they are due to the lack of mechanical wear. Currently, this is the main reason for the increasing number of applications in turbomachinery;
- The cost structure of an AMB is that of a typical mechatronics product. The costs for developing a prototype, mainly because of the demanding software, can be rather high. On the other side, a series production will lower the costs considerably because of the portability of that software.

Active Magnetic Bearings can be classified as a typical mechatronic product due to its nature which involves mechanical, electrical and control aspects, merging them in a single system. Rotordynamic field offer several examples of application areas [1], [6] : (a) Turbomachinery, (b) Vibration isolation, (c) Machine tools and electric drives, (d) Energy storing flywheels, (e) Instruments in space and physics, (f) Non-contacting suspensions for micro-techniques, (g) Identification and test equipment in rotordynamics, (f) Microapplications such as gyroscopic sensors [7], [8].The attractive potential of active magnetic suspensions motivated a considerable research effort for the past decade focused mostly on electrical actuation subsystem and control strategies [3], [9], [10], [11], [12], [13], [14].

This chapter illustrates the design, the modeling, the experimental tests and validation of all subsystems of a rotor on a five-axes active magnetic suspension. The mechanical, electrical, electronic and control strategies aspects are explained with a mechatronic approach evaluating all the interactions between them. The main goals of the manuscript are: a) Illustrate the design and the modeling phases of a five-axes active magnetic suspension; b) Discuss the design steps and the practical implementation of a standard suspension control strategy; c) Introduce an off-line technique of electrical centering of the actuators. The experimental test rig is a shaft (Weight: 5.3 kg. Length: 0.5 m) supported by two radial and one axial cylindrical active magnetic bearings and powered by an asynchronous high frequency electric motor.The chapter starts on an overview of the most common technologies used to support rotors with a deep analysis of their advantages and drawbacks with respect to active magnetic bearings. Furthermore a discussion on magnetic suspensions state of the art is carried out highlighting the research efforts directions and the goals reached in the last years.In the central sections, a detailed description of each subsystem is performed along with the modeling steps. In particular the rotor is modeled with a FE code while the actuators are considered in a linearized model.

The last sections of the chapter are focused on the control strategies design and the experimental tests.An off-line technique of actuators electrical centering is explained and its advantages are described in the control design context. This strategy can be summarized as follows. Knowing that: a) each actuation axis is composed by two electromagnets; b) each electromagnet needs a current closed-loop control; c) the bandwidth of this control is depending on the mechanical Airgap,then the technique allows obtaining the same value of the closed-loop bandwidth of the current control of both the electromagnets on the same actuation axis. This approach improves performance and gives more steadiness to the control behavior.The decentralized approach of the control strategy allowing the full suspensions on five axes is illustrated from the design steps to the practical implementation on the control unit.Finally, the experimental tests are carried out on the rotor to validate the suspension control and the off-line electrical centering. The numerical and experimental results are superimposed and compared to prove the effectiveness of the modeling approach.

2. System Architecture

The rig used for the modeling, the design and the experimental tests is an electrical spindle (picture reported in Figure 1) consisting of a shaft supported by two radial and one axial active magnetic bearings with cylindrical geometry and powered by an asynchronous high frequency electric motor. Two mechanical ball bearings, with radial and axial airgaps equal to half of the levitation ones, are positioned at the ends of the shaft to guarantee a safely touch-down of the shaft for anomalous working conditions with excessive whirling amplitude. The rotation axis is horizontal and the weight has the direction of each bearing.

Figure 1. Picture of the rotor.

Parameter	Symbol	Value	Unit
Rotor mass	m	5.31	kg
Rotor transversal Inertia	$J_x=J_y$	1.153•10-1	Kg/m2
Rotor polar Inertia	J_z	1.826•10-3	Kg/m2
Bearing rad. 1 location	a	214.5	mm A/V
Bearing rad. 2 location	b	212.6	mm A/V
Axial/Radial Airgap	g	0.75e-3	mm
Isotropic support stiffness	f / x	2.5•10-5	N/m

Table 1. Rotor mechanical and geometric parameters.

Table 1 reports the main parameters of the rotor. Figure 2 illustrates the section view of the system showing the layout of sensors, actuators, motor and rotor.

Active Magnetic Bearings applied to rotating machines can be considered as a typical mechatronic application, since it involves the control of mechanical system (the rotor) by means of an electronic control unit which elaborates the commands to feed electrical power drivers regulating the electromechanical actuators. The information to perform closed loop control architecture is given by displacement and current sensors.

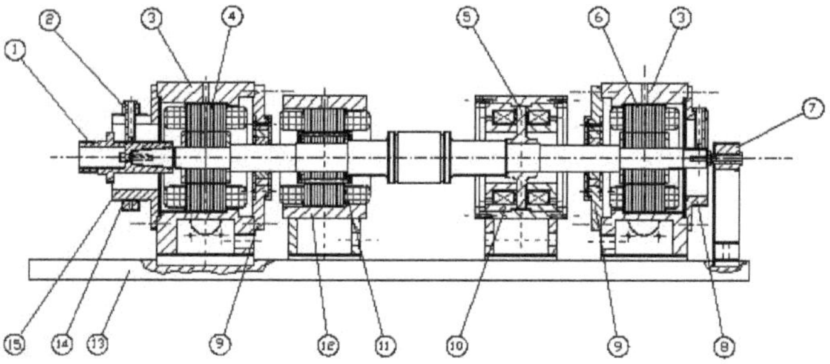

Figure 2. System section view: 1) tang, 2) radial sensor, 3) radial AMB support, 4) radial AMB laminated stator, 5) axial AMB disc, 6) radial AMB laminated stator, 7) axial sensor, 8) sensor cap, 9) bushing cap, 10) axial AMB electromagnet, 11) electric motor, 12) motor support, 13) foundation, 14) threaded ring, 15) sensor cap.

The interactions and the main functions of these subsystems are highlighted in Figure 3.The reported scheme is a standard representation of the system. However each block can be of different nature depending on the application. A short summary of the technologies typically used for each subsystem and the technologies used in the rig described in this chapter are reported in the following sections.

2.1. Control

Two main families of control architecture can be listed for active magnetic bearings:

- Decentralized SISO control: the action of each actuator is independent from the others and exploits a dedicated control law and sensors information;
- Centralized MIMO control: actuators are coupled as well as sensors information. A single control action is devoted to feed power drivers.

Several control strategies have been implemented and tested on rotors equipped with active magnetic bearings:

- Gain scheduled control [15];
- Adaptive control [16], [17];

- Robust H∞ control [18];
- Robust sliding mode control [19];
- Robust control via eigenstructure assignment dynamical compensation [20];
- Optimal control [21];
- Dynamic programming control [22];
- Genetic algorithm control [23];
- Fuzzy logic control [24];
- Feedback linearization control [25];
- Time-delay control [26];
- Control by transfer function approach [27];
- μ-synthesis control [28].

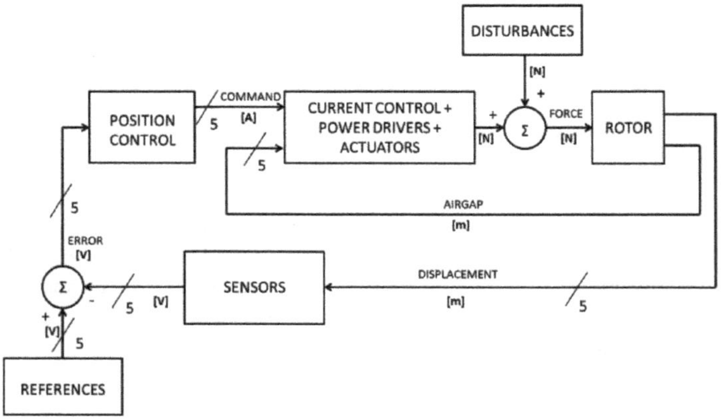

Figure 3. Overall system architecture.

In this chapter a decentralized PID strategy is implemented on a control module equipped with a DSP/FPGA–based digital control unit (EKU2.1). This digital platform allows the rapid reconfigurations of the overall system throw up to 108 (from FPGA) and 46 (from DSP) configurable digital I/O lines for input/output, event, PWM, capture/generation and user functions.Both DSP and FPGA have a dedicated Hard Real-Time Operating System (HRTOS) based on a non-pre-emptive scheduler (DSP side), involving ISR time or event triggered. EKU2.1 uses a single-master (DSP) multi-slave (FPGA) point-to-point communication protocol, based on Wishbone format; a system bus manages data exchange between the two cores. Software code is developed using the target-dependent tools Texas Instruments® Code Composer Studio.

2.2. Power drivers

The electronic circuits of power amplification stage to convert low power controller output signal to a high power stator input signal is chosen according to the kind of application. Basically, three main families of electronic circuits can be identified:

1. Linear analogue amplifiers have push–pull transistors at the output stage. They allow to enhance the current capability and to integrate a high-gain linear amplifier, such as a power operational amplifier. The linear amplifiers have the advantage of precise current and voltage regulation as well as low noise and they have a current rating of less than 10 A. Operation at the rated current is available only with effective cooling with heat sinks. Therefore the amplifier dimensions are large, resulting in high cost. The efficiency is low because of high losses in the push–pull transistor.

2. Switched-mode amplifiers enhance efficiency. Since the losses in the power devices are reduced, the heat sinks are much smaller and, as a result, switched mode amplifiers are compact in dimension so that the cost is low. Switched-mode operation of power devices is widely used in industry, e.g., for general-purpose inverters in ac drives and computer power supplies. This category of amplifiers is dominant in magnetic bearing drivers.

Hybrid amplifiers take advantage of linear and switched-mode amplifiers. At low current the push–pull transistors operate as a linear amplifier but at high current they operate in switched mode. To take advantage of a hybrid amplifier, it is quite important to modify the winding structure in magnetic bearings.

The rig object of study in this chapter is equipped with an H-bridge switching amplifier for each actuator.The power stage consists of an Embedded Isolated Power Module Board with four fully independent MOSFET/IGBT legs that supports up to 25 amperes with 100 volt of DC Bus. Also a maximum PWM switching frequency is 80kHz making this module suitable for high performance driving applications where control loop bandwidth and current ripple are important factors.The scheme used to feed power drivers is reported in Figure 4.

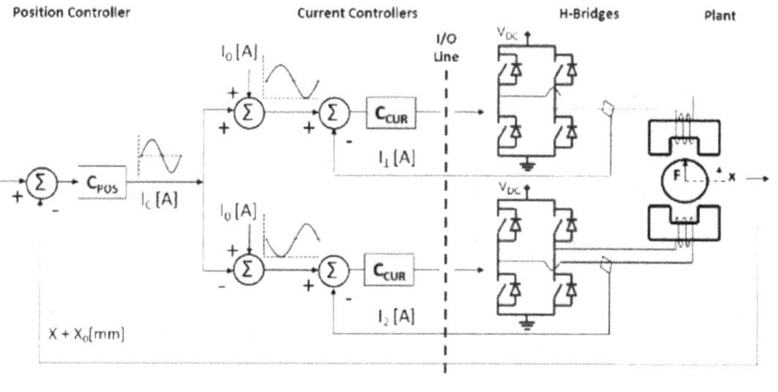

Figure 4. Power driver scheme.

Standard AMBs equipment for rotors suspensions are realized with five couples of cylindrical shape electromagnets to perform five dof active control. Conical shape of magnetic bearings exerting forces both in axial and radial direction simultaneously allow to save one couple of electromagnets and hence to reduce the size although the bearing design results more complex than standard cylindrical solution. This geometry permits to reachhigher rotation speed, limited in cylindrical solution by the strains growing in axial bearing disc.

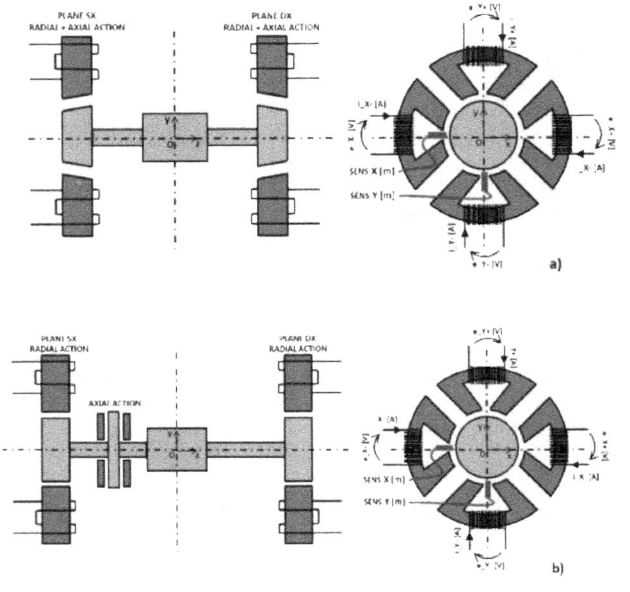

Figure 5. Geometry of actuation stage. a) Conical profile. b) Cylindrical profile.

2.3. Actuators

Actuators geometry and configuration depends on the electromagnets profile and on the number of actuators per actuation stage.

In this work classical configuration with four cylindrical actuators per actuation stage is dealt with. The disposition of the ten electromagnets is illustrated in Figure 6.

The main electrical and geometrical actuators parameters are listed in Table 2.

Parameter	Symbol	Value	Unit
Vacuum permeability	μ_0	1.26e-006	H/m
Voltage supply	V_{DC}	50	V

Parameter	Symbol	Value	Unit
	AXIAL Actuator		
Number of turns	N_{AX}	120	-
Circuitation length	l_{AX}	48e-3	m
Active section on airgap	S_{AX}	1210e-6	m^2
Nominal airgap	$g0_{AX}$	0.75e-3	m
Resistance	R_{AX}	0.5	Ω
Nominal inductance	$L\ 0_{AX}$	0.0146	H
	RADIAL actuator		
Number of turns	N_{RAD}	110	-
Circuitation length	l_{RAD}	135.2e-3	m
Active section on airgap	S_{RAD}	480e-6	m^2
Nominalairgap	$g0_{RAD}$	0.75e-3	M
Resistance	R_{RAD}	0.5	Ω
Nominal Inductance	$L\ 0_{RAD}$	0.0049	H

Table 2. Actuators parameters.

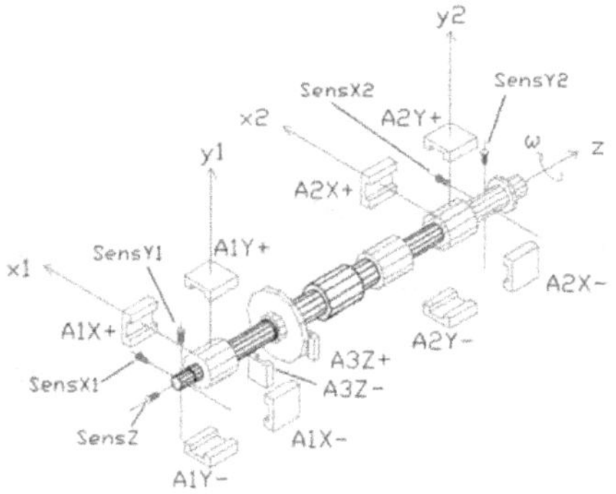

Figure 6. Actuators configuration.

2.4. Sensors

An important part of the performance of a magnetic bearing depends on the characteristics of the displacement sensors used. In order to measure the position of a moving rotor, contact-free sensors must be used which, moreover, must be able to measure on a rotating surface. Consequently, the geometry of the rotor, i.e. its surface quality, and the homogeneity of the material at the sensor will also influence the measuring results. A bad surface will thus produce noise disturbances, and geometry errors may cause disturbances with the rotational frequency or with multiples thereof.

In addition, depending on the application, speeds, currents, flux densities and temperatures are to be measured in magnetic bearing systems.

When selecting the displacement sensors, depending on the application of the magnetic bearing, measuring range, linearity, sensitivity, resolution, and frequency range are to be taken into account as well as:

- Temperature range, temperature drift of the zero point and sensitivity;
- Noise immunity against other sensors, magnetic alternating fields of the electromagnets, electromagnetic disturbances from switched amplifiers;
- Environmental factors such as dust, aggressive media, vacuum, or radiation;
- Mechanical factors such as shock and vibration;
- Electrical factors such as grounding issues associated with capacitive sensors.

The most important displacement sensors technologies are:

- Inductive sensors;
- Eddy-current sensors;
- Eddy Current Radial Displacement Sensor on a PCB (Transverse Flux Sensor)
- Capacitive sensors;
- Magnetic sensors.

The rig described in this chapter is equipped with five eddy current displacement sensors: high-frequency alternating current runs through the air-coil embedded in a housing. The electromagnetic coil section induces eddy currents in the conductive object whose position is to be measured, thus absorbing energy from the oscillating circuit. Depending on the clearance, the inductance of the coil varies, and external electronic circuitry converts this variation into an output signal. The usual modulation frequencies lie in a range of 1 - 2 MHz, resulting in useful measuring frequency ranges of 0 Hz up to approximately 20 kHz.

3. Modeling

The dynamic behavior of the rotor can be described by means of different models. This section describes the modeling techniques and the adopted assumptions, how acting forces and displacements are ordered and selected, the equation of motion used for the dynamic description and the variable used to describe the models.

Many modeling techniques can be adopted; an analytical rigid approach (based on the 4 d.o.f. rotor modeling) is here presented beside the most common Finite Element (FE) approach. The discretization for FEM software is reported in Figure 7.b.

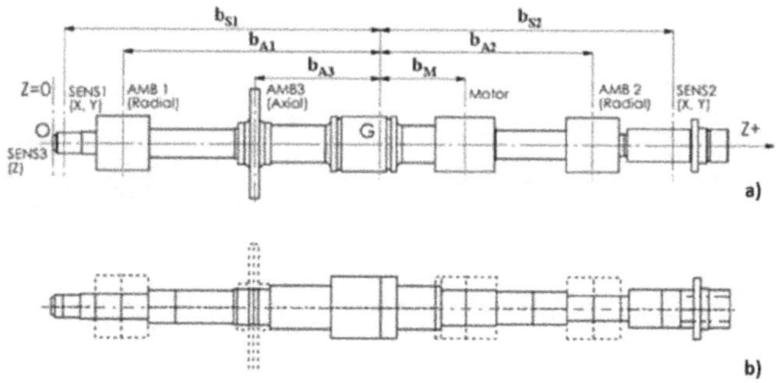

Figure 7. Rotor section view. a) View with dimensions. b) Discretization for FEM modeling.

The main hypothesis here adopted is to consider a constant spin speed. In this case the rotor behavior on the X-Y plane (known as flexural) is not coupled with the behavior on the Z-direction (axial). Other important assumption is that any rotation (except for spinning rotation) should be small.

3.1. Model Block Diagram

A simple description of the rotor model block diagram is presented in Figure 8. Forces (due to AMBs, motor and external) acting on the rotor are un-grouped for the X-Y and for the Z behavior, these signals are fed to the block describing the dynamic behaviors. Outputs of these blocks are the states (displacements and velocities) of the systems, reported as displacements and speeds to sensors, AMBs and motor. The constant spin speed Ω is used in the X-Y model for the gyroscopic behavior and reported as output.Spin speed and displacement on the sensors are physical entities measured by specific sensors, and signal are reported to the Sensor block; the other displacements and relative velocities (to AMBs and motor) should be used for intrinsic feedback such as back electromotive force in the motor or magnetic bearings.

3.1.1. Model inputs / outputs

Model inputs are the forces acting on the rotor, while outputs are typically displacements either on sensors or on AMBs and motor (Figure 9). The rotor is suspended by two radial magnetic bearing (AMB1 and AMB2) which generate four forces oriented as the reference plant reference frame and acting in the center of the relative AMB; these forces act the behavior on X-Y plane. A further magnetic bearing (AMB3) is used to constrain displacements along Z axis (axial). The five forces due to AMBs are collected in vector fAMB. The electric motor, used to generate rotation torque (not included in this kind of model where the spin speed is assumed to be constant), can also act two forces, in radial direction, while is not capable to generate a force in the Z direction. According to this, vector fMOT has two components acting in the center of the motor.

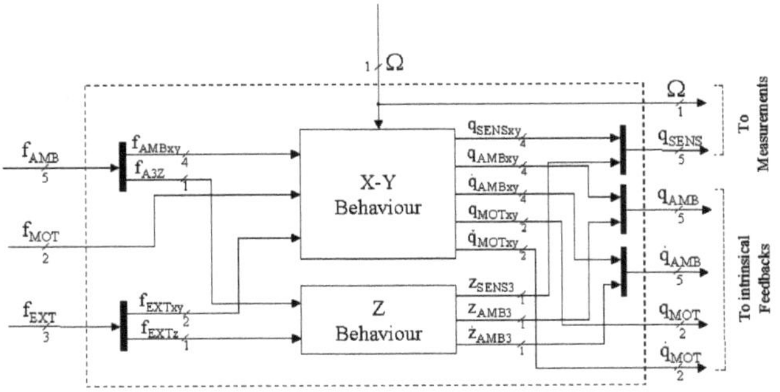

Figure 8. Rotor model block diagram.

In order to simplify the description of the system the generic external force are supposed to act directly to the center of mass of the rotor; these three forces are oriented as the plant reference frame. These components are the resultant of any external force, such as impact forces. While the X-Y behaviour is uncoupled from the Z behavior, acting forces due to AMBs and external can be rewritten dividing the forces acting in X-Y plane from forces acting on Z axis. Table 3 reports acting forces (inputs) on the rotor.

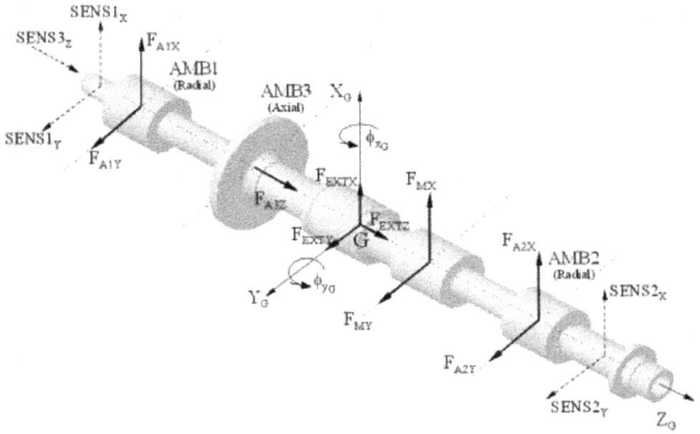

Figure 9. Actuation forces and sensors position.

AMB	Motor	External	
$\mathbf{f}_{AMB} = \begin{Bmatrix} F_{A1X} \\ F_{A1Y} \\ F_{A2X} \\ F_{A2Y} \\ F_{A3Z} \end{Bmatrix} = \begin{Bmatrix} \mathbf{f}_{AMBxy} \\ F_{A3Z} \end{Bmatrix}$	$\mathbf{f}_{MOT} = \begin{Bmatrix} F_{MX} \\ F_{MY} \end{Bmatrix}$	$\mathbf{f}_{EXT} = \begin{Bmatrix} F_{EXTx} \\ F_{EXTy} \\ F_{EXTz} \end{Bmatrix} = \begin{Bmatrix} \mathbf{f}_{EXTxy} \\ F_{EXTz} \end{Bmatrix}$	(1)

Table 3. Rotor Inputs.

Referring to Figure 8, a set of outputs is used for measurements (the spin speed Ω and the displacements on the sensor qSENS) and another set is used for intrinsical feedback (displacements and velocities on AMBs qAMB and $\dot{q}$ AMB, and on the motor qMOT and $\dot{q}$ MOT).

Table 4 reports displacements and velocities (outputs) on the rotor.

Sensors	AMB	Motor	
$\mathbf{q}_{SENS} = \begin{Bmatrix} x_{SENS1} \\ y_{SENS1} \\ x_{SENS2} \\ y_{SENS2} \\ z_{SENS3} \end{Bmatrix} = \begin{Bmatrix} \mathbf{q}_{SENSxy} \\ z_{SENS3} \end{Bmatrix}$	$\mathbf{q}_{AMB} = \begin{Bmatrix} x_{AMB1} \\ y_{AMB1} \\ x_{AMB2} \\ y_{AMB2} \\ z_{AMB3} \end{Bmatrix} = \begin{Bmatrix} \mathbf{q}_{AMBxy} \\ z_{AMB3} \end{Bmatrix}$	$\mathbf{q}_{MOT} = \begin{Bmatrix} x_{MOT} \\ y_{MOT} \\ z_{MOT} \end{Bmatrix}$	(2)

Table 4. Rotor outputs.

Input		Output		
X-Y Behaviour	Z Behaviour	X-Y Behaviour	Z Behaviour	
$\mathbf{f}_{xy} = \begin{Bmatrix} F_{A1X} \\ F_{EXTx} \\ F_{MX} \\ F_{A2X} \\ F_{A1Y} \\ F_{EXTx} \\ F_{MY} \\ F_{A2Y} \end{Bmatrix}$	$\mathbf{f}_z = \begin{Bmatrix} F_{A3Z} \\ F_{EXTz} \end{Bmatrix}$	$\mathbf{y}_{xy} = \begin{Bmatrix} x_{SENS1} \\ x_{AMB1} \\ x_{MOT} \\ x_{AMB2} \\ x_{SENS2} \\ y_{SENS1} \\ y_{AMB1} \\ y_{MOT} \\ y_{AMB2} \\ y_{SENS2} \end{Bmatrix}$	$\mathbf{y}_z = \begin{Bmatrix} z_{SENS3} \\ z_{AMB3} \end{Bmatrix}$	(3)

Table 5. Inputs / Outputs vector orders.

Input/Output vector reported in Table 5. and ordered to be compliant with blocks that generate such forces. In order to address the modeling technique (especially FE based), input/output vector should be reordered in the way reported in Table 5.

The dynamic behavior of the rotor can be described by mean of the equations of motion (EoM). In the following section the equations used for the model is described. The typical equations are reported for a generic rotor model and then applied to a rigid analytical model and to the FE model.While the spin speed is constant the X-Y behavior is uncoupled from Z behavior in the same way the equations can be separated.

X-Y Behavior

Equation of Motion:

$$\begin{aligned}&\mathbf{M}_{xy}\ddot{\mathbf{q}}_{xy}(t)+\left(\mathbf{L}_{xy}+\Omega\mathbf{G}_{xy}\right)\dot{\mathbf{q}}_{xy}(t)+\left(\mathbf{K}_{\Omega 0xy}+\Omega^2\mathbf{K}_{\Omega 2xy}+\Omega\mathbf{H}_{xy}\right)\mathbf{q}_{xy}(t)=\\&=\mathbf{f}_{sxy}+\Omega^2\mathbf{f}_{umb}\begin{Bmatrix}\sin(\Omega t)\\ \cos(\Omega t)\end{Bmatrix}+\mathbf{S}_{ixy}\mathbf{f}_{xy}(t)\end{aligned} \tag{1}$$

Measure Equation:

$$\begin{Bmatrix}\mathbf{y}_{xy}(t)\\ \dot{\mathbf{y}}_{xy}(t)\end{Bmatrix}=\mathbf{S}_{oxy}\begin{Bmatrix}\mathbf{q}_{xy}(t)\\ \dot{\mathbf{q}}_{xy}(t)\end{Bmatrix} \tag{2}$$

Z Behavior

Equation of Motion:

$$\mathbf{M}_z\ddot{\mathbf{q}}_z(t)+\mathbf{L}_z\dot{\mathbf{q}}_z(t)+\left(\mathbf{K}_{\Omega 0z}+\Omega^2\mathbf{K}_{\Omega 2z}\right)\mathbf{q}_z(t)=\mathbf{f}_{sz}+\mathbf{S}_{iz}\mathbf{f}_z(t) \tag{3}$$

Measure Equation:

$$\begin{Bmatrix}\mathbf{y}_z(t)\\ \dot{\mathbf{y}}_z(t)\end{Bmatrix}=\mathbf{S}_{oz}\begin{Bmatrix}\mathbf{q}_z(t)\\ \dot{\mathbf{q}}_z(t)\end{Bmatrix} \tag{4}$$

Refer to Table 6 for matrices naming and description.

Name	Description
Ω	Spin Speed
$\boldsymbol{q}(t)$	Generalized displacements

Name	Description
$\boldsymbol{M}$	Mass (symmetric) matrix
$\boldsymbol{L}$	Damping (symmetric) matrix
$\boldsymbol{G}$	Gyroscopic (skew-symmetric) matrix
$\boldsymbol{K}_{\Omega 0}$	Stiffness (symmetric) matrix: spin speed independent
$\boldsymbol{K}_{\Omega 2}$	Stiffness (symmetric) matrix: spin speed dependent
$\boldsymbol{H}$	Circulatory (skew-symmetric) matrix
$\boldsymbol{f}_s$	Static forces
$\boldsymbol{R}$	Rotation Matrix
$\boldsymbol{f}_{umb}$	Unbalance forces
$\boldsymbol{f}(t)$	External forces
$\boldsymbol{S}_i$	Input selection matrix
$\boldsymbol{y}(t)$	Output displacements
$\boldsymbol{S}_o$	Output selection matrix
$\boldsymbol{\Phi}$	Selected eigenvector for modal (MK) reduction

Table 6. Matrices names and description.

3.2. 4dof model

Generic Eom for rotors previously described, can be applied to a rigid analytical model based on the 4 d.o.f theory (for X-Y behavior), with an additional d.o.f. for the Z behavior. In this model the equation of motion are develop in a center of mass coordinate system as reported in Figure 10.

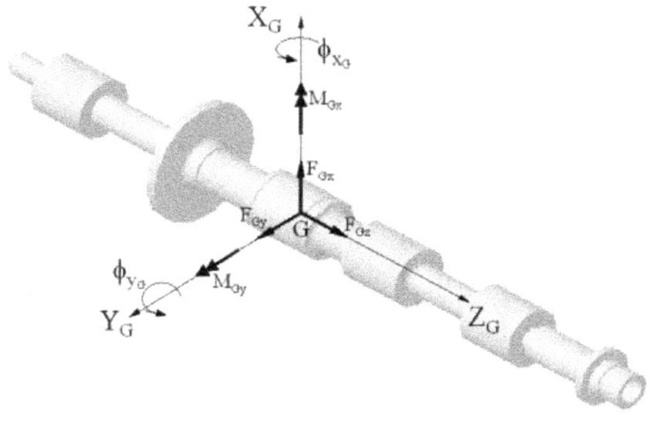

Figure 10. d.o.f. model with generalized displacements and forces.

The physical properties used in the model are:

Name	Description
m	Mass of the rotor [kg]
J_t	Transversal moment of inertia about any axis in the rotation plane [kgm2]
J_p	Transversal moment of inertia about any axis in the rotation plane [kgm2]
g_x, g_y	Gravity along x and y direction [m/s2]
$\varepsilon_x, \varepsilon_y$	Static unbalance eccentricity along x and y direction [m]
χ	Torque unbalance: angular error [rad]
b_{A1}	AMB1 distance from center of mass [m]
b_{A2}	AMB2 distance from center of mass [m]
b_M	Motor distance from center of mass [m]
b_{S1}	Sensor1 distance from center of mass [m]
b_{S2}	Sensor2 distance from center of mass [m]

Table 7. Physical properties of rigid analytical model.

X-Y Behavior

The generalized displacements vector is composed by the center of mass coordinates:

$$\mathbf{q}_{xy}(t) = \begin{Bmatrix} x_G \\ y_G \\ \varphi_{x_G} \\ \varphi_{y_G} \end{Bmatrix} \tag{5}$$

If the model has no damping the EoM (4) becomes:

$$\begin{bmatrix} m & 0 & 0 & 0 \\ 0 & m & 0 & 0 \\ 0 & 0 & J_t & 0 \\ 0 & 0 & 0 & J_t \end{bmatrix} \ddot{\mathbf{q}}_{xy}(t) + \Omega \begin{bmatrix} 0 & 0 & 0 & 0 \\ 0 & 0 & 0 & 0 \\ 0 & 0 & 0 & J_p \\ 0 & 0 & -J_p & 0 \end{bmatrix} \dot{\mathbf{q}}_{xy}(t) = \\ = \begin{Bmatrix} mg_x \\ mg_y \\ 0 \\ 0 \end{Bmatrix} + \Omega^2 \begin{bmatrix} -m\varepsilon_y & m\varepsilon_x \\ m\varepsilon_x & m\varepsilon_y \\ -\chi\left(J_t - J_p\right) & 0 \\ 0 & \chi\left(J_t - J_p\right) \end{bmatrix} \begin{Bmatrix} \sin(\Omega t) \\ \cos(\Omega t) \end{Bmatrix} + \mathbf{S}_{txy} \mathbf{f}_{xy}(t) \tag{6}$$

The measure equation has the same structure reported in (5).

Z Behavior

Under the same assumptions equation (6) becomes:

$$m\ddot{z}(t) = \mathbf{f}_{sz} + \mathbf{S}_{iz}\mathbf{f}_z(t) \qquad (7)$$

The measure equation has the same structure reported in (7).

In order to be compliant to input/output vector described in Table 8 the selection matrices of equation (9) and (10) with their relative measure equation should be:

X-Y Behaviour	Z Behaviour	
$\mathbf{S}_{ixy} = \begin{bmatrix} 1 & 1 & 1 & 1 & 0 & 0 & 0 & 0 \\ 0 & 0 & 0 & 0 & 1 & 1 & 1 & 1 \\ 0 & 0 & 0 & 0 & b_{A1} & 0 & -b_M & -b_{A2} \\ -b_{A1} & 0 & b_M & b_{A2} & 0 & 0 & 0 & 0 \end{bmatrix}_{4x8}$	$\mathbf{S}_{iz} = \begin{bmatrix} 1 & 1 \end{bmatrix}_{1x2}$	(11)
$\mathbf{S}_{oxy} = \begin{bmatrix} \mathbf{S}'_{oxy} & \mathbf{0} \\ \mathbf{0} & \mathbf{S}'_{oxy} \end{bmatrix}_{20x8}, \mathbf{S}'_{oxy} = \begin{bmatrix} 1 & 0 & 0 & -b_{S1} \\ 1 & 0 & 0 & -b_{A1} \\ 1 & 0 & 0 & b_M \\ 1 & 0 & 0 & b_{A2} \\ 1 & 0 & 0 & b_{S2} \\ 0 & 1 & b_{S1} & 0 \\ 0 & 1 & b_{A1} & 0 \\ 0 & 1 & -b_M & 0 \\ 0 & 1 & -b_{A2} & 0 \\ 0 & 1 & -b_{S2} & 0 \end{bmatrix}$	$\mathbf{S}_{oz} = \begin{bmatrix} 1 & 0 \\ 1 & 0 \\ 0 & 1 \\ 0 & 1 \end{bmatrix}_{4x2}$	

Table 8. Input/Output selection matrices for rigid analytical model.

3.3. Flexible Rotor Model (FE)

A flexible model is here described. This model is generated by using a finite element code especially designed for rotating machines (Dynrot). The outputs of this code are the matrices reported in equation (4) to (7).

3.3.1. Full Model

In the case all the nodes displacements are used the full dynamic behavior is described. The displacements vector q collects the translation displacement of the model nodes in the following order:

3.3.1. Reduced Model

Usually a reduced model is used. A typical reduction method is the modal (MK) reduction where only some modes of vibration are selected.

The displacements to modal transformation are reported in Table 10:

X-Y Behaviour	Z Behaviour	
$\mathbf{q}_{xy}(t)=\begin{Bmatrix} x_1 \\ x_2 \\ \dots \\ x_n \\ y_1 \\ y_2 \\ \dots \\ y_n \end{Bmatrix}$	$\mathbf{q}_z(t)=\begin{Bmatrix} z_1 \\ z_2 \\ \dots \\ z_n \end{Bmatrix}$	(12)

Table 9. Generalized coordinates for full FE model.

X-Y Behaviour	Z Behaviour	
$\boldsymbol{\xi}_{xy}(t)=\boldsymbol{\Phi}_{xy}\boldsymbol{q}_{xy}(t)$	$\xi_z(t)=\boldsymbol{\Phi}_z\boldsymbol{q}_z(t)$	(13)

Table 10. Nodal to Modal transformation.

The equations of motion for the reduced model are formerly the same reported in equations (4) to (7), where the nodal displacements is substituted by the modal displacements and matrices and vector are reported in their modal forms.

Input/output selection matrices should be transformed, starting from FEM matrices, as indicated in:

X-Y Behaviour	Z Behaviour	
$\boldsymbol{S}_{\xi ixy}=\boldsymbol{\Phi}_{xy}^T\boldsymbol{S}_{ixy}$	$\boldsymbol{S}_{\xi i}=\boldsymbol{\Phi}_z^T\boldsymbol{S}_{iz}$	(14)
$\boldsymbol{S}_{\xi oxy}=\boldsymbol{S}_{oxy}\boldsymbol{\Phi}_{xy}$	$\boldsymbol{S}_{\xi oz}=\boldsymbol{S}_{oz}\boldsymbol{\Phi}_z$	

Table 11. Input / Output matrices conversion from Nodal to Modal.

3.4. State Space representationofrotordynamicequations

Dynamic equations (4) to (7) can be reported, with explicit spin speed, in the state space representation in the following way, either for X-Y or Z behavior:

$$\begin{aligned}\dot{\mathbf{x}}(t) &= \left(\mathbf{A}_{\Omega 0} + \Omega \mathbf{A}_{\Omega 1} + \Omega^2 \mathbf{A}_{\Omega 2}\right)\mathbf{x}(t)\mathbf{u}(t) + \mathbf{B}\mathbf{u}(t) \\ \mathbf{y}(t) &= \mathbf{C}\mathbf{x}(t) + \mathbf{D}\mathbf{u}(t)\end{aligned} \tag{8}$$

Where:

Name	X-Y Behaviour	Z Behaviour	
$\boldsymbol{x}(t)$	$\begin{Bmatrix} \boldsymbol{q}_{xy}(t) \\ \dot{\boldsymbol{q}}_{xy}(t) \end{Bmatrix}$	$\begin{Bmatrix} \boldsymbol{q}_{z}(t) \\ \dot{\boldsymbol{q}}_{z}(t) \end{Bmatrix}$	(16)
$\boldsymbol{A}_{\Omega 0}$	$\begin{bmatrix} \mathbf{0} & \boldsymbol{I} \\ -\boldsymbol{M}_{xy}^{-1}\boldsymbol{K}_{\Omega 0xy} & -\boldsymbol{M}_{xy}^{-1}\boldsymbol{L}_{xy} \end{bmatrix}$	$\begin{bmatrix} \mathbf{0} & \boldsymbol{I} \\ -\boldsymbol{M}_{z}^{-1}\boldsymbol{K}_{\Omega 0z} & -\boldsymbol{M}_{z}^{-1}\boldsymbol{L}_{z} \end{bmatrix}$	
$\boldsymbol{A}_{\Omega 1}$	$\begin{bmatrix} \mathbf{0} & \mathbf{0} \\ -\boldsymbol{M}_{xy}^{-1}\boldsymbol{H}_{xy} & -\boldsymbol{M}_{xy}^{-1}\boldsymbol{G}_{xy} \end{bmatrix}$	$[\mathbf{0}]$	
$\boldsymbol{A}_{\Omega 2}$	$\begin{bmatrix} \mathbf{0} & \mathbf{0} \\ -\boldsymbol{M}_{xy}^{-1}\boldsymbol{K}_{\Omega 2xy} & \mathbf{0} \end{bmatrix}$	$\begin{bmatrix} \mathbf{0} & \mathbf{0} \\ -\boldsymbol{M}_{z}^{-1}\boldsymbol{K}_{\Omega 2z} & \mathbf{0} \end{bmatrix}$	
$\boldsymbol{B}$	$\begin{bmatrix} \mathbf{0} & \mathbf{0} & \mathbf{0} \\ \boldsymbol{I} & \boldsymbol{I} & \boldsymbol{S}_{ixy} \end{bmatrix}$	$\begin{bmatrix} \mathbf{0} & \mathbf{0} \\ \boldsymbol{I} & \boldsymbol{S}_{iz} \end{bmatrix}$	
$\boldsymbol{u}(t)$	$\begin{Bmatrix} \boldsymbol{f}_{sxy} \\ \boldsymbol{f}_{umb} \\ \boldsymbol{f}_{xy} \end{Bmatrix}$	$\begin{Bmatrix} \boldsymbol{f}_{sz} \\ \boldsymbol{f}_{z} \end{Bmatrix}$	
$\boldsymbol{y}(t)$	$\begin{Bmatrix} \boldsymbol{y}_{xy}(t) \\ \dot{\boldsymbol{y}}_{xy}(t) \end{Bmatrix}$	$\begin{Bmatrix} \boldsymbol{y}_{z}(t) \\ \dot{\boldsymbol{y}}_{z}(t) \end{Bmatrix}$	
$\boldsymbol{C}$	$\boldsymbol{S}_{oxy}$	$\boldsymbol{S}_{oz}$	
$\boldsymbol{D}$	$[\mathbf{0}]$	$[\mathbf{0}]$	

Table 12. State Space Representation variables.

4. Control design and results

The aim of this section is to explain the steps followed to perform the suspension control design. A conventional decentralized control strategy is illustrated with two nested loops, the inner for current control and the outer for the position.The detailed description of this strategy is followed by the exposition of an off-line electrical centering strategy which is used equalize electrical parameters of the electromagnets on each actuation stage and allows to get a steady and balanced control action.

4.1. ActiveMagnetic Suspension Control

Figure 4 shows the classical control strategy for one axis AMB. The system is characterized with a nested control structure, where the inner loops describe the current loops used to achieve a direct actuator's effort drive (force) and the outer loop is used to compensate the rotor position error from the nominal air-gap. Generally, the same strategy is applied for each axis; so they are managed independently from each other and control is called decentralized. The driving of one axis is performed with two separate H-bridges. To exert a positive force on the rotor, current in the upper coil is increased by the control current while the current in the lower coil is decreased by control current and vice versa for negative forces. Also, to linearize the current to force characteristic of an electromagnet a constant bias current is applied to both coils respectively.Position control is performed by using five decentralized PID. The design and tuning of control laws parameters are well described in [1] and [2]. Here Bode diagram transfer function of position control law is reported (Figure11).

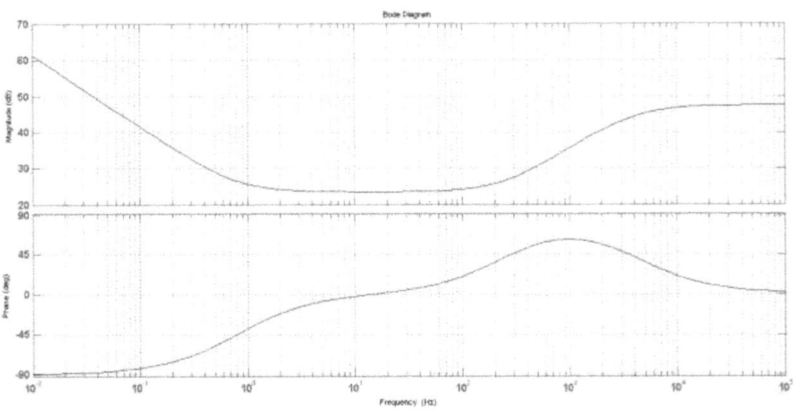

Figure 11. Control position Bode diagram.

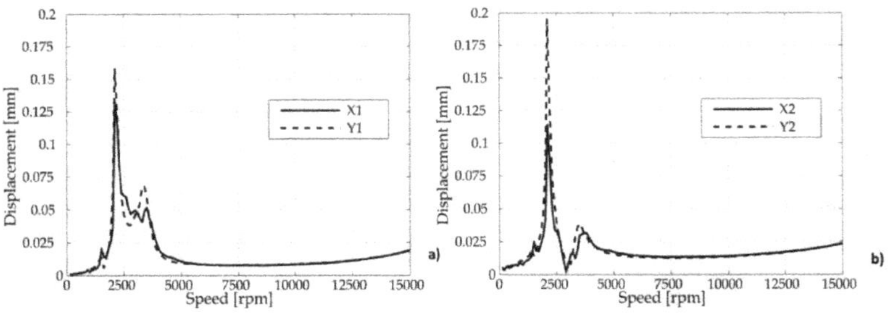

Figure 12. Unbalance response. a) Left actuation stage. b) Right actuation stage.

The experimental characterization has been performed by using classical tools of rotordynamic analysis: Unbalance responses (Figure 12), Waterfalls (Figure 13), orbital tube and orbital view (Figure 14, Figure 15). Theoretical notes on rotodynamic analysis can be found in [29].

4.2. Off-lineElectricalCentering

Active Magnetic Bearings offer several technological advantages that make their use mandatory in some particular applications, typically when clean environment is required or maintenance is expensive or difficult to manage. On the other hand, the main drawbacks are mostly the costs, higher than classical ball bearings due to the introduction of sensors and electronic equipment, and the complexity in the design phases of electrical and electronics subsystems and control strategies.

One of the aspects where this difficulty is more evident is the centering of the rotating part respect to the stator and in particular to the sensors. Sensors are indeed designed to detect microns of displacements and little inaccuracies in measurements lead to bad working or to system instability in the worst cases.

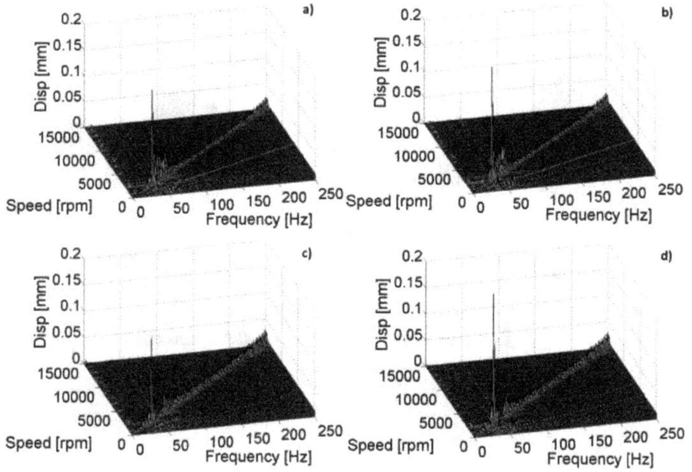

Figure 13. Waterfall plot. a) X1; b) Y1; c) X2; d) Y2.

The designer can choice to perform either a geometrical or an electrical centering, depending on the priorities of the application requirements. The first (Figure 16.a) consists in putting the rotor at the mechanical center neglecting the electrical differences in electromagnets coils parameters, inductance above all. The latter (Figure 16.b), on the contrary, leads to an equalization of electrical parameters, even if the rotor is not spinning around the geometrical center of the actuators.

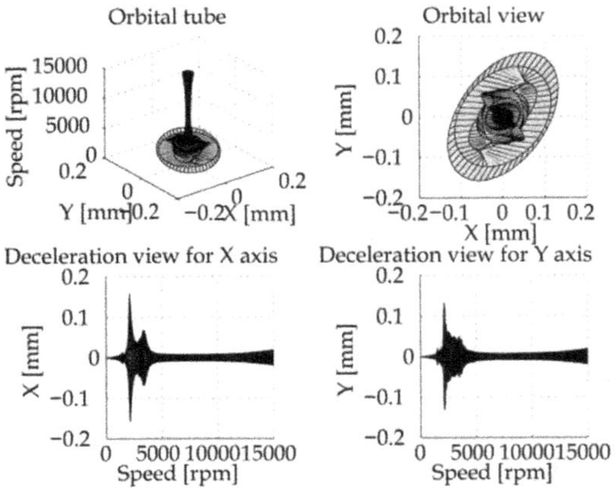

Figure 14. Orbital tube and orbital view representation on X1Y1 - plane. a) three dimensional view of the tube; b) Projection on the xy-plane; c) and d) projection on the Ωx and Ωy –planes.

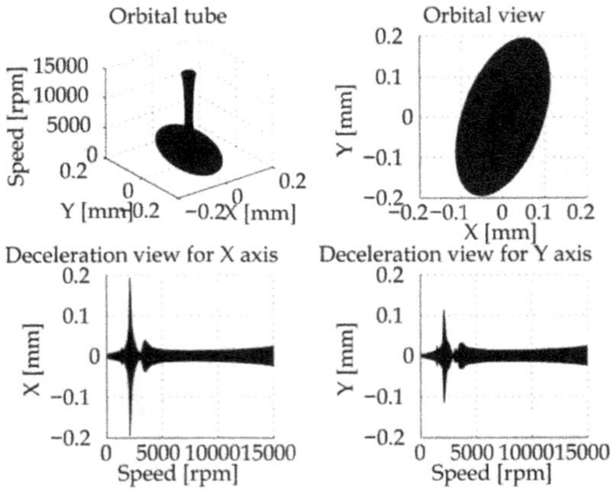

Figure 15. Orbital tube and orbital view on X2Y2 - plane. a) three dimensional view of the tube; b) Projection on the xy-plane; c) and d) projection on the Ωx and Ωy –planes.

In this section an off-line electrical centering technique is exposed. The goal is to make the rotor spins around a point which is not compulsorily coincident with the geometrical center of the actuators but grants the symmetry of the electrical parameters of them. It is well known that the inductance value of an electromagnet is depending on the distance between the ferromagnetic target (the rotor in this case) and the electromagnet itself.

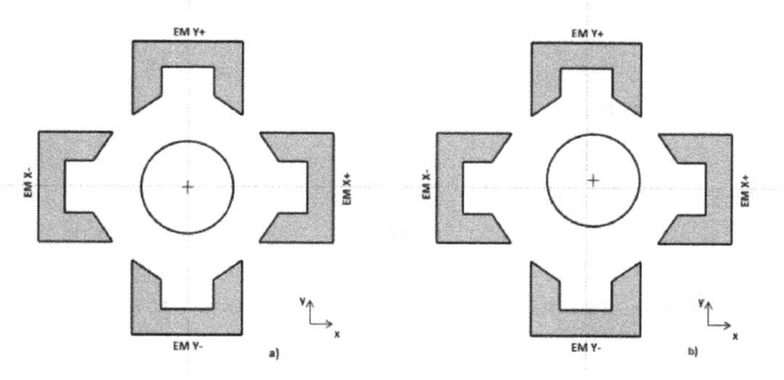

Figure 16. Mechanical centering (a) vs. Electrical centering (b).

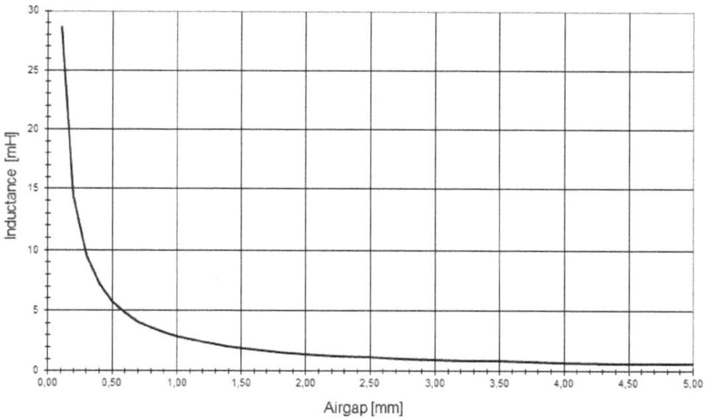

Figure 17. Inductance vs. Position.

This behavior is nonlinear as illustrated in Figure 17 and little variation of position and hence on inductance lead to big variation of actuator electrical pole and current control dynamics.Since the electrical dynamics of an electromagnet is depending on supply voltage, resistance and inductance values (Table 2) as in Eq. 17:

$$E = \frac{VDC}{R + Ls} \tag{9}$$

the electrical pole of the electromagnet is strictly dependent on the distance between the target and the actuator as reported in Figure 18. This issue has as consequence a different behavior of closed loop current control as illustrated in Figure 19 (a and b). It can be noticed that the same current control applied to the two opposite electromagnets of the same actuation axis without electrical centering generates two different closed loop responses. Few microns of Airgap generates differences of hundreds of Hertz on current control bandwidth.

Considering that this behavior is generated by a difference of inductance value of the two electromagnets, by acting on the position reference with offset corrections of the outer position control, the rotor can be set to spin around a point where electrical parameters are equalized and current loop bandwidths of both the electromagnets are the same (Figure 19 (c and d)). Further studies and research are being conducted on this strategy since this process can be performed in an on-line automatic routine with an adaptive technique, able to change the control parameters of the inner current loop while the Airgap is changing, i.e. when the rotor is oscillating.

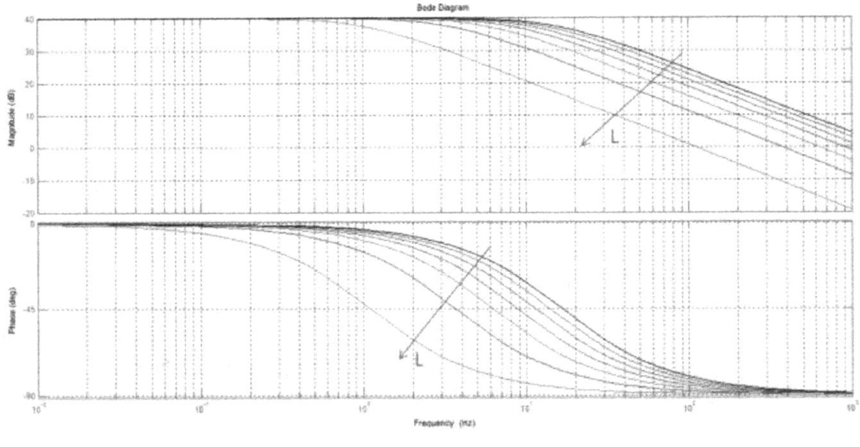

Figure 18. Electrical pole trend at varying of inductance value.

5. Conclusions

In this chapter the modeling, the design and the experimental tests phases of a rotor equipped with active magnetic bearings have been described. The description deals with rotordynamic aspects as well as electrical, electronic and control strategies subsystem. The control design of a standard decentralized SISO strategy and the details of an innovative off-line electrical centering technique have been exposed.Experimental results have been exposed highlighting mainly rotordynamics and control aspects.

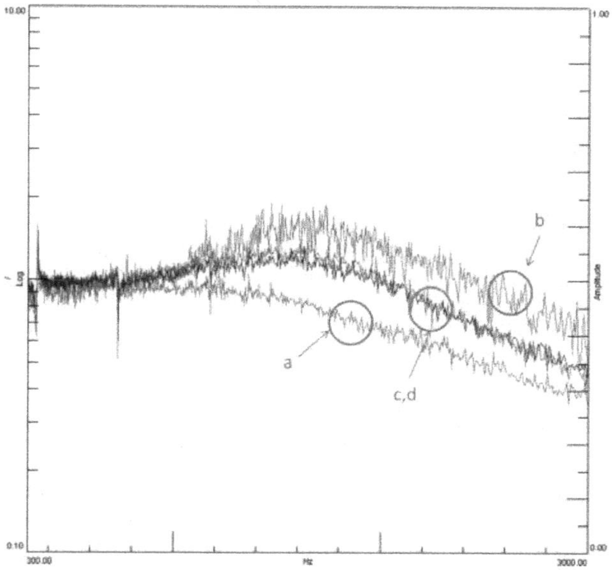

Figure 19. Electrical centering results. a) X- current loop Bode diagram before centering; b) X+ current loop Bode diagram before centering; c, d) X-/X+ current loop Bode diagrams after centering.

Author details

Andrea Tonoli, Angelo Bonfitto*, Mario Silvagni and Lester D. Suarez

*Address all correspondence to: angelo.bonfitto@polito.it

Mechanics Department, Mechatronics Laboratory – Politecnico di Torino, Italy

References

[1] Bleuler, H., Cole, M., Keogh, P., Larsonneur, R., Maslen, E., Nordmann, R., Okada, Y., Schweitzer, G., & Traxler, A. (2009). *Magnetic Bearings Theory, Design, and Application to Rotating Machinery*, Springer, Berlin Heidelberg.

[2] Chiba, A., Fukao, T., Ichikawa, O., Oshima, M., Takemoto, M., & Dorrell, D. G. (2005). *Magnetic Bearings and Bearingless drives*, Elsevier, Oxfor.

[3] Tonoli, A., Bonfitto, A., De Lépine, X., & Silvagni, M. (2009). Self-sensing Active Magnetic Dampers for vibration control. *Journal of Dynamic Systems, Measurement and Control*, 131.

[4] Tonoli, A., Amati, N., Impinna, F., Detoni, J. G., Bleuler, H., & Sandtner, J. (2010). Dynamic Modeling and Experimental Validation of Axial Electrodynamic Bearings. Wuhan, China. *Proceedings of the 12th International Symposium on Magnetic Bearings*, 978-7-90050-306-0.

[5] Genta, G., De Lépine, X., Impinna, F., Detoni, J. G., Amati, N., & Tonoli, A. (2009). Sensitivity Analysis of the Design Parameters of Electrodynamic Bearings. New Delhi, India. *IUTAM Symposium on Emerging Trends in Rotor Dynamics*, 287-296, DOI: 10.1007/978-94-007-0020-8, 1875-3507.

[6] Bonfitto, A. (2012). *Active Magnetic Control of Rotors*, LAP LAMBERT Academic publishing AG & Co. KG, 978-3-84842-104-6.

[7] Bosgiraud, T. (2008). Two Degrees of Freedom Miniaturized Gyroscope based on Active Magnetic Bearings. *PhD Thesis EPFL*, Lausanne, Switzerland.

[8] Bleuler, H., Bonfitto, A., Tonoli, A., & Amati, N. (2010). Miniaturized gyroscope with bearingless configuration. *In: Twelfth International Symposium on Magnetic Bearings (ISMB12)*, Wuhan, China, August 22-25.

[9] Schammass, A., Herzog, R., Buhler, P., & Bleuler, H. (2005). New results for self-sensing active magnetic bearings using modulation approach. *IEEE Trans. Contr. Syst. Tech.*, 13(4), 509-516.

[10] Araki, Mizuno T., , K., & Bleuler, H. (1996). Stability analysis of self-sensing magnetic bearing controllers. *IEEE Trans. Contr. Syst. Tech.*, 4(5), 572-579.

[11] Maslen, E. H., Montie, D. T., & Iwasaki, T. (2006). Robustness limitations in self-sensing magnetic bearings. *Journal of Dynamic Systems, Measurement and Control*, 128, 197-203.

[12] Vischer, D., & Bleuler, H. (1990, 12-14 July). A new approach to sensorless and voltage controlled AMBs based on network theory concepts. Tokyo, Japan. *Second International Symposium on Magnetic bearings*, 301-306.

[13] Vischer, D., & Bleuler, H. (1993). Self-sensing magnetic levitation. *IEEE Trans. on Magn.*, 29(2), 1276-1281.

[14] Noh, M. D., & Maslen, E. H. (1997). Self-sensing magnetic bearings using parameter estimation. *IEEE Trans. Instr. Meas.*, 46(1), 45-50.

[15] Betschon, F., & Knospe, C. R. (2001). Reducing magnetic bearing currents via gain scheduled adaptive control. *IEEE/ASME Transactions on Mechatronics*, 6(4), 437-443.

[16] Lum, K. Y., Coppola, V. T., & Bernstein, D. S. (1996). Adaptive autocentering control for an active magnetic bearing supporting a rotor with unknown mass imbalance. *IEEE Trans. Contr. Syst. Technol.*, 4(5), 587-597.

[17] Betschon, F., & Schob, R. (1998). On-line adapted vibration control. Cambridge, MA. *in Proc. 6th Int. Symp. Magnetic Bearings*, Mass. Inst. Technol., 362-371.

[18] Hirata, M., Nonami, K., & Ohno, T. (1998). Robust control of a magnetic bearing system using constantly scaled H-infinity control. Cambridge, MA. *in Proc. 6th Int. Symp. Magnetic Bearings*, Mass. Inst. Technol., 713-722.

[19] Kanemitsu, Y., Ohsawa, M., & Marui, E. (1994). Comparison of control laws for magnetic levitation. Zuerich. *in Fourth International Symposium on Magnetic Bearings, ETH*, 13-18.

[20] Duan, G. R., & Howe, D. (2003). Robust magnetic bearing control via eigenstructure assignment dynamical compensation. *IEEE Trans. Contr. Syst. Technol.*, 11(2), 204-215.

[21] Zhuravlyov, Y. N. (2002). On LQ-control of magnetic bearing. *IEEE Trans. Contr. Syst. Technol.*, 8, 344-355.

[22] Steffani, H. F., Hofmann, W., & Cebulski, B. (1998). A controller for a magnetic bearing using the dynamic programming method of Bellman. Cambridge, MA. *in Proc. 6th Int. Symp. Magnetic Bearings*, Mass. Inst. Technol., 569-576.

[23] Li, L. (1998). On-line tuning of AMB controllers using genetic algorithms. Cambridge, MA. *in Proc. 6th Int. Symp. Magnetic Bearings*, Mass. Inst. Technol., 372-382.

[24] Hung, J. Y. (1995). Magnetic bearing control using fuzzy logic. *IEEE Trans. Ind. Applicat.*, 31, 1492-1497.

[25] Lindlau, J. D., & Knospe, C. R. (2002). Feedback linearization of an active magnetic bearing with voltage control. *IEEE Trans. Contr. Syst. Technol.*, 10, 21-30.

[26] Youcef-Toumi, K., & Reddy, S. (1992). Dynamic analysis and control of high speed and high precision active magnetic bearings. *ASME J. Dyn. Syst., Meas., Contr.*, 114, 623-633.

[27] Mizuno, T. (2001). Analysis on the fundamental properties of active magnetic bearing control systems by a transfer function approach. *JSME Int. J. Series C. Mech. Syst. Mach. Elements Manufact.*, 44, 367-373.

[28] Losch, F., Gahler, C., & Herzog, R. (1998, Aug. 5-7). synthesis controller design for a 3 MW pump running in AMBs. Cambridge, MA. *in Proc. 6th Int. Symp. Magnetic Bearings*, Mass. Inst. Technol., 415-428.

[29] Genta, G. (2005). *Dynamics of Rotating Systems*, Springer, New York, 1-658, 978-0-38720-936-4.

Chapter 3

Estimation and Active Damping of Unbalance Forces in Jeffcott-Like Rotor-Bearing Systems

Francisco Beltran-Carbajal,
Gerardo Silva-Navarro and Manuel Arias-Montiel

Additional information is available at the end of the chapter

1. Introduction

Nowadays, rotordynamics is a technological field in which research is very active because in spite of basic phenomena have been widely studied, there are many aspects that still need theoretical and practical work in order to construct and analyze models to represent with more precision the dynamic behavior of real machines. Dynamic studies in rotordynamics usually are preformed by numerical simulations using the mathematical models reported in literature. The mathematical description of rotating systems allows the possibility to predict their dynamic behavior and to use this information for the design of control algorithms in order to preserve the desired stability and dynamic performance. The accuracy of a model is determined by comparing its response and the response of the real system to the same input signal [19]. Rotor systems are subjected, in an intrinsic way, to endogenous disturbances, centrifugal forces by the inevitable unbalance phenomenon. Magnitude of these unbalance forces depends on the rotor mass, angular speed and distance between geometric center and center mass of rotor [10, 11, 29]. This last parameter is known as eccentricity and represents one of the most difficult parameter to measure or to estimate in a rotor system and consequently, it is an important aspect for the accuracy model.

The recent trends in rotordynamic systems are moving to higher speeds, higher powers, lighter and more compact machinery, which has resulted in machines operating above one or more critical speeds and increasing the vibration problems [5, 31]. In literature, the unbalance phenomenon has been widely reported as the main source of undesired vibration in rotating machinery [5, 10, 11, 29]. An unacceptable level of vibration can cause failure in the bearings, high levels of noise, wearing in the mechanical components and eventually, catastrophic failures in machines [10, 29], hence, control algorithms are needed to reduce the unbalance effects and to take vibrations amplitudes to acceptable values for a safe machine operation.

In literature, some different approaches to solve the problem of the model accuracy have been proposed. An estimation procedure to identify the distributed eccentricity along the shaft of a Jeffcott-like rotor was presented by Yang and Lin (2002). The eccentricity distributions of the shaft are assumed as polynomial of certain grade and the disk eccentricity is considered as lumped. Measurements in different locations along the shaft and to different rotational speeds are needed. By some numerical simulations, they found the method efficiency depends on the number of sensors available and on the number of operating speeds. Some other authors have proposed the problem solution in a lumped parameters approach. Maslen et al. (2002) developed a method to analytically adjust the models of rotor systems to make them consistent with experimental data under the assumption that the predominant uncertainties in the models occur at discrete points, from effects like seal coefficients or foundation interactions. The purpose of the method is to modify the engineering model such that the output of the model matches the experimental data in frequency domain. They show some examples to identify lumped stiffness at the supports and seal coefficients, but not the associated unbalance parameters and the results are presented through numerical simulations. De Queiroz (2009) presented a relatively simple feedback method to identify the unknown unbalance parameters of a Jeffcott rotor based on a dynamic robust control technique, in which the disturbance forces are estimated and then, from these forces, the magnitude and phase of the unbalance are obtained. This strategy is proved by numerical simulations and the rotational speed of the machine has to satisfy the persistency of excitation condition in order to guarantee the convergence of the method. Using curve fitting techniques and optimization procedures based on least-squares methods, Mahfoud et al. (2009) proposed a method to identify the matrices of a rotordynamic model expressed in state variables, measuring the full state vector (displacement, speed and acceleration) in three steps. The impulse response for a null rotational speed is used to identify the speed non-dependent matrix, the control matrix is identified using the steady-state response and the dependent dynamic matrix is calculated from the permanent time response of the system at an operational speed. Finally, the external forces can be found proposing an inverse problem from the model with the three matrices previously determined. Recently, some results in this issue have been published in specialized literature, Sudhakar and Sekhar (2011) estimated the unbalance faults in a Jeffcott-like rotor system with fault identification approach, obtaining good results in both numerical and experimental ways, showing the need of new methods and techniques to solve the unbalance forces estimation problem.

The developments in the fields of electronics, computing and control systems have changed the approach to reduce the level of vibration amplitudes in mechanical systems. In the traditional approach, changes in stiffness or damping system parameters cause changes in the dynamic system behavior. Nowadays, control systems can adapt dynamically these stiffness or damping parameters depending on the requirements or apply force directly to reduce the vibration effects. This trend is increasingly applied in rotating machinery and other fields of structural mechanics [11]. For control purposes, many passive, semi-active and active devices have been proposed [31]. Active Magnetic Bearings (AMB) have found an important field of application in rotor systems because the advantages over other devices. The absence of contact between an AMB and rotor avoids wearing and the need of lubrication, in addition, AMB dynamics is relatively easy to control [21]. For this, many researchers have reported results about showing the viability for AMB application in vibration control of rotating machinery since the 1990's [16, 26] until recent days [1, 13, 20, 28]. Piezoelectric actuators are other devices with an increasing application in rotor systems, they represent an alternative because their main characteristics: very precise movement, compact, high force, low energy consumption, quick response time, no electromagnetic interference. Some researchers have presented numerical and experimental results in the active control of unbalance response in rotors using piezoelectric actuators showing that it is possible to control vibration amplitudes by these devices [15, 17, 22]. Finally, another alternative to vibration control in rotor systems are the semi-active devices. A semi-active vibration control system replaces the actuators to apply directly force for devices which can change the stiffness and/or damping system parameters. Due to their low energy consumption, their application in theoretical and practical issues in mechanical systems tends to increase in the last years. Magnetorheological dampers represent the most used semi-active device in rotor systems [3, 9, 12].

Generally, a control scheme to vibration attenuation is designed using a system model, so that it is very important that the model represents to the real system behavior with good accuracy. As we mentioned above, in the models used to describe the dynamic behavior of rotor systems, the amount and location of unbalance are some of the most difficult parameters to be measured and, therefore, estimation techniques are needed to establish these and other parameters to get the required accuracy in the model. Observers (or estimators) can be designed, from measurements of the input and of the response of the system to provide an approximation of system states or disturbances that can not be directly measured [14]. Observers can be considered as subsystems that combine sensed signals with other knowledge of the control system to produce estimated signals and offer important advantages: they can remove sensors, which reduces cost and improves reliability, and improve the quality of signals that come from the sensors, allowing performance enhancement. However, observers have disadvantages: they can be complicated to implement and they expend computational resources. Also, because observers form software control loops, they can become unstable under certain conditions. Observers can also provide observed disturbance signals, which can be used to improve disturbance response. In spite of observers add complexity to the system and require computational resources, an observer applied with skill can bring substantial performance benefits and do so, in many cases, while reducing cost or increasing reliability [6].

This chapter deals with the active control problem of unbalance-induced synchronous vibrations in variable-speed Jeffcott-like non-isotropic rotor-bearing systems using only measurements of the radial displacement close to the disk. In this study, the rotor-bearing system is supported by a conventional bearing at its left end and by an active control device at the right one, which is used to provide the control forces. A robust and efficient active unbalance control scheme based on on-line compensation of rotor unbalance-induced perturbation force signals is proposed to suppress the undesirable vibrations affecting the rotor-bearing system dynamics. The methodology presented by Sira-Ramirez et al. (2008) is applied to design a Luenberger linear state observer to estimate the unbalance force signals and velocities of the coordinates of the rotor center, which are required to implement the proposed control scheme. The designed state observer is called the Generalized Proportional Integral (GPI) observer because its design approach is the dual counterpart of the so-called GPI controller [7]. A state-space based extended linear mathematical model is developed to locally describe the dynamics of the perturbed rotor-bearing system for design purposes of the disturbance observer. The modelling approach of disturbance signals through a family of Taylor time-polynomials of fourth degree described in [23] is used to locally reconstruct such unknown signals. A similar approach to reconstruct disturbance signals based also on its Taylor time-polynomial expansion has been previously proposed by Sira-Ramirez et al. (2007) using the on-line algebraic parameter identification methodology described in [8]. Additionally, a Proportional-Integral (PI) control law is designed to perform robust tracking tasks of smooth rotor speed reference profiles described by Bézier interpolation polynomials. Simulation results are provided to show the efficient and robust performance of the active vibration control scheme, estimation of the unbalance forces and rotor speed controller for the tracking of a speed reference profile that takes the rotor system from a rest initial speed to an operation speed above its first critical speeds.

2. Rotor system model

The rotor system in a Jeffcott configuration is shown in Fig. 1. The rotor is supported by a conventional bearing at its left end and by an active suspension at the right one.

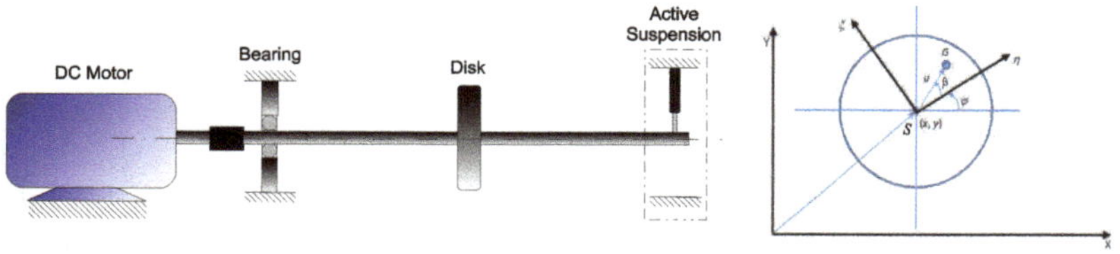

Figure 1. Jeffcott like rotor system with active suspension.

In this study, the active suspension presented by Arias-Montiel and Silva-Navarro (2010b) is considered. This control device is based on two linear electromechanical actuators and helicoidal compression springs. Electromechanical actuators provide the control forces in two perpendicular directions in order to compensate actively the unbalance effects and to get reductions in vibration amplitudes. The active suspension is depicted in Fig. 2.

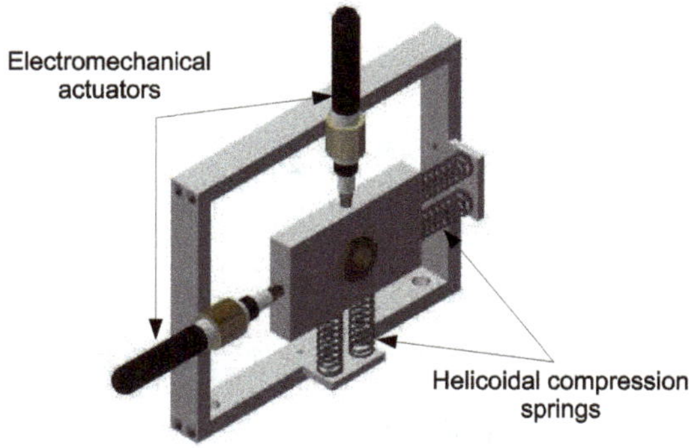

Figure 2. Active suspension with linear actuators.

The Jeffcott like rotor system consists of a disk with mass m mounted at the mid span of a flexible shaft. In Fig. 1, x and y denote the orthogonal coordinates of rotor geometric center, u is the distance between the gravity center G and the geometric center S, which is known as rotor eccentricity. Moreover, k_x, k_y, c_x and c_y are the shaft stiffness and viscous damping coefficients in x and y directions, respectively, φ and ω are the angular displacement and velocity, respectively, J is the polar moment of inertia of the rotor, c_φ is the rotational viscous damping coefficient and τ is the control torque to smoothly regulate the rotor speed through an traditional PID driver. Considering u_x and u_y as the radial control forces provided by the active suspension used to compensate the unbalance effects on the rotor in each movement plane and τ as the control torque provided by the motor, the rotor system model can be obtained by Euler-Lagrange formulation. Defining the coordinates of rotor gravity center x_G and y_G as

$$\begin{aligned} x_G &= x + u\cos(\varphi + \beta) \\ y_G &= y + u\sin(\varphi + \beta) \end{aligned} \tag{1}$$

and their time derivatives

$$\begin{aligned} \dot{x}_G &= \dot{x} - u\dot{\phi}(\sin\varphi + \beta) \\ \dot{y}_G &= \dot{y} + u\dot{\phi}(\cos\varphi + \beta) \end{aligned} \tag{2}$$

We can obtain the system kinetic energy as

$$T = \frac{1}{2}m\dot{x}_G^2 + \frac{1}{2}m\dot{y}_G^2 + \frac{1}{2}J\dot{\phi}^2 \tag{3}$$

and the potential energy as

$$V = \frac{1}{2}k_x x^2 + \frac{1}{2}k_y y^2 \tag{4}$$

So, the system Lagrangian is given by

$$L = T - V = \frac{1}{2}m\dot{x}_G^2 + \frac{1}{2}m\dot{y}_G^2 + \frac{1}{2}J\dot{\phi}^2 - \frac{1}{2}k_x x^2 - \frac{1}{2}k_y y^2 \tag{5}$$

and proposing the dissipation function of Rayleigh of the form

$$D = \frac{1}{2}c_x\dot{x}^2 + \frac{1}{2}c_y\dot{y}^2 + \frac{1}{2}c_\varphi\dot{\phi}^2 \tag{6}$$

Then, considering $\dot{\phi}=\omega$ the dynamics equations for the system can be achieved from

$$\begin{aligned}
\frac{d}{dt}\left[\frac{\partial L}{\partial \dot{x}}\right] - \frac{\partial L}{\partial x} + \frac{\partial D}{\partial \dot{x}} &= u_x \\
\frac{d}{dt}\left[\frac{\partial L}{\partial \dot{y}}\right] - \frac{\partial L}{\partial y} + \frac{\partial D}{\partial \dot{y}} &= u_y \\
\frac{d}{dt}\left[\frac{\partial L}{\partial \omega}\right] - \frac{\partial L}{\partial \phi} + \frac{\partial D}{\partial \omega} &= \tau
\end{aligned} \tag{7}$$

From equations (7), one obtains

$$\begin{aligned}
m\ddot{x} + c_x\dot{x} + k_x x - m\omega^2 u \cos(\varphi + \beta) - m\dot{\omega} u \sin(\varphi + \beta) &= u_x \\
m\ddot{y} + c_y\dot{y} + k_y y - m\omega^2 u \sin(\varphi + \beta) + m\dot{\omega} u \cos(\varphi + \beta) &= u_y \\
(J + mu^2)\dot{\omega} + c_\varphi \omega - m\ddot{x} u \sin(\varphi + \beta) + m\ddot{y} u \cos(\varphi + \beta) &= \tau
\end{aligned} \tag{8}$$

and rewriting these last equations

$$\begin{aligned}
m\ddot{x} + c_x\dot{x} + k_x x &= u_x + \xi_x \\
m\ddot{y} + c_y\dot{y} + k_y y &= u_y + \xi_y \\
J_e\dot{\omega} + c_\varphi \omega &= \tau + \xi_w \\
\dot{\varphi} &= \omega
\end{aligned} \tag{9}$$

with

$$\begin{aligned}
\xi_x &= mu\left[\dot{\omega}\sin(\varphi + \beta) + \omega^2 \cos(\varphi + \beta)\right] \\
\xi_y &= mu\left[-\dot{\omega}\cos(\varphi + \beta) + \omega^2 \sin(\varphi + \beta)\right] \\
\xi_w &= mu\left[\ddot{x}\sin(\varphi + \beta) - \ddot{y}\cos(\varphi + \beta)\right] \\
J_e &= J + mu^2
\end{aligned} \tag{10}$$

In the above, ξ_x, ξ_y and ξ_w are the centrifugal forces and perturbation torque, respectively, induced by the rotor unbalance.

Defining the state space variables as $z_1 = x$, $z_2 = \dot{x}$, $z_3 = y$, $z_4 = \dot{y}$, $z_5 = \varphi$, $z_6 = \dot{\varphi}$, the generalized forces as $u_1 = u_x$, $u_2 = u_y$ and $u_3 = \tau$ and the total unbalance amplitude y_u as system output, one obtains the following state-space description of system (8):

$$\begin{aligned}
\dot{z}_1 &= z_2 \\
\dot{z}_2 &= \frac{1}{\Delta} f_1 + \frac{1}{\Delta} g_1 \\
\dot{z}_3 &= z_4 \\
\dot{z}_4 &= \frac{1}{\Delta} f_2 + \frac{1}{\Delta} g_2 \\
\dot{z}_5 &= z_6 \\
\dot{z}_6 &= \frac{1}{\Delta} f_3 + \frac{1}{\Delta} g_3 \\
y_u &= \sqrt{z_1^2 + z_3^2}
\end{aligned} \tag{11}$$

where

$$\begin{aligned}
f_1 = & -k_x \left[\frac{1}{m} - \frac{u^2}{J_e} \cos^2 (z_5 + \beta) \right] z_1 - c_x \left[\frac{1}{m} - \frac{u^2}{J_e} \cos^2 (z_5 + \beta) \right] z_2 \\
& + \frac{u^2 k_y}{J_e} \cos (z_5 + \beta) \sin (z_5 + \beta) z_3 \\
& + \frac{c_y u^2}{J_e} \cos (z_5 + \beta) \sin (z_5 + \beta) z_4 - \frac{c_\varphi u}{J_e} z_6 \sin (z_5 + \beta) \\
& + \left[mu \left(\frac{1}{m} - \frac{u^2}{J_e} \cos^2 (z_5 + \beta) \right) \cos (z_5 + \beta) - \frac{mu^3}{J_e} \cos (z_5 + \beta) \sin^2 (z_5 + \beta) \right] z_6^2
\end{aligned}$$

$$g_1 = \left[\frac{1}{m} - \frac{u^2}{J_e} \cos^2 (z_5 + \beta) \right] u_x - \frac{u^2}{J_e} \cos (z_5 + \beta) \sin^2 (z_5 + \beta) u_y + \frac{u}{J_e} \sin (z_5 + \beta) \tau$$

$$\begin{aligned}
f_2 = & \frac{u^2 k_x}{J_e} \cos (z_5 + \beta) \sin (z_5 + \beta) z_1 + \frac{c_x u^2}{J_e} \cos (z_5 + \beta) \sin (z_5 + \beta) z_2 \\
& -k_y \left[\frac{1}{m} - \frac{u^2}{J_e} \sin^2 (z_5 + \beta) \right] z_3 - c_y \left[\frac{1}{m} - \frac{u^2}{J_e} \sin^2 (z_5 + \beta) \right] z_4 \\
& + \frac{c_\varphi u}{J_e} z_6 \cos (z_5 + \beta) + mu \left[\frac{1}{m} - \frac{u^2}{J_e} \sin^2 (z_5 + \beta) \right] \sin (z_5 + \beta) z_6^2 \\
& - \frac{mu^3}{J_e} \cos^2 (z_5 + \beta) \sin (z_5 + \beta) z_6^2
\end{aligned}$$

$$g_2 = -\frac{u^2}{J_e}\cos(z_5+\beta)\sin(z_5+\beta)\,u_x + \left[\frac{1}{m} - \frac{u^2}{J_e}\sin^2(z_5+\beta)\right]u_y$$
$$-\frac{u}{J_e}\cos(z_5+\beta)\,\tau$$

$$f_3 = -\frac{uk_x}{J_e}\sin(z_5+\beta)\,z_1 - \frac{c_x u}{J_e}\sin(z_5+\beta)\,z_2$$
$$+\frac{uk_y}{J_e}\cos(z_5+\beta)\,z_3 + \frac{c_y u}{J_e}\cos(z_5+\beta)\,z_4 - \frac{c_\varphi}{J_e}z_6$$

$$g_3 = \frac{u}{J_e}\sin(z_5+\beta)\,u_x - \frac{u}{J_e}\cos(z_5+\beta)\,u_y + \frac{\tau}{J_e}$$

$$\Delta = \frac{1}{J_e}\left(J_e - mu^2\right)$$

3. Active unbalance control

For the design of the active unbalance control scheme proposed in this chapter, consider the nonlinear ordinary differential equations that describe the dynamics of the rotor center, where only the position coordinates are available for measurement

$$\begin{aligned} m\ddot{x} + c_x\dot{x} + k_x x &= u_x + \xi_x \\ m\ddot{y} + c_y\dot{y} + k_y y &= u_y + \xi_y \end{aligned} \tag{12}$$

where

$$\begin{aligned} \xi_x &= mu\left[\dot{\omega}\sin(\varphi+\beta) + \omega^2\cos(\varphi+\beta)\right] \\ \xi_y &= mu\left[-\dot{\omega}\cos(\varphi+\beta) + \omega^2\sin(\varphi+\beta)\right] \end{aligned} \tag{13}$$

In our design approach, the unbalance forces ξ_x and ξ_y will be considered as unknown disturbance signals.

For the active unbalance suppression, Proportional-Derivative (PD) controllers with compensation of the rotor unbalance-induced disturbance signals are proposed

$$\begin{aligned} u_x &= -\alpha_{1,x}\widehat{\dot{x}} - \alpha_{0,x}x - \widehat{\xi}_x(t) \\ u_y &= -\alpha_{1,y}\widehat{\dot{y}} - \alpha_{0,y}y - \widehat{\xi}_y(t) \end{aligned} \tag{14}$$

where $\widehat{\xi}_x(t)$ and $\widehat{\xi}_y(t)$ are estimated perturbation signals of the actual time-varying unbalance forces $\xi_x(t)$ and $\xi_y(t)$, respectively, and $\widehat{\dot{x}}$ and $\widehat{\dot{y}}$ are estimates of the velocities of the rotor center in x and y directions, respectively.

In this chapter, an on-line estimation approach based on Luenberger linear estate observers is proposed to estimate the disturbance and velocity signals, using measurements of the rotor center coordinates (x, y), which could be obtained employing proximity sensors or accelerometers in practical applications. A state-space based extended linear mathematical model will be developed to locally describe the dynamics of the perturbed rotor-bearing system for design purposes of the disturbance observer. A family of Taylor time polynomials of fourth degree will be used to locally describe the unknown disturbance signals.

The use of the controllers (14) in the rotor-bearing system (12) yields the following closed-loop dynamics for the rotor center coordinates

$$\begin{aligned} \ddot{x} + \frac{1}{m}\left(c_x + \alpha_{1,x}\right)\dot{x} + \frac{1}{m}\left(k_x + \alpha_{0,x}\right)x = 0 \\ \ddot{y} + \frac{1}{m}\left(c_y + \alpha_{1,y}\right)\dot{y} + \frac{1}{m}\left(k_y + \alpha_{0,y}\right)y = 0 \end{aligned} \tag{15}$$

whose characteristic polynomials are given by

$$\begin{aligned} p_x(s) &= s^2 + \frac{1}{m}\left(c_x + \alpha_{1,x}\right)s + \frac{1}{m}\left(k_x + \alpha_{0,x}\right) \\ p_y(s) &= s^2 + \frac{1}{m}\left(c_y + \alpha_{1,y}\right)s + \frac{1}{m}\left(k_y + \alpha_{0,y}\right) \end{aligned} \tag{16}$$

Therefore, by selecting the controller gains $\alpha_{i,j}$, $i = 0, 1$, $j = x, y$, so that the characteristic polynomials (16) be Hurwitz, one can guarantee that the dynamics of the rotor center coordinates be globally asymptotically stable, i.e., $\lim_{t\to\infty} x(t) = 0$ and $\lim_{t\to\infty} y(t) = 0$. It can be observed that a consequence of the unbalance cancellation is that the rotor unbalance-induced perturbation torque signal ξ_ω affecting additively the rotor velocity dynamics is also suppressed, i.e., $\lim_{t\to\infty} \xi_\omega(t) = 0$.

For the closed-loop dynamics of the coordinates of the rotor center, the following reference system is proposed

$$\begin{aligned} \ddot{x} + 2\zeta_x\omega_{nx}\dot{x} + \omega_{nx}^2 x &= 0 \\ \ddot{y} + 2\zeta_x\omega_{nx}\dot{y} + \omega_{nx}^2 y &= 0 \end{aligned} \tag{17}$$

where ζ_i, $\omega_{ni} > 0$, $i = x, y$, are the viscous damping ratios and natural frequencies for the rotor center dynamics. Then, the gains of the controllers (14) are calculated as

$$\begin{aligned} \alpha_{1,x} &= 2m\zeta_x\omega_{nx} - c_x \\ \alpha_{1,y} &= 2m\zeta_y\omega_{ny} - c_y \\ \alpha_{0,x} &= m\omega_{nx}^2 - k_x \\ \alpha_{0,y} &= m\omega_{ny}^2 - k_y \end{aligned}$$

Otherwise, a Proportional-Integral (PI) control law is proposed for tracking tasks of an angular speed profile $\omega^*(t)$ specified for the rotor system

$$\begin{aligned} \tau &= J_e v + c_\varphi \omega \\ v &= \dot{\omega}^*(t) - \alpha_{1,\omega}\left[\omega - \omega^*(t)\right] - \alpha_{0,\omega}\int_0^t \left[\omega - \omega^*(t)\right] dt \end{aligned} \tag{18}$$

By replacing the control law (18) into the rotor-bearing system (9), it is obtained the homogenous differential equation that describes the dynamics of the angular speed tracking error $e_\omega = \omega - \omega^*(t)$, under the assumption that the unbalance was canceled by the action of the PD control forces (14),

$$\ddot{e}_\omega + \alpha_{1,\omega}\dot{e}_\omega + \alpha_{0,\omega}e_\omega = 0 \tag{19}$$

Then, the asymptotic convergence of the tracking error e_ω to zero can be achieved selecting the design parameters $\alpha_{0,\omega}$ and $\alpha_{1,\omega}$ such as the characteristic polynomial associated to tracking error dynamics in closed loop (19) given by

$$p_\omega(s) = s^2 + \alpha_{1,\omega}s + \alpha_{0,\omega} \tag{20}$$

be a Hurwitz polynomial. In this case, the asymptotic tracking of the specified angular speed profile can be verified, i.e.,

$$\lim_{t\to\infty} e_\omega(t) = 0 \Rightarrow \lim_{t\to\infty} \omega(t) = \omega^*(t)$$

In this chapter, the following Hurwitz polynomial is proposed for the closed-loop rotor angular speed dynamics

$$p_{\omega d}(s) = s^2 + 2\zeta_r\omega_{nr}s + \omega_{nr}^2 \quad (21)$$

where ζ_r and $\omega_{nr} > 0$ are the viscous damping ratio and natural frequency for rotor angular speed dynamics. Then, the gains of the controller (18) are calculated as

$$\begin{aligned} \alpha_{0,\omega} &= \omega_{nr}^2 \\ \alpha_{1,\omega} &= 2\zeta_r\omega_{nr} \end{aligned}$$

4. Asymptotic estimation of unbalance forces

In the design process of the disturbance observer, it is assumed that the perturbation force signals ξ_x and ξ_y can be locally approximated by a family of fourth degree Taylor time-polynomials [23]:

$$\xi_i(t) = \sum_{j=0}^{4} p_{j,i}t^j, \quad i = x, y \quad (22)$$

where the coefficients $p_{j,i}$ are completely unknown.

The perturbation signals can then be locally described by the following state-space based linear mathematical model:

$$\begin{aligned} \dot{\xi}_{1,i} &= \xi_{2,i} \\ \dot{\xi}_{2,i} &= \xi_{3,i} \\ \dot{\xi}_{3,i} &= \xi_{4,i} \\ \dot{\xi}_{4,i} &= \xi_{5,i} \\ \dot{\xi}_{5,i} &= 0 \end{aligned} \quad (23)$$

where $\xi_{1,i} = \xi_i$, $\xi_{2,i} = \dot{\xi}_i$, $\xi_{3,i} = \ddot{\xi}_i$, $\xi_{4,i} = \xi_i^{(3)}$, $\xi_{5,i} = \xi_i^{(4)}$, $i = x, y$.

Therefore, an extended state space model for the perturbed rotor center dynamics is given by

$$
\begin{aligned}
\dot{\eta}_{1,i} &= \eta_{2,i} \\
\dot{\eta}_{2,i} &= -\frac{k_i}{m}\eta_{1,i} - \frac{c_i}{m}\eta_{2,i} + \frac{1}{m}u_i + \frac{1}{m}\xi_{1,i} \\
\dot{\xi}_{1,i} &= \xi_{2,i} \\
\dot{\xi}_{2,i} &= \xi_{3,i} \\
\dot{\xi}_{3,i} &= \xi_{4,i} \\
\dot{\xi}_{4,i} &= \xi_{5,i} \\
\dot{\xi}_{5,i} &= 0
\end{aligned}
\tag{24}
$$

where $\eta_{1,i} = i$, $\eta_{2,i} = \dot{\eta}_{1,i}$, $i = x, y$.

From system (24), the following Luenberger linear state observer is proposed to estimate the disturbance and rotor center velocity signals

$$
\begin{aligned}
\dot{\widehat{\eta}}_{1,i} &= \widehat{\eta}_{2,i} + \beta_{6,i}\left(\eta_{1,i} - \widehat{\eta}_{1,i}\right) \\
\dot{\widehat{\eta}}_{2,i} &= -\frac{k_i}{m}\widehat{\eta}_{1,i} - \frac{c_i}{m}\widehat{\eta}_{2,i} + \frac{1}{m}u_i + \frac{1}{m}\widehat{\xi}_{1,i} + \beta_{5,i}\left(\eta_{1,i} - \widehat{\eta}_{1,i}\right) \\
\dot{\widehat{\xi}}_{1,i} &= \widehat{\xi}_{2,i} + \beta_{4,i}\left(\eta_{1,i} - \widehat{\eta}_{1,i}\right) \\
\dot{\widehat{\xi}}_{2,i} &= \widehat{\xi}_{3,i} + \beta_{3,i}\left(\eta_{1,i} - \widehat{\eta}_{1,i}\right) \\
\dot{\widehat{\xi}}_{3,i} &= \widehat{\xi}_{4,i} + \beta_{2,i}\left(\eta_{1,i} - \widehat{\eta}_{1,i}\right) \\
\dot{\widehat{\xi}}_{4,i} &= \widehat{\xi}_{5,i} + \beta_{1,i}\left(\eta_{1,i} - \widehat{\eta}_{1,i}\right) \\
\dot{\widehat{\xi}}_{5,i} &= \beta_{0,i}\left(\eta_{1,i} - \widehat{\eta}_{1,i}\right)
\end{aligned}
\tag{25}
$$

The dynamics of the estimation errors, $e_{1,i} = \eta_{1,i} - \widehat{\eta}_{1,i}$, $e_{2,i} = \eta_{2,i} - \widehat{\eta}_{2,i}$, $e_{p_k,i} = \xi_{k,i} - \widehat{\xi}_{k,i}$, $k = 1, 2, \cdots, 5$, $i = x, y$, are then given by

$$
\begin{aligned}
\dot{e}_{1,i} &= -\beta_{6,i}e_{1,i} + e_{2,i} \\
\dot{e}_{2,i} &= -\beta_{5,i}e_{1,i} - \frac{k_i}{m}e_{1,i} - \frac{c_i}{m}e_{2,i} + \frac{1}{m}e_{p_1,i} \\
\dot{e}_{p_1,i} &= -\beta_{4,i}e_{1,i} + e_{p_2,i} \\
\dot{e}_{p_2,i} &= -\beta_{3,i}e_{1,i} + e_{p_3,i} \\
\dot{e}_{p_3,i} &= -\beta_{2,i}e_{1,i} + e_{p_4,i} \\
\dot{e}_{p_4,i} &= -\beta_{1,i}e_{1,i} + e_{p_5,i} \\
\dot{e}_{p_5,i} &= -\beta_{0,i}e_{1,i}
\end{aligned}
\tag{26}
$$

Thus, the characteristic polynomials of the dynamics of the observation errors (26) are

$$p_{o,i}(s) = s^7 + \left(\beta_{6,i} + \frac{c_i}{m}\right)s^6 + \left(\beta_{5,i} + \frac{k_i}{m} + \frac{c_i}{m}\beta_{6,i}\right)s^5 + \frac{1}{m}\beta_{4,i}s^4 + \frac{1}{m}\beta_{3,i}s^3 + \frac{1}{m}\beta_{2,i}s^2 + \frac{1}{m}\beta_{1,i}s + \frac{1}{m}\beta_{0,i} \tag{27}$$

which are completely independents of any coefficients $p_{j,i}$ of the Taylor polynomial expansions of disturbance signals $\xi_i(t)$.

The design parameter for the state observer (25) are selected so that the characteristic polynomials (27) be Hurwitz polynomials. Particularly, these polynomials are proposed of the form

$$p_{o,i}(s) = (s + p_{o,i})\left(s^2 + 2\zeta_{o,i}\omega_{o,i}s + \omega_{o,i}^2\right)^3, \quad i = x, y. \tag{28}$$

with $p_{o,i}, \zeta_{o,i}, \omega_{o,i} > 0$.

Equating term by term the coefficients of both polynomials (28) and (27), one obtains that

$$\begin{aligned}
\beta_{0,i} &= m\omega_{o,i}^6 p_{o,i} \\
\beta_{1,i} &= m\left(\omega_{o,i}^6 + 6\zeta_{o,i}p_{o,i}\omega_{o,i}^5\right) \\
\beta_{2,i} &= m\left(6\omega_{o,i}^5\zeta_{o,i} + 12p_{o,i}\omega_{o,i}^4\zeta_{o,i}^2 + 3p_{o,i}\omega_{o,i}^4\right) \\
\beta_{3,i} &= m\left(12\omega_{o,i}^4\zeta_{o,i}^2 + 3\omega_{o,i}^4 + 8p_{o,i}\omega_{o,i}^3\zeta_{o,i}^3 + 12p_{o,i}\omega_{o,i}^3\zeta_{o,i}\right) \\
\beta_{4,i} &= m\left(8\omega_{o,i}^3\zeta_{o,i}^3 + 12\omega_{o,i}^3\zeta_{o,i} + 12p_{o,i}\omega_{o,i}^2\zeta_{o,i}^2 + 3p_{o,i}\omega_{o,i}^2\right) \\
\beta_{5,i} &= 12\omega_{o,i}^2\zeta_{o,i}^2 + 3\omega_{o,i}^2 + 6p_{o,i}\omega_{o,i}\zeta_{o,i} - \frac{c_i}{m}\beta_{6,i} - \frac{k_i}{m} \\
\beta_{6,i} &= p_{o,i} + 6\omega_{o,i}\zeta_{o,i} - \frac{c_i}{m}
\end{aligned}$$

5. Simulation results

In order to verify the dynamic behavior of the rotor speed controller, active unbalance control scheme and estimation of the unbalance forces, some numerical simulations were carried out using the numerical parameters shown in Table 1.

The performance of the rotor speed controller (18) was evaluated for the tracking of the smooth speed reference profile $\omega^*(t)$ shown in Fig. 3, which allows to take the rotor from

$m = 3.85$ kg	$c_y = 14$ N s/m	$u = 222\ \mu$m
$k_x = 1.9276 \times 10^5$ N/m	$d = 0.020$ m	$\beta = \frac{\pi}{4}$ rad
$c_x = 12$ N s/m	$r_{disk} = 0.076$ m	$c_\varphi = 1.5 \times 10^{-3}$ Nm s/rad
$k_y = 2.0507 \times 10^5$ N/m	$l = 0.7293$ m	

Table 1. Rotor System Parameters.

an initial speed $\bar{\omega}_1$ for $t \leq T_1$ to the desired final operation speed $\bar{\omega}_2$ for $t \geq T_2$. In general, the unbalance response has more interest when the rotor is running above its first critical speeds $\omega_{cr1x} = \sqrt{k_x/m} = 223.76$ rad/s $= 2136.8$ rpm and $\omega_{cr1y} = \sqrt{k_y/m} = 230.79$ rad/s $=$ 2203.9 rpm.

The speed profile specified for the rotor system is described by

$$\omega^*(t) = \begin{cases} \bar{\omega}_1 \text{ for } 0 \leq t < T_1 \\ \bar{\omega}_1 + (\bar{\omega}_2 - \bar{\omega}_1)\,\psi(t, T_1, T_2) \text{ for } T_1 \leq t \leq T_2 \\ \bar{\omega}_2 \text{ for } t > T_2 \end{cases} \tag{29}$$

where $\bar{\omega}_1 = 0$ rad/s, $\bar{\omega}_2 = 300$ rad/s $= 2864.8$ rpm, $T_1 = 0$ s, $T_2 = 10$ s and $\psi(t, T_1, T_2)$ is a Bézier polynomial defined as

$$\psi(t) = \left(\frac{t - T_1}{T_2 - T_1}\right)^5 \left[r_1 - r_2\left(\frac{t - T_1}{T_2 - T_1}\right) + r_3\left(\frac{t - T_1}{T_2 - T_1}\right)^2 - \ldots - r_6\left(\frac{t - T_1}{T_2 - T_1}\right)^5\right]$$

with constants $r_1 = 252$, $r_2 = 1050$, $r_3 = 1800$, $r_4 = 1575$, $r_5 = 700$ and $r_6 = 126$.

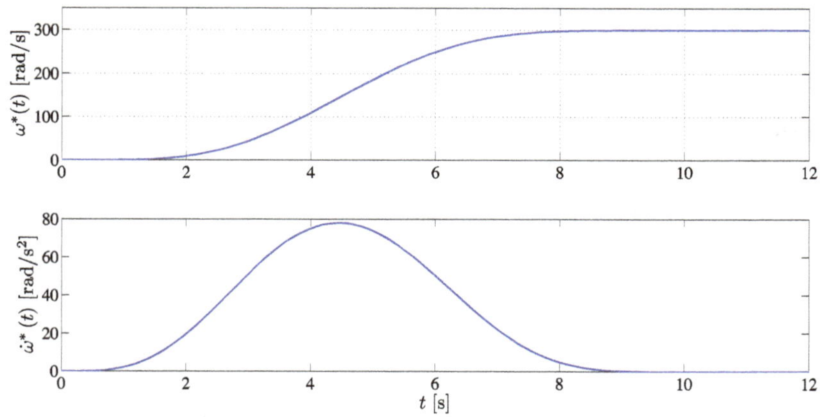

Figure 3. Smooth reference profiles for the rotor speed and acceleration, $\omega^*(t)$ and $\dot{\omega}^*(t)$.

In Fig. 4 the robust and efficient performance of the PI speed controller (18) is shown. Here, the active unbalance control scheme (14) is not performed, i.e., $u_x = u_y \equiv 0$. Therefore, some irregularities of the control torque action can be observed when the rotor passes through its first critical speeds. Fortunately, the presented speed controller results quite robust against the bounded torque perturbation input signal ξ_w induced by the rotor unbalance.

It is important to note that the control gains were selected to get a closed-loop rotor speed dynamics having a Hurwitz characteristic polynomial:

$$P_\omega(s) = s^2 + 2\zeta_r \omega_{nr} s + \omega_{nr}^2$$

with $\omega_{nr} = 15$ rad/s and $\zeta_r = 0.7071$.

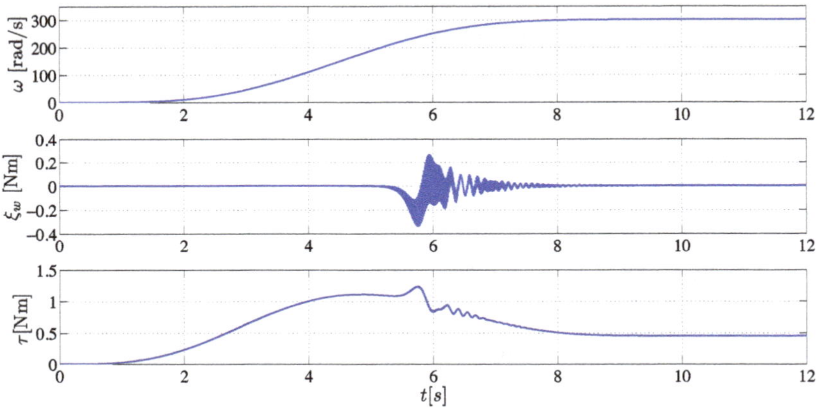

Figure 4. Rotor system response using local PI speed controller without active unbalance control.

Fig. 5 depicts the open-loop rotor unbalance response while the rotor is taken from the rest initial speed ($\bar{\omega}_1 = 0$ rad/s) to the operating speed ($\bar{\omega}_2$= 300 rad/s) above its first critical speeds by using the PI rotor speed controller (18). The presence of high vibration amplitude levels (above 9 mm) at the resonant peaks can be observed. Note in Fig. 6 that the centrifugal forces induced by the rotor unbalance are quite significant. Thus, the active rotor balancing controllers (14) should be actively compensate those perturbation forces in real time.

On the other hand, Figs. 7-9 depict the closed-loop rotor-bearing system response by using simultaneously the disturbance observer-based active unbalance control scheme (14), PI rotor speed controller (18) and disturbance observer (25). One can see in Fig. 7 the robust performance of the PI speed controller (18), achieving an effective tracking of the smooth speed reference profile (29). Since the rotor unbalance-induced torque perturbation input signal ξ_w is canceled by the active balancing controllers (14), a smooth curve of the control torque is accomplished, eliminating the irregularities presented in the control torque response without active unbalance control (4).

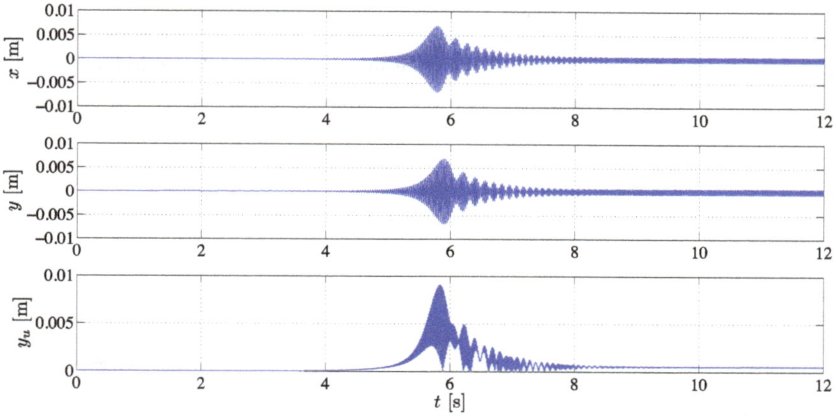

Figure 5. Open-loop rotor unbalance response with local PI rotor speed controller.

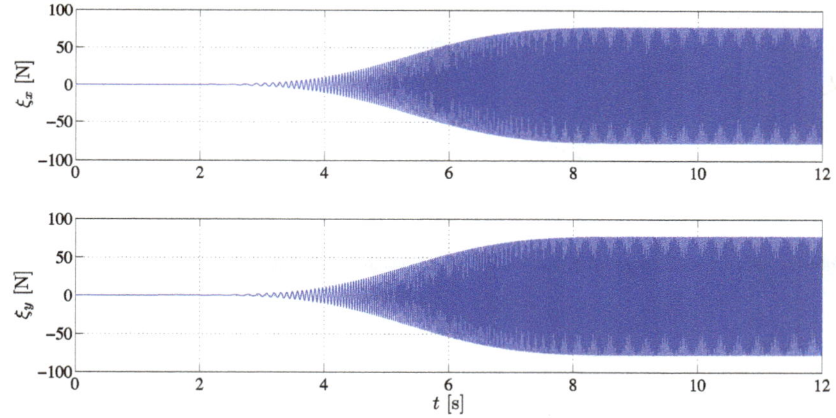

Figure 6. Rotor unbalance forces without active unbalance control.

The closed-loop rotor unbalance response is described in Fig. 8. The active unbalance suppression can be clearly noted. In this case, the gains of the active unbalance controllers were selected to get a closed-loop system dynamics having the Hurwitz characteristic polynomials:

$$P_x(s) = s^2 + 2\zeta_x\omega_{nx}s + \omega_{nx}^2$$
$$P_y(s) = s^2 + 2\zeta_y\omega_{ny}s + \omega_{ny}^2$$

with $\omega_{nx} = \omega_{ny} = 10$ rad/s and $\zeta_x = \zeta_x = 0.7071$.

Fig. (9) describes the active vibration control scheme response, which applies the active compensation of the estimated unbalance force signals $\widehat{\xi}_x$ and $\widehat{\xi}_y$ shown in Fig. (10).

The characteristic polynomials, assigned to the observation error dynamics, must be faster than the rotordynamics and, therefore, are specified as

$$P_{o,i}(s) = (s + p_{o,i}) \left(s^2 + 2\zeta_{o,i}\omega_{o,i}s + \omega_{o,i}^2\right)^3, \quad i = x, y.$$

with desired parameters $p_{o,i} = \omega_{o,i} = 1200$ rad/s and $\zeta_{o,i} = 100$.

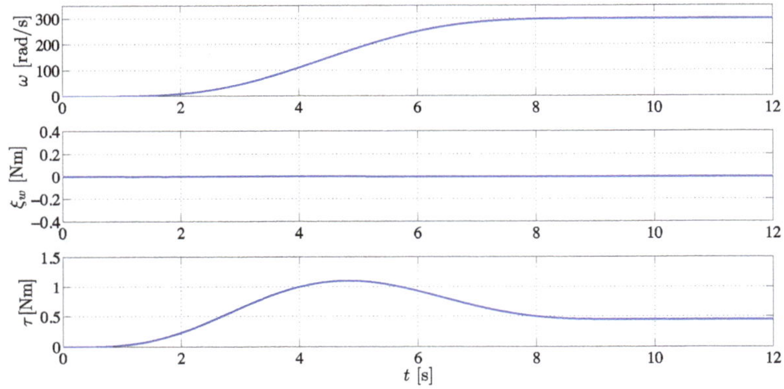

Figure 7. Rotor system response using local PI speed controller with active unbalance control.

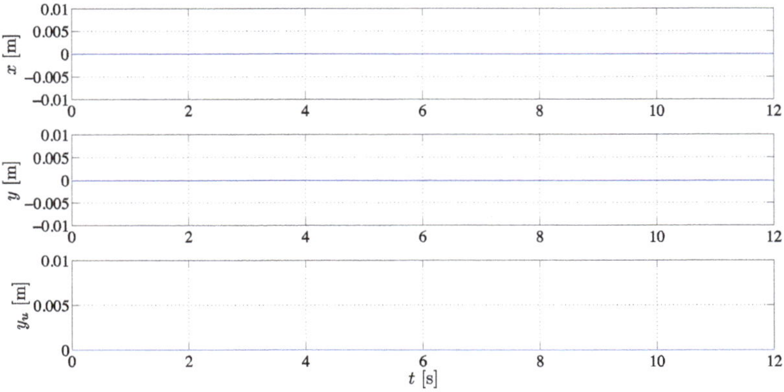

Figure 8. Closed-loop rotor unbalance response with local PI rotor speed controller.

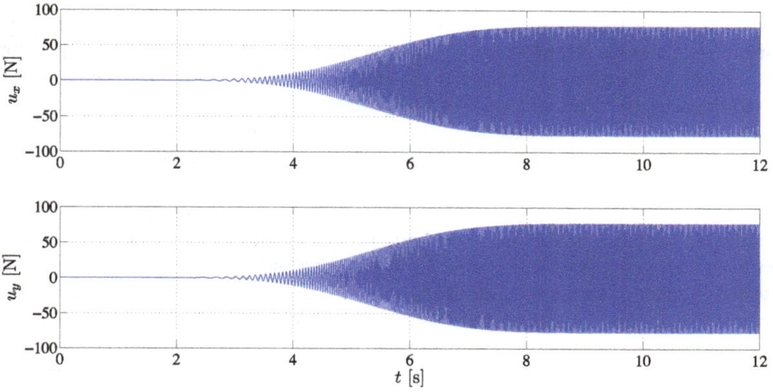

Figure 9. Response of active unbalance Controllers.

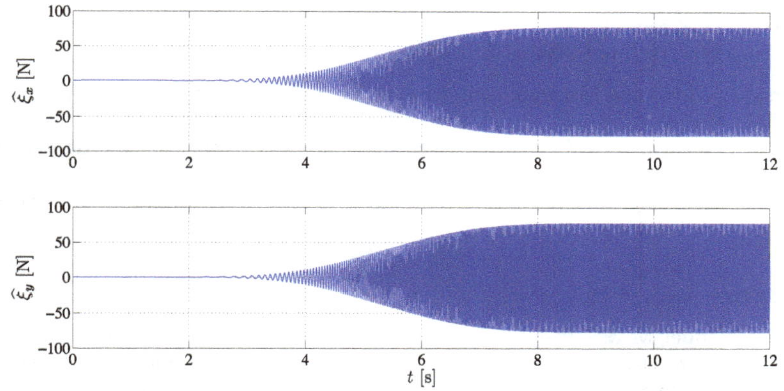

Figure 10. Estimates of the closed-loop rotor unbalance force signals using disturbance observer (25).

6. Conclusions

In this chapter, we have proposed a PD-like active vibration control scheme for robust and efficient suppression of unbalance-induced synchronous vibrations in variable-speed Jeffcott-like non-isotropic rotor-bearing systems of three degrees of freedom using only measurements of the radial displacement close to the disk. In this study, we have considered the application of an active suspension device, which is based on two linear electromechanical actuators and helicoidal compression springs, to provide the control forces required for on-line balance of the rotor system. The presented control approach is mainly based on the compensation of bounded perturbation force signals induced by the rotor unbalance, and the specification of the desired closed-loop rotor-bearing system dynamics (viscous damping ratios and natural frequencies). A robust and fast estimation scheme of the perturbation force signals and velocities of the rotor center coordinates based on Luenberger linear state observers has also been proposed. In the state observer design process, the perturbation force signals were locally approximated by a family of Taylor time-polynomials of fourth degree. Therefore, each perturbation signal was locally described by a state space-based linear mathematical model of fifth order. Then, an extended lineal mathematical model was obtained to locally describe the dynamics of the perturbed rotor system to be used in the design of the disturbance and state observer. In addition, a PI rotor speed controller was proposed to perform robust tracking tasks of smooth rotor speed reference profiles described by Bézier interpolation polynomials. Simulations results show the robust and efficient performance of the active vibration control scheme and rotor speed controller proposed in this chapter, as well as the fast and effective estimation of the perturbation force signals, when the rotor system is taken from a rest initial speed to an operation speed above its first critical velocities. The proposed methodology can be applied for more complex and realistic rotor-bearing systems (e.g., more disks, turbines, shaft geometries), finite element models, monitoring and fault diagnosis quite common in industrial rotating machinery.

Author details

Francisco Beltran-Carbajal[1],
Gerardo Silva-Navarro[2] and Manuel Arias-Montiel[3]

1 Universidad Autonoma Metropolitana, Unidad Azcapotzalco, Departamento de Energia, Mexico, D.F., Mexico
2 Centro de Investigacion y de Estudios Avanzados del I.P.N., Departamento de Ingenieria Electrica, Seccion de Mecatronica, Mexico, D.F., Mexico
3 Universidad Tecnologica de la Mixteca, Instituto de Electronica y Mecatronica, Huajuapan de Leon, Oaxaca, Mexico

References

[1] Arias-Montiel, M. & Silva-Navarro, G. (2010). Active Unbalance Control in a Two Disks Rotor System Using Lateral Force Actuators, *Proceeding of 7th International Conference on Electrical Engineering, Computing Science and Automatic Control*, pp. 440-445, Tuxtla Gutierrez, México.

[2] Arias-Montiel, M. & Silva-Navarro, G. (2010b). Finite Element Modelling and Unbalance Compensation for an Asymmetrical Rotor-Bearing System with Two Disks, *In: New Trends in Electrical Engineering, Automatic Control, Computing and Communication Sciences, Edited by C.A. Coello-Coello, Alex Pozniak, José A. Moreno-Cadenas and Vadim Azhmyakov*, pp. 127-141, Logos Verlag Berlin GmbH, Germany.

[3] Cabrera-Amado, M. & Silva-Navarro, G. (2010). Semiactive Control for the Unbalance Compensation in a Rotor-Bearing System, *In: New Trends in Electrical Engineering, Automatic Control, Computing and Communication Sciences, Edited by C.A. Coello-Coello, Alex Pozniak, José A. Moreno-Cadenas and Vadim Azhmyakov*, pp. 143-158, Logos Verlag Berlin GmbH, Germany.

[4] De Queiroz, M. S. (2009). An Active Identification Method of Rotor Unbalance Parameters, *Journal of Vibration and Control*, Vol. 15, pp. 1365-1374.

[5] De Silva, C. W. (2007). *Vibration Damping, Control and Design*, CRC Press, USA.

[6] Ellis, G. (2002). *Observers in Control Systems*, Academic Press, USA.

[7] Fliess, M., Marquez, R., Delaleau, E.& Sira-Ramirez, H. (2002). Correcteurs Proportionnels-Integraux Généralisés, *ESAIM Control, Optimisation and Calculus of Variations*, Vol. 7, No. 2, pp. 23-41.

[8] Fliess, M. & Sira-Ramirez, H. (2003). An algebraic framework for linear identification, *ESAIM: Control, Optimization and Calculus of Variations*, Vol. 9, pp. 151-168.

[9] Forte, P., Paterno, M. & Rustighi, E. (2004). A Magnetorheological Fluid Damper for Rotor Applications, *International Journal of Rotating Machinery*, Vol. 10, No. 3, pp. 175-182.

[10] Friswell, M. I., Penny, J. E. T., Garvey, S. D. & Lees, A. W. (2010). *Dynamics of Rotating Machines*, Cambridge University Press, USA.

[11] Genta, G. (2009). *Vibration Dynamics and Control*, Springer, Berlin, Germany.

[12] Guldbakke, J. M. & Hesselbach, J. (2006). Development of Bearings and a Damper Based on Magnetically Controllable Fluids, *Journal of Physics: Condensed Matter*, Vol. 18, pp. S2959-S2972.

[13] Gosiewski, Z. & Mystkowsky, A. (2008). Robust Control of Active Magnetic Suspension: Analytical and Experimental Results, *Mechanical Systems and Signal Processing*, Vol. 22, No. 6, pp. 1297-1303.

[14] Inman, D. J. (2006). *Vibration with Control*, John Wiley & Sons, England.

[15] Jung-Ho, P., Young-Bog, H., So-Nam, Y. & Hu-Seung, L. (2010). Development of a Hybryd Bearing Using Permanent Magnets and Piezoelectric Actuators, *Journal of Korean Physical Society*, Vol. 57, No. 4, pp. 907-912.

[16] Knospe, C. R, Hope, R. W., Fedigan, S. J. & Williams, R. D. (1995). Experiments in the Control of Unbalance Response Using Magnetic Bearings, *Mechatronics*, Vol. 5, No. 4, pp. 385-400.

[17] Lihua, Y., Yanhua, S. & Lie, Y. (2012). Active Control of Unbalance Response of Rotor Systems Supported by Tilting-Pad Gas Bearings, *Proceedings of the Institution of Mechanical Engineers, Part J: Journal of Engineering Tribology*, Vol. 226, No. 2, pp. 87-98.

[18] Mahfoud J., Der Hagopian J., Levecque N. & Steffen Jr. V. (2009), Experimental model to control and monitor rotating machines, *Mechanism and Machine Theory*, Vol. 44, pp. 761-771 .

[19] Maslen, E. H., Vázquez, J. A. & Sortore, C. K. (2002). Reconciliation of Rotordynamic Models with Experimental Data, *Journal of Engineering for Gas Turbines and Power*, Vol. 124, pp. 351-356.

[20] Pi-Cheng, T., Mong-Tao, T, Kuan-Yu, C., Yi-Hua, F. & Fu-Chu, C. (2011). Design of Model-Based Unbalance Compensator with Fuzzy Gain Tuning Mechanism for an Active Magnetic Bearing System, *Expert Systems with Applications*, Vol. 38, No. 10, pp. 12861-12868.

[21] Schweitzer, G. & Maslen, E. H. (Eds) (2010). *Magnetic Bearings - Theory, Design and Application to Rotating Machinery*, Springer, Germany.

[22] Simoes, R. C., Steffen Jr., V., Der Hagopian, J. & Mahfoud, J. (2007). Modal Active Vibration Control of a Rotor Using Piezoelectric Stack Actuators, *Journal of Vibration and Control*, Vol. 13, No. 1, pp. 45-64.

[23] Sira-Ramirez, H., Beltran-Carbajal, F. & Blanco-Ortega, A. (2008). A Generalized Proportional Integral Output Feedback Controller for the Robust Perturbation Rejection in a Mechanical System, *e-STA*, Vol. 5, No. 4, pp. 24-32.

[24] Sira-Ramirez, H., Feliu-Batlle, V., Beltran-Carbajal, F. & Blanco-Ortega, A. (2008). Sigma-Delta modulation sliding mode observers for linear systems subject to locally unstable inputs, *16th Mediterranean Conference on Control and Automation*, pp. 344-349, Ajaccio, France, June 25-27.

[25] Sira-Ramirez, H., Silva-Navarro, G. & Beltran-Carbajal, F. (2007). On the GPI Balancing Control of an Uncertain Jeffcott Rotor Model, *Proceeding of 4th International Conference on Electrical and Electronics Engineering (ICEEE 2007)*, pp. 306-309, Mexico city, Mexico.

[26] Sivrioglu, S. & Nonami, K. (1998). Sliding Mode Control with Time Varying Hyperplene for AMB Systems, *IEEE/ASME Transactions on Mechatronics*, Vol. 3, No. 1, pp. 51-59.

[27] Sudhakar, G. N. D. S. & Sekhar, A. S. (2011). Identification of Unbalance in a Rotor Bearing System, *Journal of Sound and Vibration*, Vol. 330, pp. 2299-2313.

[28] Tammi, K. (2009). Active Control of Rotor Vibrations by Two Feedforward Control Algorithms, *Journal of Dynamic Systems, Measurment, and Control*, Vol. 131, pp. 1-10.

[29] Vance, J. M., *Rotordynamics of Turbomachinery* (1988). John Wiley & Sons, USA.

[30] Yang, T. & Lin, C. (2002). Estimation of Distribuited Unbalance of Rotors, *Journal of Engineering for Gas Turbines and Power*, Vol. 124, pp. 976-983.

[31] Zhou, S. & Shi, J. (2001). Active Balancing and Vibration Control of Rotating Machinery: A Survey, *The Shock and Vibration Diggest*, Vol. 33, No. 4, pp. 361-371.

Chapter 4

Vibration Control of Flexible Structures Using Semi-Active Mount : Experimental Investigation

Seung-Bok Choi, Sung Hoon Ha and Juncheol Jeon

Additional information is available at the end of the chapter

1. Introduction

In many dynamic systems such as robot and aerospace areas, flexible structures have been extremely employed to satisfy various requirements for large scale, light weight and high speed in dynamic motion. However, these flexible structures are readily susceptible to the internal/external disturbances (or excitations). Therefore, vibration control schemes should be exerted to achieve high performance and stability of flexible structure systems. Recently, in order to successfully achieve vibration control for flexible structures smart materials such as piezoelectric materials [1-2], shape memory alloys [3-4], electrorheological (ER) fluids [5-6] and magnetorheological (MR) fluids [7] are being widely utilized. Among these smart materials, ER or MR fluid exhibits reversible changes in material characteristics when subjected to electric or magnetic field. The vibration control of flexible structures using the smart ER or MR fluid can be achieved from two different methods. The first approach is to replace conventional viscoelastic materials by the ER or MR fluid. This method is very effective for shape control of flexible structures such as plate [5]. The second approach is to devise dampers or mounts and apply to vibration control of the flexible structures. This method is very useful to isolate vibration of large structural systems subjected to external excitations [6-7]. In this work, a new type of MR mount is proposed and applied to vibration control of the flexible structures.

In order to reduce unwanted vibration of the flexible structure system, three different types of mounts are normally employed: passive, semi-active and active. The passive rubber mount, which has low damping, shows efficient vibration performance at the non-resonant and high frequency excitation. Thus, the rubber mount is the most popular method applied for various vibrating systems. However, it cannot have a favorable performance due to small damping effect at the resonant frequency excitation. On the other hand, the passive hydraulic mount has been developed to utilize dynamic absorber effect or meet large damping requirement in the resonance of low frequency domain [8]. However, the high dynamic stiffness property of the hydraulic mount may deteriorate isolation performance in the non-resonant excitation domain. Thus, the damping and stiffness of the passive mounts are not simultaneously controllable to meet imposed performance criteria in a wide frequency range. The active mounts are normally operated by using external energy supplied by actuators in order to generate control forces on the system subjected to excitations [9]. The control performance of the active mount is fairly good in a wide frequency range, but its cost is expensive. Moreover, its configuration is complex and its stability may not be guaranteed in a certain operation condition. On the other hand, the semi-active mounts cannot inject mechanical energy into the structural systems. But, it can adjust damping to reduce unwanted vibration of the flexible structure systems. It is known that using the controllable yield stress of ER or MR fluid, a very effective semi-active mount can be devised for vibration control of the flexible structures. The flow operation of the ER or MR mount can be classified into three different modes: shear mode [6], flow mode [10] and squeeze mode [11].

In this article, a new type of semi-active MR mount shown in the figure 1 is proposed and applied to vibration control of flexible structures. As a first step to achieve the research goal, the configuration of a mixed-mode MR mount is devised and the mathematical model is formulated on the basis of non-dimensional Bingham number. After manufacturing an appropriate size of MR mount, the field-dependent damping force is experimentally evaluated with respect to the field intensity. The MR mount is installed on the beam structure as a semi-active actuator, while the beam structure is supported by two passive rubber mounts. The dynamic model of the structural system incorporated with the MR mount is then derived in the modal coordinate, and an optimal controller is designed in order to control unwanted vibration responses of the structural system subjected to external excitations. The controller is experimentally implemented and control performances such as acceleration of the structural systems are evaluated in frequency domain.

2. MR Mount

In this work, a new type of the mixed-mode MR mount which is operated under the flow and shear motion is proposed. The schematic configuration of the MR mount proposed in this work is shown in Figure 1 (a). The MR mount consists of rubber element and MR dash-pot. The MR dash-pot is assembled by MR fluid, piston (or plunger), electromagnet coil, flux guide, and housing. The MR fluid is filled in the gap between piston and outer cylindrical housing. The electromagnetic coil is wired inside of the cylindrical housing. The housing can be fixed to the supporting structure, and the plunger is attached to the top end of the rubber element. The rubber element has a role to support the static load and isolate the vibration transmission at the non-resonant and high frequency regions. During the relative motion of the plunger and housing, MR fluid flows through annular gap. Thus, the pressure drop due to flow resistance of MR fluid in the annular gap can be obtainable. At the same time, the MR dash-pot has additional shear resistance due to relative motion of annular gap walls. Therefore, the proposed MR dash-pot operates under both the flow and shear modes. If no magnetic field is applied, the MR dash-pot only produces a damping force caused by the fluid resistance associated with the viscosity of the MR fluid. However, if the magnetic field is applied through the annular gap, the MR mount produces a controllable damping force due to the yield stress of the MR fluid. As it can be seen from Figure 1 (a), the proposed MR mount has compact structure and operates without frictional components.

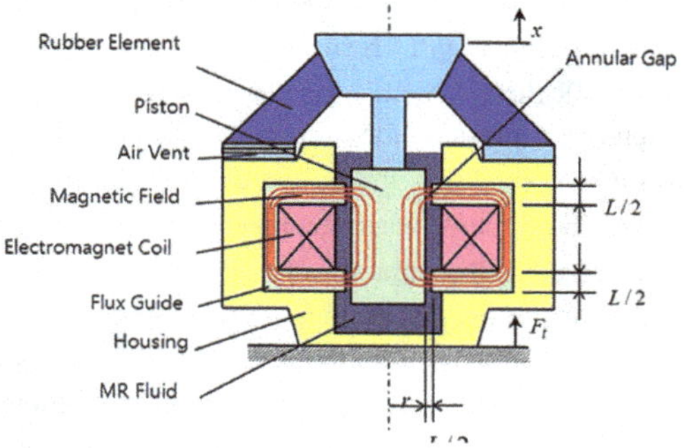

(a) configuration

(b) photograph

Figure 1. The proposed MR mount.

The transmitted force (F_t) through the proposed MR mount can be represented by

$$F_t = k_r x + c_r \dot{x} + F_{mr} \tag{1}$$

In the above, k_r and c_r are stiffness and damping of the rubber element, respectively. x is the deflection of the rubber element. The damping force (F_{mr}) due to the flow resistance of the MR fluid in the gap can be represented by considering the plug flow of the MR fluid under the mixed mode operation as follows [7].

$$\varphi_F = a\varphi_r^3 + \left(\frac{b}{6}\varphi_c + c\right)\varphi_r^2 \tag{2}$$

where,

$$\varphi_F = \frac{F_{mr}}{6\pi\eta v_p L},\ \varphi_c = \frac{\tau_y h}{\eta v_p},\ \varphi_r = \frac{r}{h} \tag{3}$$

In the above, φ_F is the dimensionless damping force, φ_c is the Bingham number, and φ_r is the dimensionless geometric parameter. η is the post-yield plastic viscosity, v_p is the relative velocity between piston and housing, L is the annular gap length, τ_y is the field-dependent yield shear stress of the MR fluid, h is the annular gap size, and r is the piston radius. The dimensionless parameter φ_c represents the ratio of the dynamic yield shear stress to the viscous shear stress. Moreover, φ_c shows the influence of the magnetic field on the damping force of the MR mount. The dimensionless geometric parameter φ_r is characterized by the piston radius r and gap size h . The dimensionless damping force φ_F increases as the Bingham number φ_c and dimensionless geometric parameter φ_r increase. This implies that high yield stress and large piston area are required to generate high damping force of the MR mount. On the other hand, the gap length L depends on the dimensionless damping force φ_F only. Therefore, the damping force F_{mr} can be directly scaled by the gap length L . It is noted that the damping force F_{mr} can be expressed by the velocity of rubber element and the field-dependent yield shear stress of the MR fluid. The parameters a and c of equation (2) are chosen to be 1 by considering Newtonian flow behavior of the MR fluid in the absence of magnetic field. And then, the parameter b , which reflects the effect of Bingham flow of the MR fluid, is chosen to be 2.47 using a least square error criterion.

The field-dependent yield stress τ_y of the MR fluid (MRF-132LD, Lord Corporation) employed in this study has been experimentally obtained by $0.13H^{1.3}$ kPa. Here the unit of magnetic field H is kA/m. The post-yield plastic viscosity η was also experimentally evaluated by 0.59 Pasec. The rheometer (MCR300, Physica, Germany) was used for obtaining the value of the yield stress and the post-yield plastic viscosity. By considering both the nondimensional damping force equations (2,3) and the size of the structural system, an appropriate size of MR mount was designed and manufactured as shown in Figure 1 (b). The piston radius r , gap size h , and gap length L are designed to be 8.5 mm, 1.5 mm, and 10 mm, respectively. The field-dependent damping force is predicted and experimentally measured by exciting the MR mount with the sinusoidal signal which has frequency of 15 Hz and amplitude of 0.07m/sec. Figure 2 presents controllable damping force characteristic of the proposed MR mount. In order to measure the field-dependent damping force of the MR mount, the MR mount is placed between the load cell and electromagnetic exciter. When the shaker table moves up and down by a command signal generated from the exciter controller, the MR mount produces the damping force and it is measured by the load cell. As expected, as the input current increases, the damping force also increases. It is also observed that the predicted field-dependent damping force shows small difference from the measured force in the post-yield velocity region. On the other hand, the damping forces show much large difference in the pre-yield region. This is because the model (2,3) is discontinuous and cannot represent the hysteretic behavior of the MR mount in the transition from pre-yield to post-yield.

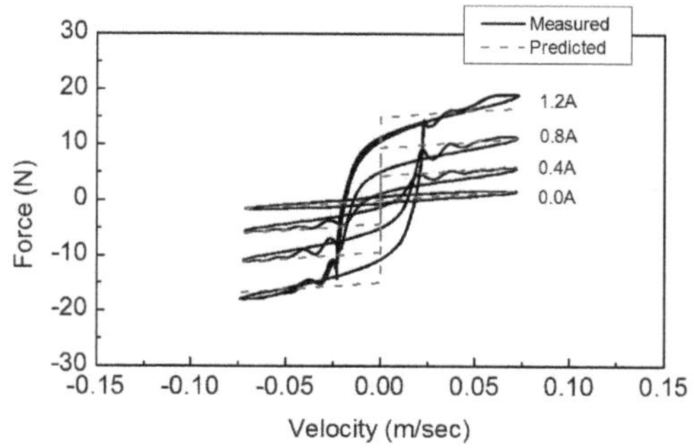

Figure 2. The field-dependent damping force of the MR mount.

The nondimensional form (2,3) of MR mount under mixed mode can be transformed to the Bingham plastic model as follows [12]. This simple model is widely used for the controller implementations.

$$F_{mr} = c_f(I)v_p + F_y(I)\mathrm{sgn}(v_p) \tag{4}$$

In the above, c_f and F_y are the post-yield damping constant and the yield force, respectively. I is the current applied to the MR mount. In order to identify the parameters of the models, a constrained least-mean-squared (LMS) error minimization procedure is used. The cost function J for Bingham plastic model is defined by

$$J(c_f, F_y) = \sum_{k=1}^{N} [f(t_k) - \hat{f}(t_k)]^2 \tag{5}$$

where $\hat{f}(t_k)$ is the force calculated using the model given by equation (4), $f(t_k)$ is the measured force shown in figure 2, and t_k is the time at which the kth sample has been taken. Parameters c_f and F_y of the Bingham plastic model are estimated so as to minimize the cost function J . The parameter optimization is performed on a test case of the identification data set. It is noted that the optimization procedure is applied to the data of the post-yield regions for Bingham plastic model. Based on the measured response in Figure 2, the field-dependent damping constant c_f and yield force F_y are identified by $(23+25I)$ Nsec/m and

$8.28I^{1.85}$ N, respectively. Here, the unit of current I is A. These values will be used as system parameters in the controller implementation for the structural system.

3. Structural System

In order to investigate the applicability of the proposed MR mount to vibration control, a flexible structure system is established as shown in Figure 3. The MR mount is placed between the exciting mass and steel beam structure. When the mass is excited by external disturbance, the force transmitting through the MR mount excites the beam structure. Thus, the vibration of the beam structure can be controlled by activating the MR mount. $y(p_j)$ is the displacement at position p_j of the structural system. The MR mount is placed at position p_2 of the beam structure, while two passive rubber mounts are placed at p_1 and p_3 to support the beam structure. The governing equation of the motion of the proposed structural system can be obtained using the mode summation method as follows [13]:

$$\ddot{q}_i(t) + 2\zeta_i\omega_i\dot{q}_i(t) + \omega_i^2 q_i(t) = \frac{Q_i(t)}{I_i} + \frac{Q_{exi}(t)}{I_i}, \; i=1,\,2,\,\cdots,\,\infty \tag{6}$$

where,

$$Q_i(t) = (\varphi_i(p_4) - \varphi_i(p_2))(-F_{mr}(t)) \tag{7}$$

$$Q_{ex_i}(t) = \varphi_i(p_4)F_{ex}(t) \tag{8}$$

In the above, $q_i(t)$ is the generalized modal coordinate, $\varphi_i(x)$ is the mode shape value at position x , ω_i is the modal frequency, ζ_i is the damping ratio, and I_i is the generalized mass of the i th mode. $Q_i(t)$ is the generalized force including the damping force of the MR mount, and $Q_{exi}(t)$ is the generalized force including the exciting force $F_{ex}(t)$. It is noted that the spring and damping effects of the rubber elements are resolved in the modal parameters of the structural system.

In order to determine system parameters such as modal frequencies and mode shapes, modal analysis is undertaken by adopting a commercial software (MSC/NASTRAN for Windows V4.0). The finite element model of the flexible structure consists of 30 beam elements, 3 spring elements, and 2 mass elements. The nodes of the structural system are constrained in the x, z directions, and the 2-node elastic beam element is used to model the beam. The geometry of the steel beam is 1500mm (length) 60mm (length) 15mm (thickness). The rubber mounts are placed on the 50mm (p_1) and 1450mm (p_3) from left end of the beam. The MR mount is placed on the 600mm (p_2) from the left end of the beam instead of the center of the beam. This is because that the MR mount located at center of the beam cannot control the rotational modes. The 2-node spring elements are used to model the rubber mounts and rubber element of the MR mount. The material property for the spring elements is set by 60kN/m which has been experimentally evaluated. To model the mass of lower part of the MR mount and vibrating mass, the 1-node lumped mass element is used. The material properties for the mass elements are 1.5kg and 3kg for the lower part of the MR mount and the vibrating mass, respectively.

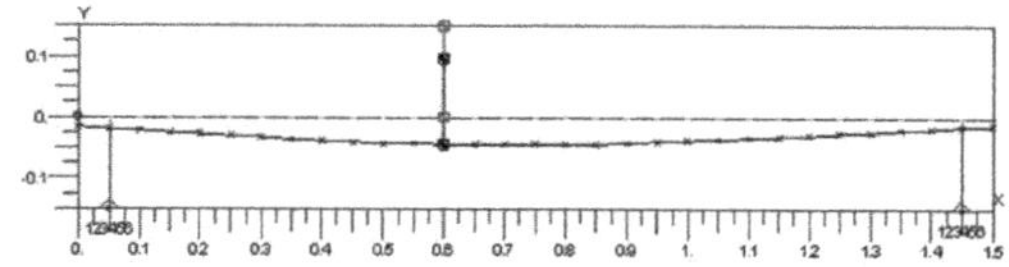

(a) 1st mode

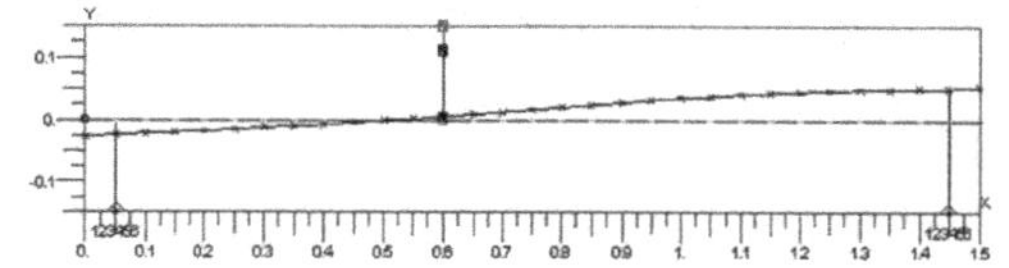

(b) 2nd mode

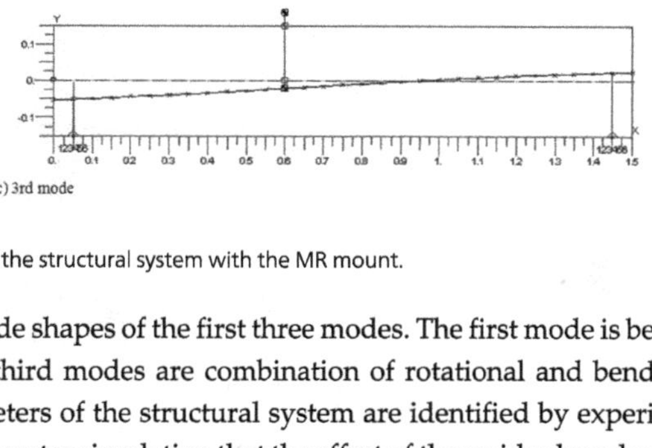

(c) 3rd mode

Figure 3. Configuration of the structural system with the MR mount.

Figure 4 presents mode shapes of the first three modes. The first mode is bending mode, while the second and the third modes are combination of rotational and bending modes. In this study, modal parameters of the structural system are identified by experimental modal test. It is evaluated by computer simulation that the effect of the residual modes is quite small compared with dominant mode. Therefore, in this paper, only dominant modes are considered.

Figure 5 shows the schematic configuration of the experimental setup for the modal parameter identification. The mass on the MR mount is excited by the electromagnetic exciter. The MR mount is removed for the modal parameter identification of structure system only. Thus, this test shows the modal test of the passive structural system in which the mass is supported by the rubber mount. The excitation force and frequency are regulated by the exciter controller. Accelerometers are attached to the mass and beam, and their positions are denoted by $x=p_1$, $x=p_2$, $x=p_3$ and $x=p_4$. From this, the following system parameters are obtained : ω_1 56.9 rad/s, ω_2 144.8 rad/s, ω_3 168.4 rad/s, ζ_1 0.0260, ζ_2 0.0497, and ζ_3 0.0630.

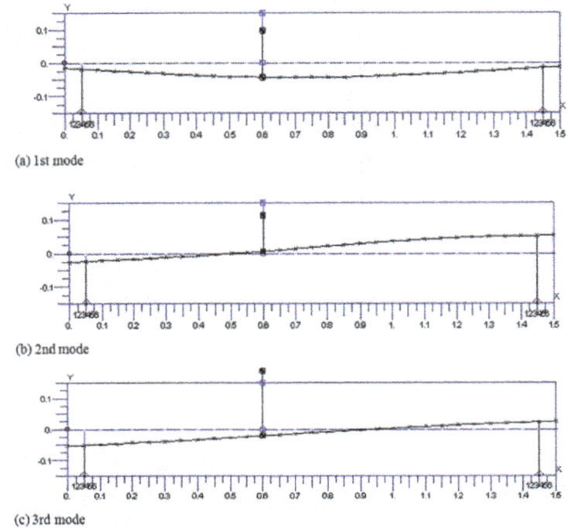

Figure 4. Mode shapes of the structural system.

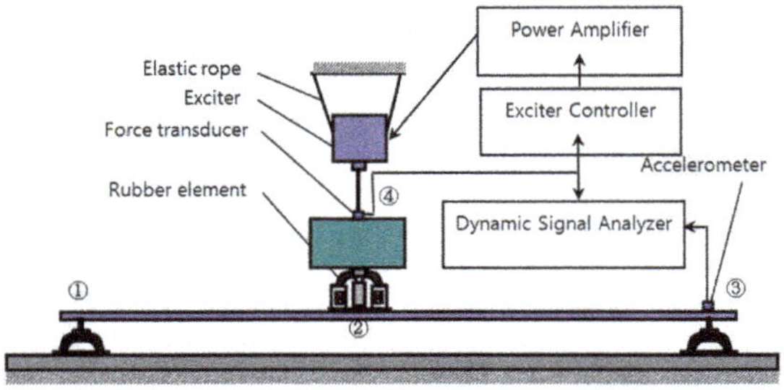

Figure 5. Experimental setup for the modal test of MR fluid mount system.

By considering first three modes as controllable modes, the dynamic model of the structural system, given by equations (6-8), can be expressed in a state space control model as follows.

$$\dot{x}(t)=Ax(t)+Bu(t)+w_1 \tag{9}$$

$$y(t)=Cx(t)+w_2 \tag{10}$$

where

$$\boldsymbol{x}(t)=[q_1(t)\dot{q}_1(t)q_2(t)\dot{q}_2(t)q_3(t)\dot{q}_3(t)]^T \tag{11}$$

$$y(t)=[\dot{y}(p_4, t)\dot{y}(p_2, t)]^T \tag{12}$$

$$u(t)=[F_{mr}(t)] \tag{13}$$

$$A=\begin{bmatrix} 0 & 1 & 0 & 0 & 0 & 0 \\ -\omega_1^2 & -2\zeta_1\omega_1 & 0 & 0 & 0 & 0 \\ 0 & 0 & 0 & 1 & 0 & 0 \\ 0 & 0 & -\omega_2^2 & -2\zeta_2\omega_2 & 0 & 0 \\ 0 & 0 & 0 & 0 & 0 & 1 \\ 0 & 0 & 0 & 0 & -\omega_3^2 & -2\zeta_3\omega_3 \end{bmatrix} \tag{14}$$

$$B=\begin{bmatrix} 0 \\ (\varphi_1(p_2)-\varphi_1(p_4))/I_1 \\ 0 \\ (\varphi_2(p_2)-\varphi_2(p_4))/I_2 \\ 0 \\ (\varphi_3(p_2)-\varphi_3(p_4))/I_3 \end{bmatrix} \tag{15}$$

$$C=\begin{bmatrix} 0 & \varphi_1(p_4) & 0 & \varphi_2(p_4) & 0 & \varphi_3(p_4) \\ 0 & \varphi_1(p_2) & 0 & \varphi_2(p_2) & 0 & \varphi_3(p_2) \end{bmatrix} \tag{16}$$

In the above, $x(t)$ is the state vector, $y(t)$ is the measured output vector, A is the system matrix, B is the control input matrix, C is the output matrix, and $u(t)$ is the control input. w_1 and w_2 are uncorrelated white noise characterized by covariance matrices V_1 and V_2 as follows [14].

$$\begin{aligned} Cov(w_1, w_1^T)&=V_1 \\ Cov(w_2, w_2^T)&=V_2 \\ Cov(w_1, w_2^T)&=0 \end{aligned} \tag{17}$$

4. Vibration Control

In order to attenuate unwanted vibration of the flexible structural system, the linear quadratic Gaussian (LQG) controller, consisting of linear quadratic regulator (LQR) and Kalman-Bucy filter (KBF), is adopted. As a first step to formulate the LQG controller the performance index (J) to be minimized is given by

$$J = \int_0^{\infty} \{ x^T(t) Q x(t) + u^T(t) R u(t) \} dt \tag{18}$$

In the above, $\boldsymbol{Q}$ is the state weighting semi-positive matirx and R is the input weighting positive constant. Since the system (A, B) in equation (9) is controllable, the LQR controller for the MR mount can be obtained as follows.

$$u(t) = -\boldsymbol{P}^{-1}\boldsymbol{B}^T \boldsymbol{x}(t) = -\boldsymbol{K}\boldsymbol{x}(t) \tag{19}$$

In the above, $\boldsymbol{K}$ is the state feedback gain matrix, and $\boldsymbol{P}$ is the solution of the following algebraic Riccati equation :

$$A^T P + PA - PBR^{-1}B^T P + Q = 0 \tag{20}$$

Thus, the control force of the MR mount can be represented as

$$\begin{aligned} u(t) &= -\boldsymbol{K}\boldsymbol{x}(t) \\ &= -(k_1 q_1(t) + k_2 \dot{q}_1(t) + k_3 q_2(t) + k_4 \dot{q}_2(t) + k_5 q_3(t) + k_6 \dot{q}_3(t)) \end{aligned} \tag{21}$$

In this work, by the tuning method the control gains of the k_1, k_2, k_3, k_4, k_5 and k_6 are selected by -536068, 8315.9, -327595, 3706.3, -15.99 and 2359.4, respectively. On the other hand, the proposed MR mount is semi-active. Therefore, the control signal needs to be applied according to the following actuating condition [15]:

$$u(t) = \begin{cases} u(t), & for\, u(t)(\dot{y}(p_4, t) - \dot{y}(p_2, t))0 \\ 0, & for\, u(t)(\dot{y}(p_4, t) - \dot{y}(p_2, t)) \leq 0 \end{cases} \tag{22}$$

This condition physically indicates that the actuating of the controller $u(t)$ assures the increment of energy dissipation of the stable system.

Since the states of $q_i(t)$ and $\dot{q}_i(t)$ of the LQR controller (21) are not available from direct measurement, the Kalman-Bucy filter (KBF) is formulated. The KBF is a state estimator which is optimally considered in the statistical sense. The estimated state, $\hat{x}(t)$, can be obtained from the following state observer [14].

$$\hat{x}(t) = A\hat{x}(t) + Bu(t) + \boldsymbol{L}\,(\boldsymbol{y}(t) - \boldsymbol{C}\hat{x}(t)) \tag{23}$$

where,

$$\boldsymbol{L} = \boldsymbol{O}\boldsymbol{C}^T \boldsymbol{V}_2^{-1} = \begin{bmatrix} l_{11} l_{12} \cdots l_{16} \\ l_{21} l_{22} \cdots l_{26} \end{bmatrix}^T \tag{24}$$

In the above, $\boldsymbol{L}$ is the observer gain matrix, and $\boldsymbol{O}$ is the solution of the following observer Riccati equation :

$$AO+\boldsymbol{O}A^{T}-\boldsymbol{O}C^{T}\boldsymbol{V}_{2}^{-1}CP+V_{1}=0 \tag{25}$$

Using the estimated states, the control force of the MR mount is obtained as follows.

$$u(t)=-(k_1\hat{q}_1(t)+k_2\hat{\dot{q}}_1(t)+k_3\hat{q}_2(t)+k_4\hat{\dot{q}}_2(t)+k_5\hat{q}_3(t)+k_6\hat{\dot{q}}_3(t)) \tag{26}$$

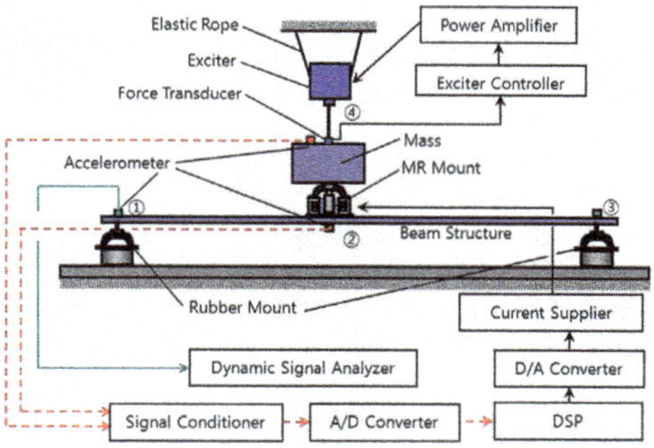

(a) schematic diagram of the experimental setup

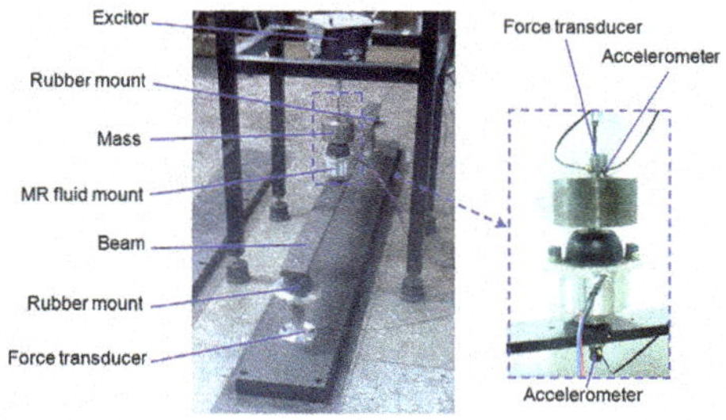

(b) photograph of the experimental setup

Figure 6. Experimental setup for vibration control.

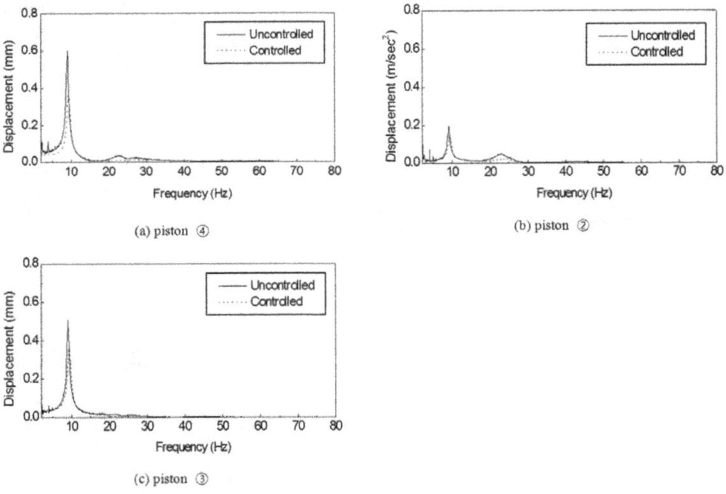

Figure 7. Displacement response of the structural system with the MR mount.

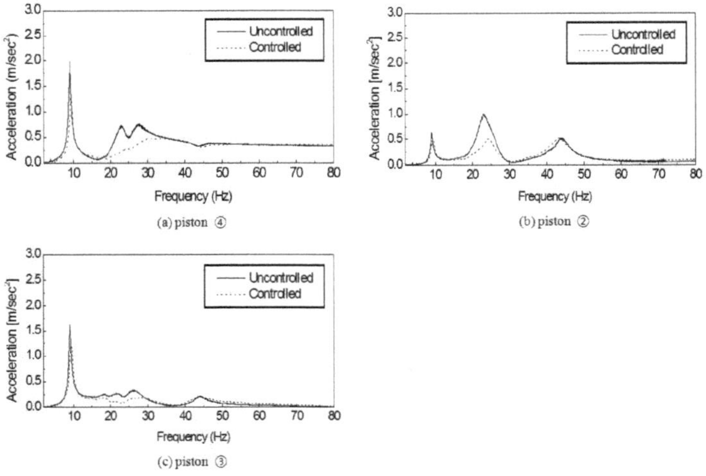

Figure 8. Acceleration response of the structural system with the MR mount.

In order to implement the LQG controller, an experimental setup is established as shown in Figure 6. The mass supported by the MR mount is excited by the electromagnetic exciter, and the excitation force and frequency are regulated by the exciter controller. Accelerometers are attached to the beam and mass, and their positions are denoted by ▪ $(x = p_1)$, ▪ $(x = p_2)$, ▪ $(x = p_3)$, and ▪ $(x = p_4)$. Two accelerometers, attached on the positions ▪ and ▪, are used for the feedback signals. Velocity signals at these positions are obtained by using integrator circuit of charge amplifier. And the other accelerometers are used to measure vibration of the structural system. The velocity signal which denoted by dashed line in Figure 6 is fed back to the DSP (digital signal processor) via an A/D(analog to digital) converter. The state variables required for the LQR (26) are then estimated by the KBF (23), and control voltage is determined by means of the LQR controller in the DSP. The control current is applied to the MR mount via a D/A (digital to analog) converter and a current supplier. The sampling rate in the controller implementation is set by 2kHz.

Figures 7 and 8 present the measured displacement and acceleration of the structural system. The amplitude of excitations force is set by 1N. The excitation frequency range for the structural system is chosen from 5 to 80Hz. The uncontrolled and controlled responses are measured at the positions of the mass (▪) and beam structure (▪ , ▪). It is observed from Figures 7 and 8 that the resonant modes of the structural system have been reduced by controlling the damping force of the MR mount. Especially, the 2nd and 3rd modes vibration at the position ▪ has been substantially reduced by activating the MR mount. It is noted that the uncontrolled response is obtained without input current.

5. Conclusion

A new type of the mixed-mode MR mount was proposed and applied to vibration control of a flexible beam structure system. On the basis of non-dimensional Bingham number, an appropriate size of the MR mount was designed and manufactured. After experimentally evaluating the field-dependent damping force of the MR mount, a structural system consisting of a flexible beam and vibrating rigid mass was established. The governing equation of motion of the system was formulated and a linear quadratic Gaussian (LQG) controller was designed to attenuate the vibration of the structural system. It has been demonstrated through experimental realization that the imposed vibrations of the structural system such as acceleration and displacement are favorably reduced by activating the proposed MR mount associated with the optimal controller. The control results presented in this study are quite self-explanatory justifying that the proposed semi-active MR mount can be effectively utilized to the vibration control of various structural systems such as flexible robot arm and satellite appendages.

Acknowledgements

This work was partially supported by the National Research Foundation of Korea (NRF) grant funded by the Korea government (MEST)(No. 2010-0015090).

Author details

Seung-Bok Choi*, Sung Hoon Ha and Juncheol Jeon

*Address all correspondence to: seungbok@inha.ac.kr

Smart Structures and Systems Laboratory, Department of Mechanical Engineering, Inha University, Incheon 402-751,, Korea

References

[1] Crawley, E. F., & de Luis, J. (1987). Use of Piezoelectric Actuators as Elements of Intelligent Structures. *AIAA Journal*, 25(10), 1373-1385.

[2] Choi, S. B., & Kim, M. S. (1997). New Discrete-Time Fuzzy-Sliding Mode Control with Application to Smart Structures. *AIAA Journal of Guidance, Control and Dynamics*, 20(5), 857-864.

[3] Roger, C. A. (1990). Active Vibration and Structural Acoustic Control of Shape Memory Alloy Hybrid Composites: Experimental Results. *Journal of Acoustical Society of America*, 88(6), 2803-2811.

[4] Choi, S. B., & Cheong, C. C. (1996). Vibration Control of a Flexible Beam Using SMA Actuators. *AIAA Journal of Guidance, Control, and Dynamics*, 19(5), 1178-1180.

[5] Choi, S. B., Park, Y. K., & Jung, S. B. (1999). Modal Characteristics of a Flexible Smart Plate Filled with Electrorheological Fluids. *AIAA Journal of Aircraft*, 36(2), 458-464.

[6] Choi, S. B. (1999). Vibration Control of a Flexible Structure Using ER Dampers. *ASME Journal of Dynamic Systems, Measurement and Control*, 121(1), 134-138.

[7] Hong, S. R., & Choi, S. B. (2003). Vibration Control of a Structural System using Magneto-rheological Fluid Mount. *Proceedings of the 14th International Conference on Adaptive Structures and Technologies*, 182-194, (Seoul, Korea).

[8] Ushijima, T. K., Takano, K., & Kojima, H. (1988). High Performance Hydraulic Mount for Improving Vehicle Noise and Vibration. *SAE Technical Paper Series. 880073.*

[9] Tanaka, N., & Kikushima, Y. (1998). Optimal Design of Active Vibration Isolation Systems. *Transactions of the ASME, Journal of Vibration, Acoustics, Stress and Reliability in Design*, 110(1), 42-48.

[10] Hong, S. R., Choi, S. B., & Han, M. S. (2002). Vibration Control of a Frame Structure using Electrorheological Fluid Mounts. *International Journal of Mechanical Sciences*, 44(10), 2027-2045.

[11] Williams, E. W., Rigby, S. G., Sproston, J. L., & Stanway, R. (1993). Electrorheological Fluids Applied to an Automotive Engine Mount. *Journal of Non-Newtonian Fluid Mechanics*, 47, 221-238.

[12] Hong, S. R., Choi, S. B., Choi, Y. T., & Wereley, N. M. (2003). Comparison of Damping Force Models for an Electrorheological Fluid Damper. *International Journal of Vehicle Design*, 33(1-3), 17-35.

[13] Meirovitch, L. (1990). Dynamics and Control of Structures. John Wiley & Sons Inc.

[14] Machejowski, J. M. (1989). Multivariable feedback design. Addison-Wesley Publishing Company.

[15] Karnopp, D. (1990). Design Principles for Vibration Control Systems using Semi-active Dampers. *Journal of Dynamic Systems, Measurement and Control*, 112(3), 448-455.

Chapter 5

Free Vibration Analysis of Spinning Spindles: A Calibrated Dynamic Stiffness Matrix Method

Seyed M. Hashemi and Omar Gaber

Additional information is available at the end of the chapter

1. Introduction

The booming aerospace industry and high levels of competition has forced companies to constantly look for ways to optimize their machining processes. Cycle time, which is the time it takes to machine a certain part, has been a major concern at various Industries dealing with manufacturing of airframe parts and subassemblies. When trying to machine a part as quick as possible, spindle speed or metal removal rates are no longer the limiting factor; it is the chatter that occurs during the machining process. Chatter is defined as self-excited vibrations between the tool and the work piece. A tight surface tolerance is usually required of a machined part. These self-excited vibrations leave wave patterns inscribed on the part and threaten to ruin it, as its surface tolerances are not met. Money lost due to the destructive nature of chatter, ruining the tools, parts and possibly the machine, has driven a lot of research into determining mathematical equations for the modeling and prediction of chatter. It is well established that chatter is directly linked to the natural frequency of the cutting system, which includes the spindle, shaft, tool and hold combination.

The first mention of chatter can be credited to Taylor [18], but it wasn't until 1946 that Arnold [3] conducted the first comprehensive investigation into it. His experiments were conducted on the turning process. He theorized that the machine could be modelled as a simple oscillator, and that the force on the tool decreased as the speed of the tool increased with relation to the work piece. In his equations, the proportionality constant of the speed of the tool to the force was subtracted from the damping value of the machine; when the proportionality constant increases beyond the damping value of the machine, negative damping occurs causing chatter. This was later challenged by Gurney and Tobias who theorized the now widely accepted belief that chatter is caused by wave patterns traced onto the surface of the work piece by preceding tool passes [9]. The phase shift of the preceding wave to the wave currently being traced determines whether there is any amplification in the tool head movement. If there exists a phase shift between the two tool passes, then the uncut chip cross-sectional area is varied over the pass. The cutting force is dependent on the chips cross sectional area, and so, a varying cutting force is produced [19]. To perform calculations on this system, they modelled a grinding machine as a mass-spring system as opposed to an oscillating system. It had a single degree of freedom, making its calculations quite simple. The spring-mass system is also the widely used modelling theory for how a vibrating tool should be characterized today.

Prior to 1961, the research papers published on the machining processes regarded them as steady state, discrete processes [8]. This erroneous idea led to the creation of machines that were overly heavy and thick walled. It was believed that this provided high damping to the forces on the tool tips that were thought to be static. To properly predict chatter, one must realize that machining is a continuous, dynamic process with tooltip forces that are in constant fluctuation. When performing calculations, the specific characteristics of each machine must also be taken into consideration. If one takes two identical tools, placed into two identical machines, and perform the same machining process on two identical parts, the lifespan of the tools will not be the same. The dynamics and response of each of the machines differs slightly due to structural imperfections, imbalances, etc. Therefore, calculations must always take the machine-tool dynamics into consideration [23]. The modes of the machines structure determine the frequency and the direction that the tool is going to vibrate at [11]. Rather than the previously used machine design philosophy of "where there's vibration, add mass", it was then stated that designers must investigate the mode-forms, weak points, bearing clearances, and self-inducing vibratory components of their machine design to try to reduce chatter [8, 15]. Certain researchers even further investigatedthe required number of structural modes to produce accurate results [6]. Since it is impractical to investigate an unlimited bandwidth of a signal, restrictions must be made. This has generally been restrained to one or two modes of vibration of the machine. Their study proves that using low order models, that only incorporate two modes, are sufficiently accurate to model the machines.

In 1981, one of the first papers documenting the non-linearity of the vibratory system occurring during chatterwas published [20]. Self-excited chatter is a phenomenon that grows, but does not grow indefinitely. There is a point in time where the vibrations stabilize because of the tool jumping out of the cut. As the vibrations amplify, the tool head displacement increases. The displacement of the tool is not linear, but occurs in all three dimensions. When the force on the tool due to chatter causes displacement away from the work piece that exceeds the depth of cut, the tool will lose contact with the work piece. When this occurs, the work piece exerted forces on the tool all go to zero. The only forces acting on it now are the structural forces that want to keep the tool on its planned route. It is impossible for chatter to amplify further past this point, and so, this is where it stabilizes. Previous reports do not account for this stabilization. Their results are accurate up to this point, but then diverge from the experimentally obtained results. Tlusty's investigation was then complemented by adding the behaviour of the tool after the onset of chatter [12]. The paper discusses the effects preceding passes of the tool have on the current state. It was generally accepted that wave patterns left on the work piece from a previous pass greatly effects the current pass, however, it is demonstrated that tool passes two or more turns prior to the current also have an effect. The phase difference and frequency of the waves etched into the surface of the prior turns interact with one another, and if the conditions are correct, interact in a critical way that produces increasing amplitude vibrations [10]. While Tlusty was able to theorize that chatter stabilizes at a certain point due to the tool leaving the work piece, [12] set off to prove this theory. They had the novel idea to turn the machine-work piece system into a circuit. Current was passed through the machine and into the work piece while turning. When chatter occurred, they noticed drops in current at the machine-tool contact point. This proves that an open circuit was being created, proving that the tool was losing contact with the work piece. They also sought out to prove why cutting becomes more stable at lower speeds. They believed that there was a resistive force caused by the tool moving forward along the cut. They found this resistive force to be inversely proportional to the cutting speed, and directly proportional to the relative velocity of the tool to the work piece. When this force was taken into account in their equations, it produces a wider region of stability while the spindle is at lower speeds. This resistive force was proven to be responsible for the large regions of stability at low spindle speeds, and is what diminishes at higher spindle speeds resulting in less stability. The majority of papers published prior to the 80's examined chatter with reference to the turning and boring processes. Milling is plagued with the same issues of chatter, but its modelling becomes more complicated. Tlusty and Ismail [21] characterized the chatter in the milling process by examining the periodicity of the forces that occur at the tool that are not present in other processes. During the milling process, cutter teeth come into and out of contact with the work piece. It is on the surface of these teeth that the force is applied. The same number of teeth are not always in contact with the work piece, and each tooth may be removing a different amount of material at a time. This leads to widely varying forces at the tool tip, creating a more challenging system to model. Forced vibrations can be attributed to periodic forces that the machine is subjected to. This can include an imbalance on rotating parts, or forces the machine transfers to the tool while moving. Chatter must be isolated from this in experiments so that the observations and calculations can be kept specific to the chatter phenomenon.

Once an accurate model of the milling process had been created, a reliable stability lobe can be constructed. Stability lobes plot the axial depth of cut vs. spindle speed. The resulting graph has a series of lobes that intersect each other at certain points. The area that is formed underneath the intersection of these lobes represents conditions that will produce stable machining. The area above these intersections represent unstable machining conditions. The concept of the stability lobe was first proposed by Tobias [24] As the mathematical modelling of the machining systems improved, so did the accuracy of the stability lobes. Prior to the paper by Tlusty *et al.* [22], most stability lobe calculations contained many simplifying assumptions, and therefore, were not very accurate; all teeth on the cutter were assumed to be oriented in the same direction, and also had a uniform pitch. They eliminated all of these assumptions and proved their math represented reality more accurately. A quarter of a centurylater, Mann *et al.* [14] discovered unstable regions in a stability lobe graph that existed underneath the stability boundary for the milling operation. They resemble islands in the fact that they are ovular areas contained within the stable regions, complicating the previously thought simple stability lobe model. It was found that stability lobes taken from modal parameters of the machine at rest (static) were not as accurate as the stability lobes produced from the dynamic modal properties. Zaghbani & Songmene [25] obtained these dynamic properties using operational modal analysis (OMA). OMA uses the autoregressive moving average method and least square complex exponential method to obtain these values, producing a dynamic stability lobe that more accurately represents stable cutting conditions. These stability lobes have proven to be an invaluable asset to machinists and machine programmers. They provide a quick and easy reference to choose machining parameters that should produce a chatter free cut [2].

Tool wear is an often-overlooked factor that contributes to chatter. With the aid of more powerful computers this variable can now be included in simulations. The cutting tool is not indestructible and will change its shape while machining, and consequently affects the stability of the system and stability lobes [7]. As the tool becomes worn, its limits of stability increase. Therefore, the axial depth of cut can be increased while maintaining the same spindle speed that would have previously created chatter. The rate of wear was incorporated into the stability lobe calculations for the tools so that it was now also a function of wear. To verify their calculations, the tools were ground to certain stages of wear and then tested experimentally. They were found to be in strong agreement. Tool wear, however, is not something that machine shops want increased. Chatter increases the rates of tool wear, shortening their lifespan, and increasing the amount of money the shops must spend on new tools. Li *et al.* [13] determined that the coherence function of two crossed accelerations in the bending vibration of the tool shank approaches unity at the onset of chatter. A threshold needs to be set [16] and then detected using simple mechanism to alert the operator to change the machining conditions and avoid increased tool wear.

In most of the previous stability prediction methods, a Frequency Response Function (FRF) is required to perform the calculations. FRF refers to how the machine's structure reacts to vibration. It is required to do an impact test to acquire the system's FRF [17]. In this case, an accelerometer is placed at the end of the top of the tool, and a hammer is used to strike the tool. The accelerometer will measure the displacement of the tool, telling the engineer how the machine reacts to vibration. This test gives crucial information about the machine, such as the damping of the structure and its natural frequencies. This method of obtaining information is impractical; because the FRF of the machine is always changing, it would require the impact test to be performed at all the different stages of machining. Also, having to do this interrupts the manufacturing processing and having machines sitting idle costs the company money. An offline method of obtaining this information could greatly benefit machining companies by eliminating the need for the impact test. Adetoro *et al.* [1] proposed that the machine, tool and work piece could be modelled using finite element analysis.A computer simulation would be able to predict the FRF during all phases of the machining process. As the part is machined and becomes thinner, its response to vibration changes dramatically. Previous research papers assume a constant FRF throughout the whole process for the sake of reducing computations. However, a constantly updated FRF would allow for accurate, real time stability calculations.

To the authors' best knowledge, the spindle decay/bearings wear over the service time and their effects on the system natural frequencies, and consequently change of the stability lobes, have not been investigated.The objective of the present study is to determine the natural frequencies/vibration characteristics ofmachine tool spindle systems by developing its Dynamic Stiffness Matrix (DSM) [4] and applying the proper boundary conditions. These results would then be compared to the experimental results obtained from testing a common cutting system to validate/tune the model developed. The Hamilton's Principle is used to derive the differential equations governing the coupled Bending-Bending (B-B) vibration of a spinning beam, which are solved for harmonic oscillations. A MATLAB® code is developed to assemble the DSM element matrices for multiple components and applying the boundary conditions (BC). The machine spindles usually contain bearings, simulated by applying spring elements at said locations. The bearings are first modeled as Simply-Supported (S-S) frictionless pins. The S-Sboundary conditions are then replaced by linear spring elements to incorporate the flexibility of bearings into the DSM model. In comparison with the manufactures' data on the spindle's fundamental frequency, the bearing stiffness coefficients, K_S, are then varied to achieve a Calibrated Dynamics Stiffness Matrix (CDSM) vibrational model. Once the non-spinning results are confirmed and the spindle model tuned to represent the real system, the formulation could then be extended to include varying rotational speeds and torsional degree-of-freedom (DOF) for further modeling purposes. The research outcomepresented in this Chapter is to be used in the next phase of the authors' ongoing research to establish the relationship between the tool/system characteristics (incorporating spindle's service time/age), and intended machining process, through the development of relevant Stability Lobes, to achieve the best results.

2. Mathematical model

Computer Numeric Control (CNC) machines are quite often found in industries where a great deal of machining occurs. These machines are generally 3-, 4- or 5-axis, depending on the number of degrees of freedom the device has. Having the tool translate in the X, Y and Z direction accounts for the first three degrees-of-freedom (DOF). Rotation about the spindle axes account for any further DOF. The spindle contains the motors that rotate the tools and all the mechanisms that hold the tool in place. Figure 1 displays a sample spindle configuration and a typical tool/holder configuration is shown in Figure 2.

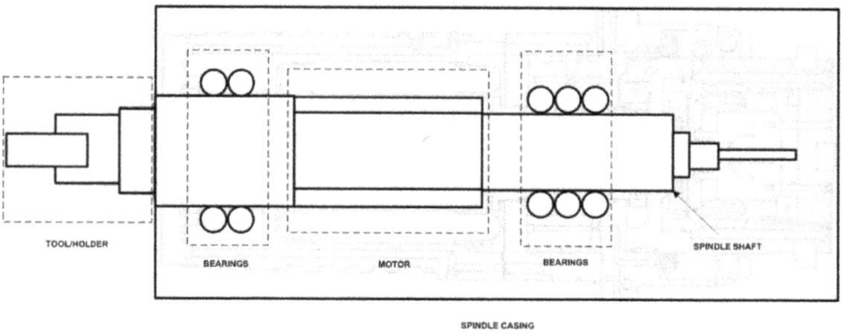

Figure 1. Typical Spindle Configuration.

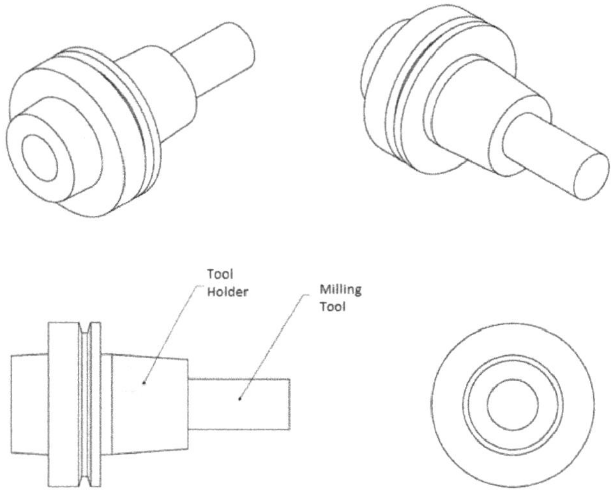

Figure 2. Typical Tool/Holder Configuration.

In this section, following the assumptions made by Banerjee & Su [5], discarding torsional vibrations, neglecting the rotary ineriaand shear deformation effects, the development of the governing differential equations of motion for coupled Bending-Bending (B-B) vibratins of a spinning beam is first briefly discussed. Then, based on the general procedure presented by Banerjee [4], the development of Dynamic Stiffness Matrix (DSM) formulation of the problem is conciselypresented. Figure 3 shows the spinnning beam, represented by a cylinder in a right-handed rectangular Cartesian coordinates system. The beam length is L, mass per unit length is $m=\rho A$, and the bending rigidities are EI_{xx} and EI_{yy}. See Figure 4 for the Degrees-of-Freedom (DOF) of the system.

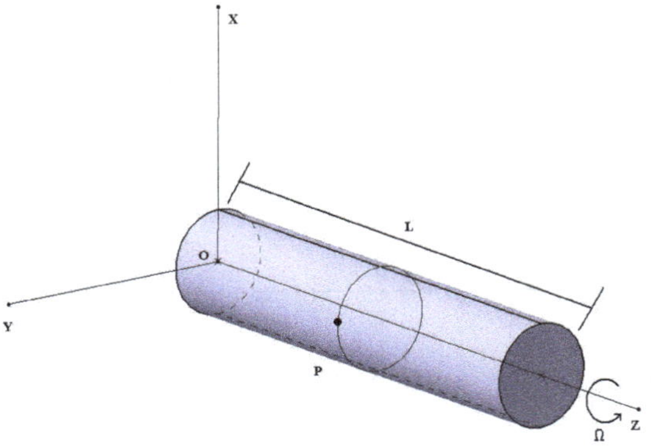

Figure 3. Spinning Beam.

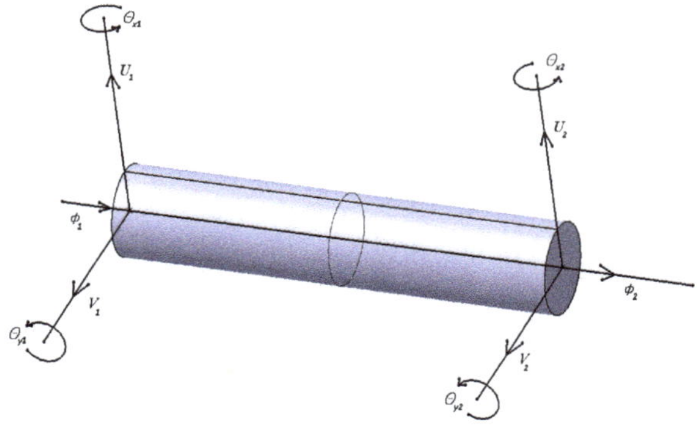

Figure 4. Degrees-of-Freedom (DOF) of the system

At an arbitrary cross section, located at *z* from *O*, *u* and *v* are lateral displacements of a point *P* in the *X* and *Y* directions, respectively. The cross section is allowed to rotate or twist about the *OZ* axis. The position vector ***r*** of the point *P* after deformation is given by

$$\boldsymbol{r} = (\mathrm{u} - \phi \mathrm{y})\boldsymbol{i} + (\mathrm{v} + \phi \mathrm{x})\boldsymbol{j} \tag{1}$$

where ***i*** and ***j*** are unit vectors in the *X* and *Y* directions, respectively. The velocity of point P is given by

$$\mathbf{v} = \dot{\mathbf{r}} + \boldsymbol{\Omega} \times \mathbf{r}, \textit{ where } \boldsymbol{\Omega} = \Omega \mathbf{k} \tag{2}$$

The kinetic and potential energies of the beam (T and U) are given by:

$$T = \frac{1}{2}\int_0^L |v|^2\, mdz = \frac{1}{2} m\int_0^L [\dot{u}^2 + \dot{v}^2 + 2\Omega(u\dot{v} - \dot{u}v) + \Omega^2(u^2 + v^2)]dz, \tag{3}$$

$$U = \frac{1}{2} EI_{xx} \int_0^L v''^2 dz + \frac{1}{2} EI_{yy} \int_0^L u''^2 dz \tag{4}$$

Using the Hamilton Principle in the usual notation state

$$\delta\int_{t_1}^{t_2} (T - U)dt = 0, \tag{5}$$

where t_1 and t_2 are the time intervals in the dynamic trajectory and δ is the variational operator. Substituting the kinetic and potential energies in the Hamilton Principle, collecting like terms and integrating by parts,leads to the following set of equations.

$$EI_{yy}u'''' - m\Omega^2 u + m\ddot{u} - 2m\Omega\dot{v} = 0, \tag{6}$$

$$-2m\Omega\dot{u} - EI_{xx}v'''' - m\ddot{v} + m\Omega^2 v = 0. \tag{7}$$

The resulting loads are then found to be in the following forms, written for Shear forces as

$$S_x = EI_{xx}u''', \text{ and } S_y = EI_{yy}v'', \tag{8}$$

and for Bending Moments as

$$M_x = EI_{xx}v'', \text{ and } M_y = -EI_{yy}u''. \tag{9}$$

Assuming the simple harmonic motion, the form of

$$u(z, t) = U(z)\sin\omega t, \text{ and } v(z, t) = V(z)\cos\omega t \tag{10}$$

where ω frequency of oscillation and U and V are the amplitudes of u and v . Substituting equations (10) into equations (6) and (7), they can be re-written as:

$$\left(EI_{yy}U'''' - m(\Omega^2+\omega^2)U\right) + 2m\Omega\omega V = 0, \tag{11}$$

$$\left(EI_{xx}V'''' - m(\Omega^2+\omega^2)V\right) + 2m\Omega\omega U = 0. \tag{12}$$

Introducing $\xi = z/L$ and $D = d/d\xi$, which are non-dimensional length and the differential operator into equations (11) and (12) leads to

$$\left[D^4 - \frac{m(\omega^2+\Omega^2)L^4}{EI_{yy}}\right]U + \frac{2m\Omega\omega L^4}{EI_{yy}}V = 0, \tag{13}$$

$$\left[D^4 - \frac{m(\omega^2+\Omega^2)L^4}{EI_{xx}}\right]V + \frac{2m\Omega\omega L^4}{EI_{xx}}U = 0. \tag{14}$$

The above equations are combined to form the following 8th-order differential equation,

$$\left[D^8 - (\lambda_x^2+\lambda_y^2)(1+\eta^2)D^4 + \lambda_x^2\lambda_y^2(1-\eta^2)^2\right]W = 0 \tag{15}$$

written in terms of W, satisfied by both U and V, where

$$\lambda_x^2 = \frac{m\omega^2L^4}{EI_{xx}},\ \lambda_y^2 = \frac{m\omega^2L^4}{EI_{yy}},\ \text{and}\ \eta^2 = \frac{\Omega^2}{\omega^2} \tag{16}$$

The solution of the differential equation is sought in the form $W = e^{r\xi}$, and when substituted into (15), leads to

$$r^8 - (\lambda_x^2+\lambda_y^2)(1+\eta^2)r^4 + \lambda_x^2\lambda_y^2(1-\eta^2)^2 \tag{17}$$

where

$$r_{1,3} = \pm\sqrt{\alpha},\ r_{2,4} = \pm\sqrt{\beta},\ r_{5,7} = \pm i\sqrt{\alpha},\ r_{6,8} = \pm i\sqrt{\beta},$$

and

$$\alpha^2 = \frac{1}{2}\left\{(\lambda_x^2+\lambda_y^2)(1+\eta^2) + \sqrt{(\lambda_x^2-\lambda_y^2)^2(1+\eta^2)^2 + 16\lambda_x^2\lambda_y^2\eta^2}\right\},$$

$$\beta^2 = \frac{1}{2}\left\{(\lambda_x^2+\lambda_y^2)(1+\eta^2) - \sqrt{(\lambda_x^2-\lambda_y^2)^2(1+\eta^2)^2 + 16\lambda_x^2\lambda_y^2\eta^2}\right\} \tag{18}$$

From the above solutions of U and V, the corresponding bending rotation about *X* and *Y* axes, Θ_x and Θ_y, respectively, are given by

$$\Theta_x = \frac{dV}{dz} = -\frac{1}{L}\frac{dV}{d\xi} = -\frac{1}{L}\begin{pmatrix} \sqrt{\alpha}B_1\cos\sqrt{\alpha}\xi - \sqrt{\alpha}B_2\sin\sqrt{\alpha}\xi + \\ \sqrt{\alpha}B_3\cosh\sqrt{\alpha}\xi + \sqrt{\alpha}B_4\sinh\sqrt{\alpha}\xi + \sqrt{\beta}B_5\cos\sqrt{\beta}\xi - \\ \sqrt{\beta}B_6\sin\sqrt{\beta}\xi + \sqrt{\beta}B_7\cosh\sqrt{\beta}\xi + \\ \sqrt{\beta}B_8\sinh\sqrt{\beta}\xi \end{pmatrix},$$

$$\Theta_y = \frac{dU}{dz} = \frac{1}{L}\frac{dU}{d\xi} = -\frac{1}{L}\begin{pmatrix} \sqrt{\alpha}A_1\cos\sqrt{\alpha}\xi - \sqrt{\alpha}A_2\sin\sqrt{\alpha}\xi + \\ \sqrt{\alpha}A_3\cosh\sqrt{\alpha}\xi + \sqrt{\alpha}A_4\sinh\sqrt{\alpha}\xi + \sqrt{\beta}A_5\cos\sqrt{\beta}\xi - \\ \sqrt{\beta}A_6\sin\sqrt{\beta}\xi + \sqrt{\beta}A_7\cosh\sqrt{\beta}\xi + \sqrt{\beta}A_8\sinh\sqrt{\beta}\xi \end{pmatrix}. \quad (19)$$

By doing similar substitutions we find

$$S_x = \left(\frac{EI_{yy}}{L^3}\right)\begin{pmatrix} -\alpha\sqrt{\alpha}A_1\cos\sqrt{\alpha}\xi + \alpha\sqrt{\alpha}A_2\sin\sqrt{\alpha}\xi + \\ \alpha\sqrt{\alpha}A_3\cosh\sqrt{\alpha}\xi + \alpha\sqrt{\alpha}A_4\sinh\sqrt{\alpha}\xi - \\ \beta\sqrt{\beta}A_5\cos\sqrt{\beta}\xi + \beta\sqrt{\beta}A_6\sin\sqrt{\beta}\xi + \beta\sqrt{\beta}A_7\cosh\sqrt{\beta}\xi \\ +\beta\sqrt{\beta}A_8\sinh\sqrt{\beta}\xi \end{pmatrix}$$

$$S_y = \left(\frac{EI_{xx}}{L^3}\right)\begin{pmatrix} -\alpha\sqrt{\alpha}B_1\cos\sqrt{\alpha}\xi + \alpha\sqrt{\alpha}B_2\sin \\ \sqrt{\alpha}\xi + \alpha\sqrt{\alpha}B_3\cosh\sqrt{\alpha}\xi + \alpha\sqrt{\alpha}B_4\sinh\sqrt{\alpha}\xi - \\ \beta\sqrt{\beta}B_5\cos\sqrt{\beta}\xi + \beta\sqrt{\beta}B_6\sin\sqrt{\beta}\xi + \\ \beta\sqrt{\beta}B_7\cosh\sqrt{\beta}\xi + \beta\sqrt{\beta}B_8\sinh\sqrt{\beta}\xi \end{pmatrix} \quad (20)$$

$$M_x = \left(\frac{EI_{yy}}{L^2}\right)\begin{pmatrix} -\alpha B_1\sin\sqrt{\alpha}\xi - \alpha B_2\cos\sqrt{\alpha}\xi + \\ \alpha B_3\sinh\sqrt{\alpha}\xi + \alpha B_4\cosh\sqrt{\alpha}\xi - \beta B_5\sin\sqrt{\beta}\xi + \\ \beta B_6\cos\sqrt{\beta}\xi + \beta B_7\sinh\sqrt{\beta}\xi + \beta B_8\cosh\sqrt{\beta}\xi \end{pmatrix}$$

$$M_y = \left(\frac{EI_{xx}}{L^2}\right)\begin{pmatrix} -\alpha A_1\sin\sqrt{\alpha}\xi - \alpha A_2\cos\sqrt{\alpha}\xi + \\ \alpha A_3\sinh\sqrt{\alpha}\xi + \alpha A_4\cosh\sqrt{\alpha}\xi - \\ \beta A_5\sin\sqrt{\beta}\xi + \beta A_6\cos\sqrt{\beta}\xi + \beta A_7\sinh\sqrt{\beta}\xi \\ +\beta A_8\cosh\sqrt{\beta}\xi \end{pmatrix},$$

where

$$A_1=k_\alpha B_1,\ A_2=k_\alpha B_2,\ A_3=k_\alpha B_3,\ A_4=k_\alpha B_4,$$
$$A_5=k_\beta B_5,\ A_6=k_\beta B_6,\ A_7=k_\beta B_7,\ A_8=k_\beta B_8$$

with

$$k_\alpha=\frac{2\lambda_y^2\eta}{\alpha^2-\lambda_y^2(1+\eta^2)},\ \text{and}\ k_\beta=\frac{2\lambda_y^2\eta}{\beta^2-\lambda_y^2(1+\eta^2)}. \tag{21}$$

To obtain the dynamic stiffness matrix (DSM) of the system the boundary conditions are then introduced into the governing equations.

For Displacements:

$$\begin{aligned} At\,z=0:\quad & U=U_1,\ V=V_1,\ \Theta_x=\Theta_{x1},\ \Theta_y=\Theta_{y1} \\ At\,z=L:\quad & U=U_2,\ V=V_2,\ \Theta_x=\Theta_{x2},\ \Theta_y=\Theta_{y2} \end{aligned} \tag{22}$$

For Forces we have

$$\begin{aligned} At\,z=0:\quad & S_x=S_{x1},\ S_y=S_{y1},\ M_x=M_{x1}=M_y=M_{y1}, \\ At\,z=L:\quad & S_x=S_{x2},\ S_y=S_{y2},\ M_x=M_{x2}=M_y=M_{y2}. \end{aligned} \tag{23}$$

Substituting the boundary conditions into the governing equations we find

$$\boxed{\boldsymbol{\delta}=\mathbf{B}\,\mathbf{R}} \tag{24}$$

where

$$\begin{aligned} \boldsymbol{\delta}&=[U_1 V_1 \Theta_{x1} \Theta_{y1} U_2 V_2 \Theta_{x2} \Theta_{y2}]^T, \\ \mathbf{R}&=[R_1 R_2 R_3 R_4 R_5 R_6 R_7 R_8]^T, \end{aligned} \tag{25}$$

and

$$\mathbf{B}=\begin{bmatrix} 0 & k_\alpha & 0 & k_\alpha & 0 & k_\beta & 0 & k_\beta \\ 0 & 1 & 0 & 1 & 0 & 1 & 0 & 1 \\ -\tau_\alpha & 0 & -\tau_\alpha & 0 & -\tau_\beta & 0 & -\tau_\beta & 0 \\ \chi_\alpha & 0 & \chi_\alpha & 0 & \chi_\beta & 0 & \chi_\beta & 0 \\ k_\alpha S_\alpha & k_\alpha C_\alpha & k_\alpha S_{h_\alpha} & k_\alpha C_{h_\alpha} & k_\beta S_\beta & k_\beta C_\beta & k_\beta S_{h_\beta} & k_\beta C_{h_\beta} \\ S_\alpha & C_\alpha & S_{h_\alpha} & C_{h_\alpha} & S_\beta & C_\beta & S_{h_\beta} & C_{h_\beta} \\ -\tau_\alpha C_\alpha & \tau_\alpha S_\alpha & -\tau_\alpha C_{h_\alpha} & -\tau_\alpha S_{h_\alpha} & -\tau_\beta C_\beta & \tau_\beta S_\beta & -\tau_\beta C_{h_\beta} & -\tau_\beta S_{h_\beta} \\ \chi_\alpha C_\alpha & -\chi_\alpha S_\alpha & \chi_\alpha C_{h_\alpha} & \chi_\alpha S_{h_\alpha} & \chi_\beta C_\beta & -\chi_\beta S_\beta & \chi_\beta C_{h_\beta} & \chi_\beta S_{h_\beta} \end{bmatrix} \quad (26)$$

with,

$$\tau_\alpha=\frac{\sqrt{\alpha}}{L},\ \tau_\beta=-\frac{\sqrt{\beta}}{L},\ \omega_\alpha=k_\alpha\tau_\alpha,\ \omega_\beta=k_\beta\tau_\beta,$$

$$\begin{aligned} &S_\alpha=\sin\sqrt{\alpha}, \quad C_\alpha=\cos\sqrt{\alpha}, \quad S_{h_\alpha}=\sinh\sqrt{\alpha}, \quad C_{h_\alpha}=\cosh\sqrt{\alpha}, \\ &S_\beta=\sin\sqrt{\beta}, \quad C_\beta=\cos\sqrt{\beta}, \quad S_{h_\beta}=\sinh\sqrt{\beta}, \quad C_{h_\beta}=\cosh\sqrt{\beta}. \end{aligned} \quad (27)$$

Substituting similarly for the force equation

$$\boxed{\mathbf{F}=\mathbf{A}\,\mathbf{R}} \quad (28)$$

where

$$\mathbf{F}=[S_{x1}S_{y1}\mathrm{M}_{x1}\mathrm{M}_{y1}S_{x2}S_{y2}\mathrm{M}_{x2}\mathrm{M}_{y2}]^T, \quad (29)$$

$$\mathbf{A}=\begin{bmatrix} -\zeta_\alpha & 0 & \zeta_\alpha & 0 & -\zeta_\beta & 0 & \zeta_\beta & 0 \\ -\varepsilon_\alpha & 0 & \varepsilon_\alpha & 0 & -\varepsilon_\beta & 0 & \varepsilon_\beta & 0 \\ 0 & -\gamma_\alpha & 0 & \gamma_\alpha & 0 & -\gamma_\beta & 0 & \gamma_\beta \\ 0 & \lambda_\alpha & 0 & -\lambda_\alpha & 0 & \lambda_\beta & 0 & -\lambda_\beta \\ \zeta_\alpha C_\alpha & -\zeta_\alpha S_\alpha & -\zeta_\alpha C_{h_\alpha} & -\zeta_\alpha S_{h_\alpha} & \zeta_\beta C_\beta & -\zeta_\beta S_\beta & -\zeta_\beta C_{h_\beta} & -\zeta_\beta S_{h_\beta} \\ \varepsilon_\alpha C_\alpha & -\varepsilon_\alpha S_\alpha & -\varepsilon_\alpha C_{h_\alpha} & -\varepsilon_\alpha S_{h_\alpha} & \varepsilon_\beta C_\beta & -\varepsilon_\beta S_\beta & -\varepsilon_\beta C_{h_\beta} & -\varepsilon_\beta S_{h_\beta} \\ \gamma_\alpha S_\alpha & \gamma_\alpha C_\alpha & -\gamma_\alpha S_{h_\alpha} & -\gamma_\alpha C_{h_\alpha} & \gamma_\beta S_\beta & \gamma_\beta C_\beta & -\gamma_\beta S_{h_\beta} & -\gamma_\beta C_{h_\beta} \\ -\lambda_\alpha S_\alpha & -\lambda_\alpha C_\alpha & \lambda_\alpha S_{h_\alpha} & \lambda_\alpha C_{h_\alpha} & -\lambda_\beta S_\beta & -\lambda_\beta C_\beta & \lambda_\beta S_{h_\beta} & \lambda_\beta C_{h_\beta} \end{bmatrix} \quad (30)$$

and

$$\zeta_\alpha = k_\alpha \alpha \sqrt{\alpha} \frac{EI_{yy}}{L^3}, \ \eta_\alpha = \alpha\sqrt{\alpha} \frac{EI_{xx}}{L^3}, \ \gamma_\alpha = \alpha \frac{EI_{xx}}{L^2}, \ \lambda_\alpha = k_\alpha \alpha \frac{EI_{yy}}{L^2},$$
$$\zeta_\beta = k_\beta \beta \sqrt{\beta} \frac{EI_{yy}}{L^3}, \ \eta_\beta = \beta\sqrt{\beta} \frac{EI_{xx}}{L^3}, \ \gamma_\beta = \beta \frac{EI_{xx}}{L^2}, \ \lambda_\beta = k_\beta \beta \frac{EI_{yy}}{L^2}. \quad (31)$$

The frequency-dependent dynamic stiffness matrix (DSM) of the spinning beam, $\mathbf{K}(\omega)$,can be derived by eliminating $\mathbf{R}$. The force amplitude is related to the displacement vector by

$$\boxed{\mathbf{F} = \mathbf{K}\boldsymbol{\delta}} \quad (32)$$

For that we find

$$\mathbf{F} = \mathbf{K}(\mathbf{B}\ \mathbf{R}) = \mathbf{A}\ \mathbf{R} \quad (33)$$

and assuming $\boldsymbol{KB} = \boldsymbol{A}$, finally leads to

$$\boxed{\mathbf{K}(\omega) = \mathbf{A}\mathbf{B}^{-1}} \quad (34)$$

Once the correctness and accuracy of the DSM code was established, a real machine spindle was then modeled, where the non-uniform spindle was idealized as a piecewise uniform (stepped) beam. Each step was modeled as a single continuous element and the above steps in the DSM formulation are repeated several times to determine the stiffness matrix for each element of the spindle. The element Dynamic Stiffness matrices are then assembled and the boundary conditions are applied. The system is simplified to contain 12 elements (13 nodes), as shown in Figure 5.

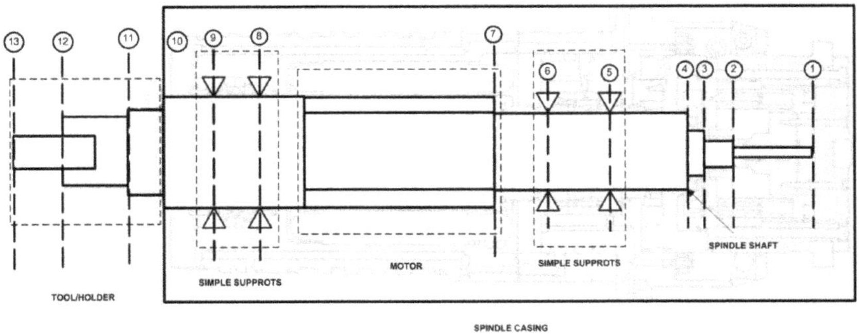

Figure 5. Simplified Spindle sections, with bearings modeled as simply-supported BC.

It is assumed that the entire system is made from the same material and the properties of tooling steel were used for all section. It was also assumed that the system is simply supported at the locations of the bearings. The simply supported boundary conditions were then modified and replaced by linear spring elements (Figure 6); the spring stiffness values were varied in an attempt to achieve a fundamental frequency equivalent to the spindle system's natural frequency reported by the manufacturer.

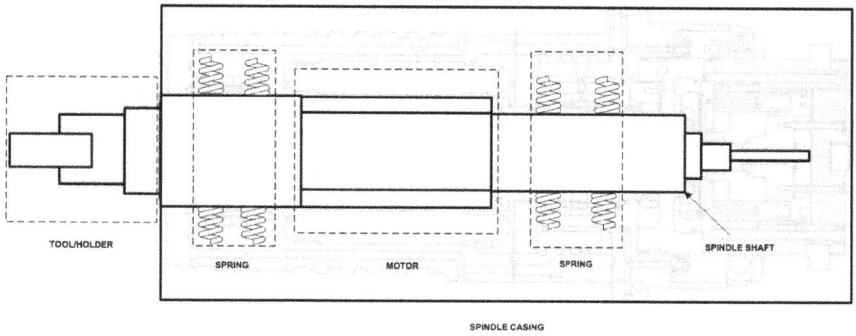

Figure 6. Spindle model, with bearings modeled as linear spring elements(modified BC).

3. Numerical tests and experimental results

3.1. DSM results

When the spindle was modeled using simply supported boundary conditions at the bearing locations, the fundamental natural frequency of the system was found to be just below 1400 Hz, i.e., higher than the nominal value provided by the manufacturer. The boundary conditions were then updated and the simple supports (bearings) were replaced by spring elements. The new spring-supported model was then updated/calibrated to achieve the spindle's nominal fundamental natural frequency by varying spring stiffness values, K_s, all assumed to be identical. It was observed that as the spring stiffness value increase the natural frequency of the system increases. The natural frequency then levels out and reaches an asymptote as the springs start behaving more like simple supports at high values of spring stiffness. It was also found that at spring stiffness value of $K_s = 2.1 \times 10^8$ N/m the system achieves the natural frequency reported by the spindle manufacturer. This value of spring stiffness will be used for any further analysis of the system. These results are shown in Figure 7.

Using the above results the natural frequency of the spindle was also found for multiple rotational speeds (Figure 8). It was observed that, as expected, as the spindle rotation speed increases the natural frequency of the system decreases. It was also found that the spindle critical spindle speed is 2.3×10^6 RPM which is well above the operating rotational speed of the spindle, i.e., 3.5×10^4 RPM.

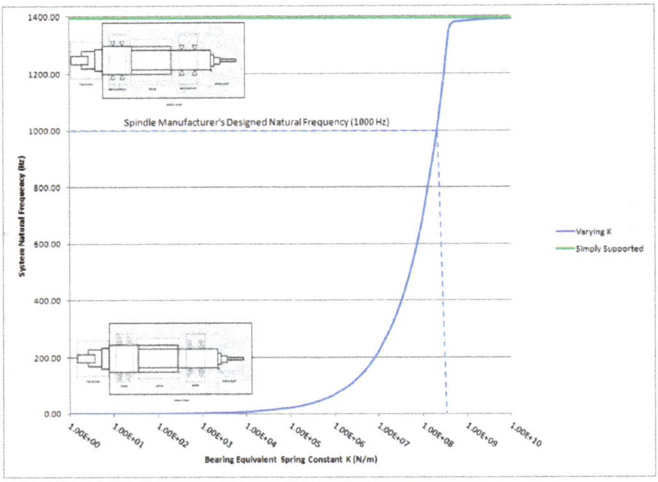

Figure 7. System Natural Frequency vs. Bearing Equivalent Spring Constant (in log scale).

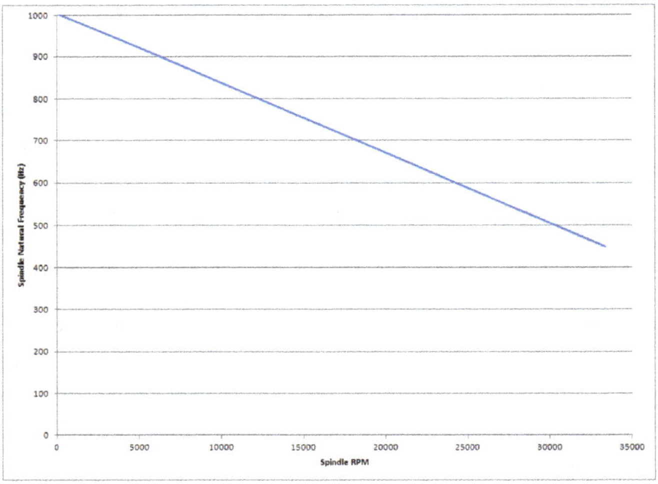

Figure 8. Spindle Natural Frequency vs. Spindle RPM.

3.2. Preliminary Experimental results

The experimentally evaluated Frequency Response Function (FRF) data were collected for a machine over the period of twelve months. A 1-inch diameter blank tool with a 2-inch protrusion was used. A typical shrink fit tool holder was also used (See Figure 9). This type of holder was selected for its rigid contact surface with the tool. Therefore, any play in the whole system was going to be attributed to the spindle. The tested machine was used to produce typical machined parts and was not restricted to one type of cut. This was done to observe the spindle decay over time while operating under normal production conditions. The tool was placed in the spindle and the spindle was returned to its neutral position as shown in Figure 9. Acceleration transducers were placed in both the X and Y direction. The tool was struck with an impulse hammer in both the X and Y directions and corresponding bending natural frequencies were evaluated over the time. Figure 10 shows the bending natural frequencies of the non-spinning spindle vs. machine hours. As can be seen, system natural frequencies in both X and Y directions reduce with spindle's life, which can be attributed to bearings decay. Further reseatrch is underway to analyze more spindles and to model the system decay by establishing a relationship between bearings stiffness, K_g, and machine hours. This, in turn, can be used to predict the optimum machining parameters as a function of spindle age.

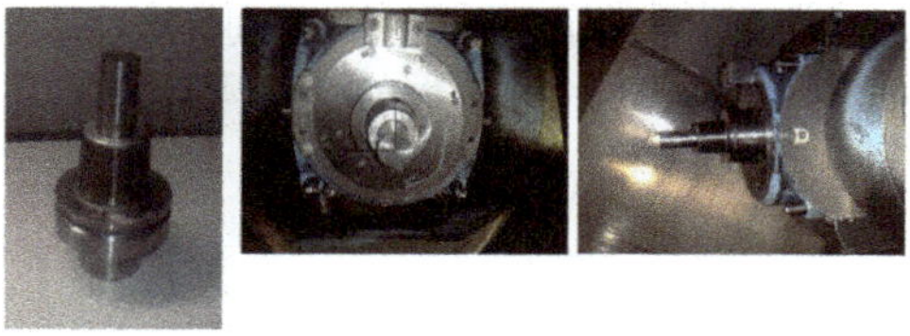

Figure 9. Blank Tool (left) and Blank Tool in Spindle (middle and right).

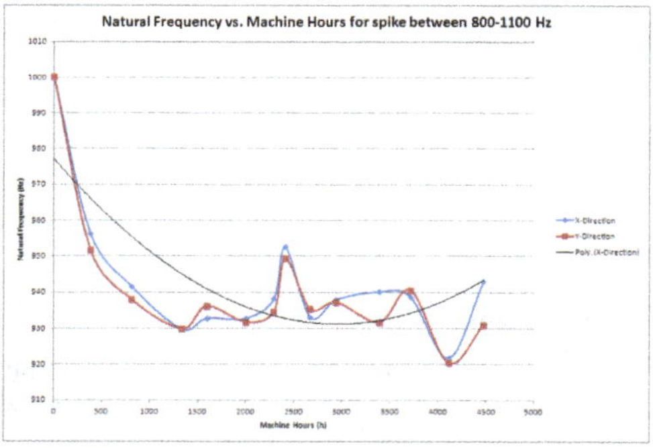

Figure 10. Natural Frequency vs. Machine Hours.

4. Conclusion

The effects of spindle system's vibrational behavior on the stability lobes, and as a result on the Chatter behavior of machine tools have been established. It has been observed that the service life changes the vibrational behavior of spindles, i.e., reduced natural frequency over the time. An analytical model of a multi-segment spinning spindle, based on the Dynamic Stiffness Matrix (DSM) formulation and exact within the limits of the Euler-Bernoulli beam bending theory, was developed. The beam exhibits coupled Bending-Bending (B-B) vibration and, as expected, its natural frequencies are found to decrease with increasing spinning speed. The bearings were included in the model using two different models; rigid, simply-supported,frictionless pins and flexible linear spring elements. The linear spring element stiffness, K_s, was then calibrated so that the fundamental frequency of the system matched the nominal data provided by the manufacturer. This step is vital to the next phase of the authors' ongoing research, where the bearing wear would be modeled in terms of spindle's service time/age, to investigate the consequent effects on the stability lobes and, in turn, on themachine Chatter.

Acknowledgements

The authors wish to acknowledge the support provided by Ryerson University and Natural Science and Engineering Research Council of Canada (NSERC).

Author details

Seyed M. Hashemi* and Omar Gaber

*Address all correspondence to: smhashem@ryerson.ca

Department of Aerospace Eng., Ryerson University, Toronto (ON), Canada

References

[1] Adetoro, O. B. , Wen, P. H., Sim, W. M., & Vepa, R. (2009, 1-3 July). Stability Lobes Prediction in Thin Wall Machining. Paper presented at World Congress on Engineering, WCE 2009, London, U.K. 520-525, 978-9-88170-125-1.

[2] Altintas, Y., & Budak, E. (1995). Analytical Prediction of Stability Lobes in Milling. *CIRP Annals- Manufacturing Technology*, 44(1), 357-362, 0007-8506.

[3] Arnold, R. N. (1946). Discussion on the Mechanism of Tool Vibration in the Cutting of Steel. *Proceedings of the Institution of Mechanical Engineers*, 154, 429-432.

[4] Banerjee, J. R. (1997, April). Dynamic Stiffness Formulation for Structural Elements: A General Approach. *Computers & Structures*, 63(1), 101-103, 0045-7949.

[5] Banerjee, J. R., & Su, H. (2004, September-October). Development of a Dynamic Stiffness Matrix for Free Vibration Analysis of Spinning Beams. *Computers and Structures*, 82(23-26), 2189-2197, 0045-7949.

[6] Butlin, T., & Woodhouse, J. (2009, November). Friction-Induced Vibration: Should Low-Order Models be Believed. *Journal of Sound and Vibration*, 328(1-2), 92-108, 0002-2460X.

[7] Clancy, B. E., & Shin, Y. C. (2002, July). A Comprehensive Chatter Prediction Model for Face Turning Operation Including Tool Wear Effect. *International Journal of Machine Tools and Manufacture*, 42(9), 1035-1044, 0890-6955.

[8] Eisele, F. (1961, December). Machine Tool Research at the Technological University of Munich. *International Journal of Machine Tool Design Research*, 1(4), 249-274, 0890-6955.

[9] Gurney, J. P., & Tobias, S. A. (1961, September). A Grphical Method for the Determination of the Dynamic Stability of Machine Tools. *International Journal of Machine Tool Design Research*, 1(1-2), 148-156, 0020-7357.

[10] Heisel, U. (1994). Vibrations and Surface Generation in Slab Milling. *CIRP Annals-Manufacturing Technology*, 43(1), 337-340, 0007-8506.

[11] Insperger, T., Mann, B. P., Stepan, G., & Bayly, P. V. (2003, April). Multiple Chatter Frequencies in Milling Processes. *Journal of Sound and Vibration*, 262(2), 333-345, 0002-2460X.

[12] Yoshitaka, K., Osame, K., & Hisayoshi, S. (1981, November). Behaviour of Self Excited Chatter Due to Multiple Regenerative Effect. *Journal of Engineering for Industry*, 103(4), 324-329, 0022-1817.

[13] Li, X. Q., Wong, Y. S., & Nee, A. Y. C. (1997, April). Tool Wear and Chatter Detection using the Coherence Function of Two Crossed Accelerations. *International Journal of Machine Tools and Manufacture*, 37(4), 425-435, 0890-6955.

[14] Mann, B. P., Edes, B. T., Easly, S. J., Young, K. A., & Ma, K. (2008, March). Chatter Vibration and Surface Location Error Prediction for Helical End Mills. *International Journal of Machine Tools and Manufacture*, 48(3-4), 350-361, 0890-6955.

[15] Peng, Z. K., Jackson, M. R., Guo, L. Z., Parkin, R. M., & Meng, G. (2010, October). Effects of Bearing Clearance on the Chatter Stability of Milling Process. *Nonlinear Analysis: Real World Applications*, 11(5), 3577-3589, 1468-1218.

[16] Rahman, M., & Ito, Y. (1986, September). Detection of the Onset of Chatter Vibration. *Journal of Sound and Vibration*, 109(2), 193-205, 0002-2460X.

[17] Solis, E., Peres, C. R., Jimenez, J. E., Alique, J. R., & Monje, J. C. (2004). A New Analytical-Experimental Method for the Identification of Stability Lobes in High-Speed

Milling. International Journal of Machine Tools and Manufacture December) 0890-6955 , 44(15), 1591-1597.

[18] Taylor, F. W. (1907). On the Art of Cutting Metal. *Trans. ASME,* 28, 31-350.

[19] Shabana, A., & Thomas, B. (1987). Chatter Vibration of Flexible Multibody Machine Tool Mechanisms. *Mechanical Machine Theory,* 22(4), 359-369, 0009-4114X.

[20] Tlusty, J., & Ismail, F. (1981). Basic Non-Linearity in Machining Chatter. *CIRP Annals-Manufacturing Technology,* 30(1), 299-304, 0007-8506.

[21] Tlusty, J., & Ismail, F. (1983, January). Special Aspects if Chatter in Milling. *Journal of Vibration, Acoustics, Stress, and Reliability in Design,* 105(1), 24-32, 0739-3717.

[22] Tlusty, J., Zaton, W., & Ismail, F. (1983). Stability Lobes in Milling. CIRP Annals-Manufacturing Technology 0007-8506 , 32(1), 309-313.

[23] Tobias, S. A., & Andrew, C. (1962, October-December). Vibration in Horizontal Milling. *International Journal of Machine Tool Design Research,* 2(4), 369-378, 0890-6955.

[24] Tobias, S. A. (1965). Machine Tool Vibration. Blackie and Sons Ltd. London, Glasgow

[25] Zaghbani, I., & Songmene, V. (2009, October). Estimation of Machine-Tool Dynamic Parameters During Machining Operation Through Operational Modal Analysis. *International Journal of Machine Tools and Manufacture,* 49(12-13), 947-957, 0890-6955.

Chapter 6

Recent Advances on Force Identification in Structural Dynamics

N. M. M. Maia, Y. E. Lage and M. M. Neves

Additional information is available at the end of the chapter

1. Introduction

This chapter presents recent advances on force identification for structural dynamics that have been developed by the authors using the concept of transmissibility for multiple degree-of-freedom (MDOF) systems.

Being applied for many years only to the single degree-of-freedom (SDOF) system or to MDOF systems in a very limited way, the transmissibility concept has been developed along the last decade or so in a consistent manner, to be applicable in a general and complete way to MDOF systems. Various applications for MDOF systems may now be found, such as evaluation of unmeasured frequency response functions (FRFs), force identification, detection of damage, etc. A review of the multiple applications of the transmissibility concept has been published recently [1].

It is the application of this generalized transmissibility concept to both the direct and inverse force identification that is described along this chapter. The direct problem is understood as the one where one knows the applied forces and wishes to estimate the reactions at the supports; the inverse force identification problem is when one wishes to determine how many forces are applied, where they are applied and which are their magnitudes.

To determine the location and magnitude of the dynamic forces that excite the system is an important issue in structural dynamics [2, 3], especially when operational forces cannot be directly measured, as it happens at inaccessible locations [4, 5]; it is often the case that transducers cannot be introduced in the structure to allow the experimental measurement of the external loads and only a limited number of sensors and positions are available. The identification of forces from vibration measurements at a few accessible locations is a very important problem in various areas, such as vibration control, fatigue life prediction and health monitoring.

Although the force identification problem may be solved from the dynamic responses by simply reversing the direct problem, this is usually ill-posed and sensitive to perturbations in the measured data.

Over the past years, the theory of inverse methods has been actively developed in many research areas presenting in common the effects of matrix ill-conditioning, reflecting the ill-posedness nature of the inverse problem itself. Those problems can often be overcome by methods such as pseudo-inversion for over-determined systems, use of Kalman filters [6, 7], Singular Value Decomposition and Tikhonov regularization [8-10].

Various research works in force identification can be found in the literature, such as those related to the identification of impact forces, implementation of prediction models based on reflected waves or simply from the dynamic responses [11-18], prediction of forces in plates for systems with time dependent properties [11] and identification of harmonic forces [13].

These methods to identify operational loads based on response measurements can be classified into three main categories: deterministic methods, stochastic methods and methods based on artificial intelligence. Two main classes of identification technique are considered in the group of deterministic methods for load identification: frequency-domain methods and time-domain methods. The force identification in time domain has been less studied than its frequency domain equivalent, therefore there are not that many force identification studies in the literature. A review on the state of the art for dynamic load identification may be found in [3, 14].

Although out of the scope of this chapter, some references are here given with respect to recent time-domain force identification developments. One interesting approach based on modal filtering [15] is the Sum of Weighted Accelerations Technique (SWAT), which allows to obtain the time-domain force reconstruction by isolating the rigid body modal accelerations. Another approach for time-domain force reconstruction is the Inverse Structural Filter (ISF) method of Kammer and Steltzner [16] that inverts the discrete-time equations of motion. A variant of this, expected to produce a stable ISF when the standard method fails was recently developed and named as Delayed Multi-step ISF (DMISF). For a more detailed description on these methods (SWAT, ISF and DMISF) see e.g. [17] and for its application to rotordynamics, see [18].

In this chapter, the authors treat the frequency-domain problem from a different perspective, which is based on the MDOF transmissibility concept. As aforementioned, usually the transmissibility of forces is defined in textbooks for SDOF systems, simply as the ratio between the modulus of the transmitted force magnitude to the support and the modulus of the applied force magnitude. For SDOF systems, the expression of either the transmissibility of motion or forces is exactly the same; however, as explained in [1], that is not the case for MDOF systems. On the one hand, the problem of extending the idea of transmissibility of motion to an MDOF system is essentially a problem of how to relate a set of unknown responses to a set of known responses associated to a given set of applied forces; on the other hand, for the transmissibility of forces the question is how to relate a set of reaction forces to a set of applied ones.

Some initial attempts on the generalization of the transmissibility concept are due toVakakis et al. [19-21], Liu et al. [22, 23] and Varoto [24]. Similar efforts can also be found in the indirect measurement of vibration excitation forces [2, 4, 5]. To the best knowledge of the authors, a general answer to the problem is due to Ribeiro [25], and in [26] the experimental evaluation of the transmissibility concept for MDOF systems is presented. The concept of transmissibility of forces for MDOF has been proposed in 2006 [27], where the authors explain the formulation of the transmissibility using both the dynamic stiffness and the receptance matrices.

The use of the transmissibility in conjunction with a two step methodology for force identification is the main novelty of this chapter. For the force identification based on the transmissibility of motion, two steps are taken, (i) firstly the number of forces and their location are obtained, and (ii) secondly the reconstruction of the load vector is performed using some of the responses obtained experimentally together with the updated numerical model. Both have been numerically developed and implemented, as well as experimentally tested in the research group during the last years to access the potential of these new methods.

In section 2, the authors review the generalized transmissibility concepts, both in terms of displacements and forces. They are introduced and deduced from two different perspectives, (i) from the frequency response functions, (ii) from the dynamic stiffness.

In section 3, a numerical model and an experimental application are presented to illustrated the transmissibility concept.

In section 4 the methodologies proposed for force identification based on the transmissibility concept are introduced.

Some simulated and experimental results are presented to show how these methodologies are able to help us identifying applied and reaction forces. The authors present a discussion on these proposed methods and on the obtained results.

2. Transmissibility in MDOF systems

The transmissibility concept may be found in any fundamental textbook on mechanical vibrations (e.g. [28]), related to SDOF systems.

The transmissibility of motion is defined as the ratio between the modulus of the response amplitude (output) and the modulus of the imposed base harmonic displacement (input). Depending on the imposed frequency, the result can vary from an amplification to an attenuation in the response amplitude relatively to the input one.

On the other hand, the transmissibility of force is defined as the ratio between the modulus of the transmitted force magnitude to the support and the modulus of the imposed force magnitude.

It happens that for SDOF systems the expression for calculating the transmissibility is the same, either referring to forces or to motion. This is not the case for MDOF systems.

The generalization of these definitions to MDOFs has been developed in the last decade, as mentioned before. In this section a brief review of these generalizations is given, introducing also the concepts and notation used for the force identification problem.

2.1. Transmissibility of motion in MDOF systems

To introduce the problem, the authors follow here as near as possible the notation used in [1]. Let K be the set of n_K co-ordinates where the displacement responses $\boldsymbol{Y}_K$ are known (measured or computed), U the set of n_U co-ordinates where the displacement responses $\boldsymbol{Y}_U$ are unknown, and A the set of co-ordinates where the forces $\boldsymbol{F}_A$ may be applied (Fig. 1).

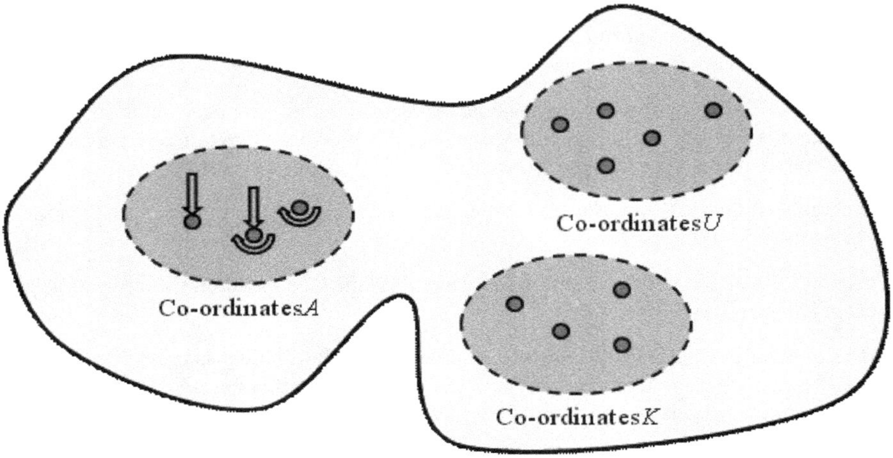

Figure 1. Illustration of an elastic body with the three sets of co-ordinates K, U and A.

To obtain the needed transmissibility of motion one may consider two distinct ways. The first is based on the frequency response function (FRF) matrices $\boldsymbol{H}(\omega)$, known as the fundamental formulation, while the second is based on the dynamic stiffness matrix $\boldsymbol{Z}(\omega)$ and is named alternative formulation.

The receptance frequency response matrix $\boldsymbol{H}(\omega)$ relates the dynamic displacement amplitudes $\boldsymbol{Y}$ with the external force amplitudes $\boldsymbol{F}$ as (using harmonic excitation, in steady-state conditions):

$$\boldsymbol{Y} = \boldsymbol{H}\,\boldsymbol{F} \quad \Leftrightarrow \quad \boldsymbol{Y} = \left(\boldsymbol{K} - \omega^2\,\boldsymbol{M} + i\omega\,\boldsymbol{C}\right)^{-1}\boldsymbol{F} \tag{1}$$

where $\boldsymbol{K}$, $\boldsymbol{M}$ and $\boldsymbol{C}$are the stiffness, mass and viscous damping matrices, respectively. $\boldsymbol{H}(\omega)$ includes all the degrees of freedom in which the system is discretized and corresponds to the inverse of the dynamic stiffness matrix $\mathbf{Z}(\omega)$. One may underline that the mass-normalized orthogonality properties are observed here:

$$\begin{aligned} \boldsymbol{\Phi}^T \boldsymbol{M} \boldsymbol{\Phi} &= \mathbf{I} \\ \boldsymbol{\Phi}^T \boldsymbol{K} \boldsymbol{\Phi} &= \mathbf{diag}(\omega_r^2) \end{aligned} \tag{2}$$

Assuming proportional damping, $\boldsymbol{C}=\alpha\boldsymbol{K}+\beta\boldsymbol{M}$ and therefore,

$$\boldsymbol{Y} = \boldsymbol{H}\,\boldsymbol{F} = \boldsymbol{\Phi}\left[\boldsymbol{diag}(\omega_r^2-\omega^2) + i\,\omega\left(\alpha\,\boldsymbol{diag}(\omega_r^2) + \beta\,\boldsymbol{I}\right)\right]^{-1}\boldsymbol{\Phi}^T\,\boldsymbol{F} \tag{3}$$

where $\boldsymbol{\Phi}$ is the mode shape matrix, ω_r is the rth natural frequency and α and β are constants.

From (1) it is easy to understand that if the responses $\boldsymbol{Y}$ at the discretization points are known, then the force reconstruction (in frequency-domain) would be given by:

$$\boldsymbol{F} = \boldsymbol{H}^{-1}\,\boldsymbol{Y} \tag{4}$$

2.1.1. Transmissibility of motion in terms of FRFs

Based on harmonically applied forces at co-ordinates *A*, one may establish that displacements at co-ordinates *U* and *K* are related to the applied forces at co-ordinates *A* by the following relationships:

$$\boldsymbol{Y}_U = \boldsymbol{H}_{UA}\,\boldsymbol{F}_A \tag{5}$$

$$\boldsymbol{Y}_K = \boldsymbol{H}_{KA}\,\boldsymbol{F}_A \tag{6}$$

Eliminating the external forces $\boldsymbol{F}_A$ between (5) and (6), one obtains

$$\boldsymbol{Y}_U = \boldsymbol{H}_{UA}\left(\boldsymbol{H}_{KA}\right)^+\boldsymbol{Y}_K = \left(\boldsymbol{T}_d\right)_{UK}^A\,\boldsymbol{Y}_K \tag{7}$$

where

$$\left(\boldsymbol{T}_d\right)_{UK}^A = \boldsymbol{H}_{UA}\left(\boldsymbol{H}_{KA}\right)^+ \tag{8}$$

is the transmissibility matrix relating both sets of displacements. $(\boldsymbol{H}_{KA})^+$ is the pseudo-inverse of the sub-matrix $\boldsymbol{H}_{KA}$. An important property of the transmissibility matrix to be used here is that it does not depend on the magnitude of the involved forces and only requires the knowledge of a set of co-ordinates that include all the co-ordinates where the forces are applied. Indeed, it is required that n_K be greater or equal to n_A. One important aspect of this definition is that sub-matrices $\boldsymbol{H}_{UA}$ and $\boldsymbol{H}_{KA}$ may be obtained experimentally.

2.1.2. Transmissibility of motion in terms of dynamic stiffness

There exists an alternative approach to obtain the transmissibility matrix for the displacements, using the dynamic stiffness matrices introduced in (1). Assuming again harmonic loading and defining two subsets, *A* and *B*, *A* being the set where the dynamic loads may be applied and *B* the set formed by the remaining co-ordinates, where no forces are applied ($\boldsymbol{F}_B = \boldsymbol{0}$), one can obtain (after grouping adequately the degrees of freedom of the problem):

$$\begin{bmatrix} \boldsymbol{Z}_{AK} & \boldsymbol{Z}_{AU} \\ \boldsymbol{Z}_{BK} & \boldsymbol{Z}_{BU} \end{bmatrix} \begin{Bmatrix} \boldsymbol{Y}_K \\ \boldsymbol{Y}_U \end{Bmatrix} = \begin{Bmatrix} \boldsymbol{F}_A \\ \boldsymbol{0} \end{Bmatrix} \qquad (9)$$

Developing eq. (9), it follows that

$$\begin{aligned} \boldsymbol{Z}_{AK}\, \boldsymbol{Y}_K + \boldsymbol{Z}_{AU}\, \boldsymbol{Y}_U &= \boldsymbol{F}_A \qquad \text{a} \\ \boldsymbol{Z}_{BK}\, \boldsymbol{Y}_K + \boldsymbol{Z}_{BU}\, \boldsymbol{Y}_U &= \boldsymbol{0} \qquad \text{b} \end{aligned} \qquad (10)$$

From (10b) one obtains the transmissibility in terms of the dynamic stiffnesses:

$$\boldsymbol{Y}_U = -\left(\boldsymbol{Z}_{BU}\right)^+ \boldsymbol{Z}_{BK} \boldsymbol{Y}_K = \left(\boldsymbol{T}_d\right)_{UK}^{A} \boldsymbol{Y}_K \qquad (11)$$

where $(\boldsymbol{Z}_{BU})^+$ is the pseudo-inverse of $\boldsymbol{Z}_{BU}$.

From (11) it is possible to obtain the response at the unknown co-ordinates, as long as the pseudo-inverse is viable, which requires that n_B is greater or equal to n_U.

Indeed, from all this resulted two conditions:

$$\left(\boldsymbol{T}_d\right)_{UK}^{A} = -\left(\boldsymbol{Z}_{BU}\right)^+ \boldsymbol{Z}_{BK} = \boldsymbol{H}_{UA}\left(\boldsymbol{H}_{KA}\right)^+ \qquad n_B \geq n_U \text{ and } n_K \geq n_A \qquad (12)$$

2.2. Transmissibility of forces in MDOF systems

To introduce the transmissibility of forces for MDOF systems, the authors follow a similar procedure to the one used in the previous sub-section. The problem consists now of relating the set of known applied forces to a set of unknown reactions (or the other way around), relating the set of known applied forces (set *K*) with a set of unknown reaction forces (set *U*), which are illustrated in Fig.2. At the set *U* it will be assumed that $\boldsymbol{Y}_U = \boldsymbol{0}$. In general, there will be other co-ordinates, where neither there are any applied forces nor there are any reactions, that shall constitute the set C.

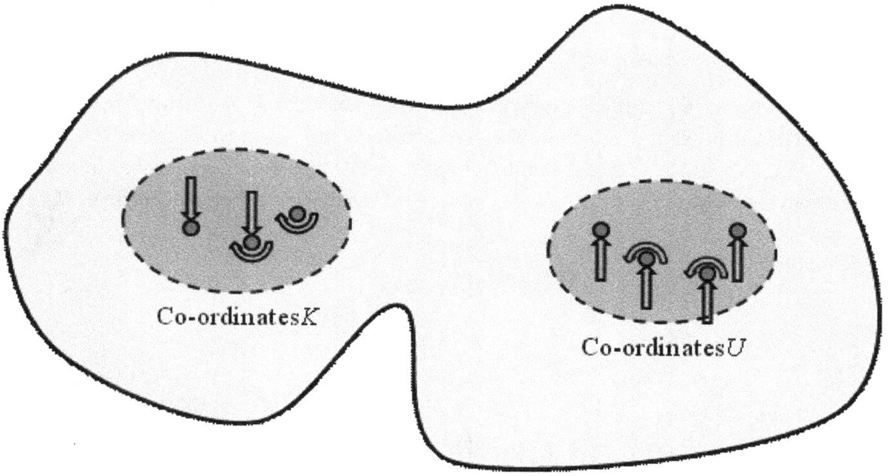

Figure 2. Illustration of both sets of co-ordinates K and U.

2.2.1. Transmissibility of forces in terms of FRFs

With the definition of the new sets K, U and C, the problem may be defined in the following way:

$$\begin{Bmatrix} \boldsymbol{Y}_K \\ \boldsymbol{Y}_U \\ \boldsymbol{Y}_C \end{Bmatrix} = \begin{bmatrix} \boldsymbol{H}_{KK} & \boldsymbol{H}_{KU} \\ \boldsymbol{H}_{UK} & \boldsymbol{H}_{UU} \\ \boldsymbol{H}_{CK} & \boldsymbol{H}_{CU} \end{bmatrix} \begin{Bmatrix} \boldsymbol{F}_K \\ \boldsymbol{F}_U \end{Bmatrix} \tag{13}$$

Imposing $\boldsymbol{Y}_U = \boldsymbol{0}$, it follows that

$$\boldsymbol{H}_{UK}\,\boldsymbol{F}_K + \boldsymbol{H}_{UU}\,\boldsymbol{F}_U = \boldsymbol{0} \tag{14}$$

and so

$$\boldsymbol{F}_U = \left(\boldsymbol{T}_f\right)_{UK}\,\boldsymbol{F}_K \tag{15}$$

where

$$\left(\boldsymbol{T}_f\right)_{UK} = -\left(\boldsymbol{H}_{UU}\right)^{-1}\boldsymbol{H}_{UK} \tag{16}$$

is the force transmissibility matrix.

This is the direct force identification method, i.e., one knows the applied forces and calculate the reactions at the supports, where the displacements are assumed as zero. The inverse problem is also possible, if one is able to measure the reaction forces and if their number is higher than the number of applied forces, in order to calculate the pseudo-inverse of $\boldsymbol{H}_{UK}$:

$$\boldsymbol{F}_K = \left(\left(\boldsymbol{T}_f\right)_{UK}\right)^{+}\boldsymbol{F}_U \tag{17}$$

where

$$\left(\left(\boldsymbol{T}_f\right)_{UK}\right)^{+} = -\left(\boldsymbol{H}_{UK}\right)^{+}\boldsymbol{H}_{UU} \tag{18}$$

Note that in spite of the fact that here the reaction forces are known, the notation U (that in principle stands for "unknown") is kept.

In the inverse problem, one may not know how many applied force exist and where they are applied. If that is the case, one must follow a different approach, as it will be explained in section 4.1

If the condition $\boldsymbol{Y}_U = \boldsymbol{0}$ is relaxed, from eq. (13) it follows that:

$$\boldsymbol{Y}_U = \boldsymbol{H}_{UK}\boldsymbol{F}_K + \boldsymbol{H}_{UU}\boldsymbol{F}_U \tag{19}$$

$$\begin{aligned} &\boldsymbol{F}_U = \left(\boldsymbol{T}_f\right)_{UK}\boldsymbol{F}_K + \left(\boldsymbol{H}_{UU}\right)^{-1}\boldsymbol{Y}_U && \text{a} \\ &\text{and} \\ &\boldsymbol{F}_K = \left(\left(\boldsymbol{T}_f\right)_{UK}\right)^{+}\boldsymbol{F}_U + \left(\boldsymbol{H}_{UK}\right)^{+}\boldsymbol{Y}_U && \text{b} \end{aligned} \tag{20}$$

2.2.2. Transmissibility of forces in terms of dynamic stiffness

Again, there is an alternative approach to obtain the force transmissibility matrix, using the dynamic stiffness matrices.

Assuming harmonic loading and the mentioned sets *K*, *U* and *C*, one can obtain (after grouping adequately the degrees of freedom of the problem) the following result:

$$\begin{bmatrix} \boldsymbol{Z}_{KK} & \boldsymbol{Z}_{KC} & \boldsymbol{Z}_{KU} \\ \boldsymbol{Z}_{CK} & \boldsymbol{Z}_{CC} & \boldsymbol{Z}_{CU} \\ \boldsymbol{Z}_{UK} & \boldsymbol{Z}_{UC} & \boldsymbol{Z}_{UU} \end{bmatrix} \begin{Bmatrix} \boldsymbol{Y}_K \\ \boldsymbol{Y}_C \\ \boldsymbol{Y}_U \end{Bmatrix} = \begin{Bmatrix} \boldsymbol{F}_K \\ \boldsymbol{F}_C \\ \boldsymbol{F}_U \end{Bmatrix} \tag{21}$$

It is worthwhile noting that joining together the sets *K* and *C* in a new set *E* makes it easier to see that imposing $\boldsymbol{Y}_U = \boldsymbol{0}$ one obtains the following relationships:

$$\begin{bmatrix} \boldsymbol{Z}_{EE} & \boldsymbol{Z}_{EU} \\ \boldsymbol{Z}_{UE} & \boldsymbol{Z}_{UU} \end{bmatrix} \begin{Bmatrix} \boldsymbol{Y}_E \\ \boldsymbol{0} \end{Bmatrix} = \begin{Bmatrix} \boldsymbol{F}_E \\ \boldsymbol{F}_U \end{Bmatrix} \tag{22}$$

from which it is clear that:

$$\begin{aligned} &\boldsymbol{Z}_{EE}\,\boldsymbol{Y}_E = \boldsymbol{F}_E && \text{a} \\ &\boldsymbol{Z}_{UE}\,\boldsymbol{Y}_E = \boldsymbol{F}_U && \text{b} \end{aligned} \tag{23}$$

Eliminating $\boldsymbol{Y}_E$ between (23a) and (23b), it turns out that

$$\boldsymbol{F}_U = \left(\boldsymbol{T}_f\right)_{UE} \boldsymbol{F}_E \tag{24}$$

where

$$\left(\boldsymbol{T}_f\right)_{UE} = \boldsymbol{Z}_{UE}\left(\boldsymbol{Z}_{EE}\right)^{-1} \tag{25}$$

The inverse problem corresponds to

$$\boldsymbol{F}_E = \left(\left(\boldsymbol{T}_f\right)_{UE}\right)^{+} \boldsymbol{F}_U \tag{26}$$

with

$$\left(\left(\boldsymbol{T}_f\right)_{UE}\right)^{+} = \boldsymbol{Z}_{EE}\left(\boldsymbol{Z}_{UE}\right)^{+} \tag{27}$$

It is important to note that only some of the co-ordinates of the set *E* have applied forces. This means that in (23) some rows of $\boldsymbol{F}_E$ are zero and only the columns (in $\boldsymbol{Z}_{EE}$) whose co-ordinates have applied forces (set *K*) are needed for the transmissibility matrix. In other words, from the set *E* only the co-ordinates corresponding to the *K* set are used.

2.3. Summary

From sections 2.2.1 and 2.2.2, one can conclude that for the direct problem of transmissibility of forces there is no restrictions in the number of co-ordinates used:

$$\begin{aligned} \left(\boldsymbol{T}_f\right)_{UK} &= -\left(\boldsymbol{H}_{UU}\right)^{-1}\boldsymbol{H}_{UK} \\ \left(\boldsymbol{T}_f\right)_{UE} &= \boldsymbol{Z}_{UE}\left(\boldsymbol{Z}_{EE}\right)^{-1} \end{aligned} \tag{28}$$

whereas in the inverse problem of transmissibility of forces there are some restrictions that can make this option not very useful in practice, especially when using the dynamic stiffnesses, since one needs to calculate the pseudo-inverse matrices:

$$\begin{aligned} \left(\left(\boldsymbol{T}_f\right)_{UK}\right)^{+} &= -\left(\boldsymbol{H}_{UK}\right)^{+}\boldsymbol{H}_{UU} \qquad && n_U \geq n_K \\ \left(\left(\boldsymbol{T}_f\right)_{UE}\right)^{+} &= \boldsymbol{Z}_{EE}\left(\boldsymbol{Z}_{UE}\right)^{+} \qquad && n_U \geq n_E \end{aligned} \tag{29}$$

3. Numerical and experimental applications

As explained before, the transmissibility matrices may be obtained from a numerical model (which should be updated for the range of frequencies involved) or from results obtained experimentally. In this section, the methodology used in each case is described and illustrated through a comparison example.

3.1. Transmissibility in terms of the numerical model

For the numerical model, one needs the knowledge of the structure within the discretization chosen, to create the receptance matrix $\boldsymbol{H}(\omega)$, which is the inverse of the corresponding dynamic stiffness matrix $\boldsymbol{Z}(\omega)$. Here, the numerical model is created using the Finite Element Method (FEM), although other alternatives may also be used. As seen before, the dynamic stiffness matrix is defined as:

$$\boldsymbol{Z}(\omega) = \boldsymbol{K} - \omega^2\boldsymbol{M} + i\omega\,\boldsymbol{C} \tag{30}$$

where C represents the viscous damping matrix, often of the proportional type, i.e., $C=\alpha K+\beta M$, where α and β are constants to be evaluated experimentally.

To build the dynamic stiffness matrix, a specific structural finite element is chosen according to the approximation considered. For example, in the case of a reasonably long and slender beam one can use the Euler-Bernoulli beam element (instead of a shell or solid structural element). Then, the global matrices are assembled for the chosen discretization of the structure.

In order to improve the accuracy of the numerical model when simulating what is obtained experimentally, concentrated masses are often added at the corresponding nodes to model the effect of the accelerometers used in the testing positions.

Although the receptance matrix $H(\omega)$ is the inverse of the corresponding dynamic stiffness matrix, one should avoid such direct numerical inversion (frequency by frequency). Instead, $H(\omega)$ is calculated from eq. (3), after a modal analysis in free vibration.

Then, using (8) or (16), one can calculate the needed transmissibility matrices $(T_d)_{UK}^A$ or $(T_f)_{UK}$. Of course, alternatively one may use the equivalent expressions (11) or (25), respectively.

3.2. Transmissibility in terms of experimental measurements

Depending on the type of transmissibility to obtain, the corresponding experimental setup should be established. Essentially, it is important to observe that for the transmissibility of motion one measures the FRFs relating co-ordinates U and K with co-ordinates A, normally using accelerometers and force transducers. For validation purposes, one may also measure the applied forces. In the examples presented next for the transmissibility of motion, a suspended (free-free) beam is always used.

For the transmissibility of forces, in the direct problem, one measures the applied forces at co-ordinates K (in the inverse problem, one measures the reaction forces at co-ordinates U). For validation purposes, one also measures the ones to be estimated. The test specimen for the transmissibility of forces is always a simply supported beam.

For the experimental setup, the following equipment is used:

- Vibration exciter (Brüel & Kjær Type 4809);
- Power amplifier (Brüel & Kjær Type 2706);
- Force transducers(PCB PIEZOTRONICS Model 208C01);
- Data acquisition equipment (Brüel & Kjær Type 3560-C).

In Fig. 3, a schematic representation of the experimental setup used for the force transmissibility tests is presented, in this case a simply supported beam with a single applied force.

The excitation signal used was a multi-sine transmitted to the exciter, with constant amplitude in the frequency. In reality the signal measured by the force transducer does not exhibits a constant amplitude along the frequency, as it depends on the dynamic response of the structure.

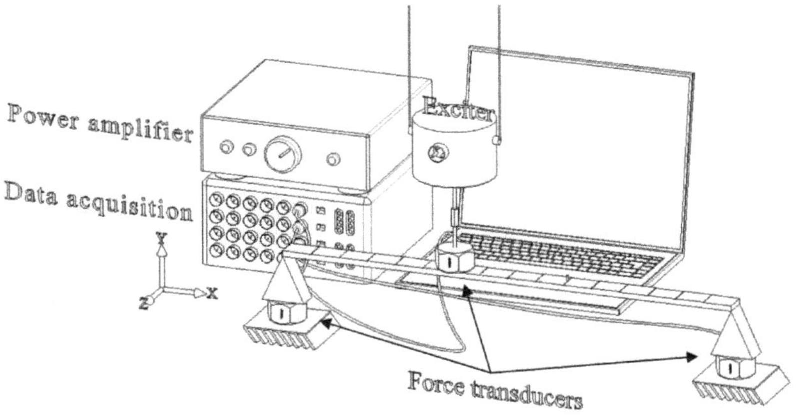

Figure 3. Example of the experimental setup developed for the transmissibility of forces.

For the transmissibility of motion, a beam suspended by nylon strings is used, to simulate free-free conditions. In order to facilitate the interchange of the available accelerometers between the measure positions without affecting the dynamics of the structure, it is important to add equivalent masses (dummies) to model the effect of the sensors.

After obtaining experimentally the needed receptances, by using (8) or (16) one can establish the transmissibility matrices $(\boldsymbol{T}_d)_{UK}^{A}$ or $(\boldsymbol{T}_f)_{UK}$.

3.3. Examples

The same steel beam was used in all the examples. With the purpose of illustrating the applicability of the presented formulations to obtain the transmissibility plots, the authors used the geometric and material parameters presented in Table 1. Note that these data correspond to the values obtained after updating the FE model.

Young's modulus – E	208 GPa
Density – ρ	7840 kg/m³
Length – L	0.8 m
Section width - b	$5.0\times10^{-3}m$
Section height - h	$20.0\times10^{-3}m$
Section area - A	$1\times10^{-4}m^2$
Second moment of area - I	$2.0883\times10^{-10}m^4$
proportional damping - α	$4s$
proportional damping - β	$2.0\times10^{-6}s^{-1}$

Table 1. Beam properties (after updating).

3.3.1. *Numerical model*

The standard two-node Euler-Bernouli bidimensional finite element is used here to build the needed numerical model of the beam.

The beam was discretized into sixteen finite elements, which correspond to $N = 17$ nodes, ordered from 1 up to 17. As the analysis and the model are limited to the plane xOy, each node has three degrees of freedom (which are u_x, u_y, θ). Hence, the matrices of the numerical model have an order of $3xN$ for the free-free beam. In what the measurements are concerned, only the displacements and applied forces along the y direction are used and therefore the numbering of nodes and co-ordinates y coincide.

The model was updated using E, ρ and I as updating parameters and a proportional damping model is included using α and β as updating parameters (Table 1).

3.3.2. *Example 1 — Transmissibility of motion*

In this example, the free-free beam has only one applied force at node 11 along the y direction, denoted as F_{11}, and the measurements are taken at nodes 3, 7, 13 and 17 also along the y direction, denoted as Y_3, Y_7, Y_{13}, and Y_{17} (Fig. 4).

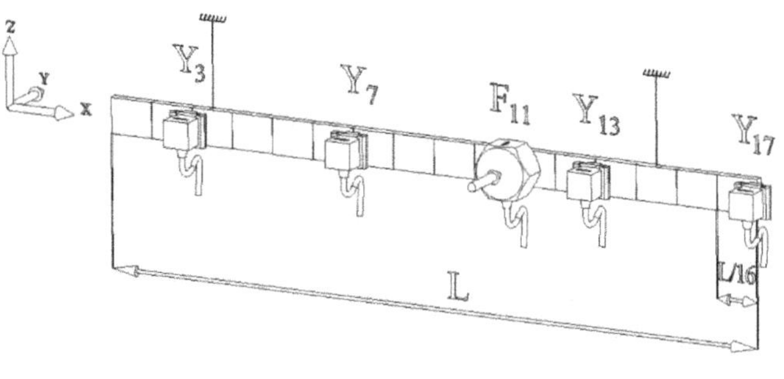

Figure 4. Schematic representation of the accelerometers and force transducer positions.

The transmissibility of motion between the pair of nodes (3,7) and the pair of nodes (13,17) may be expressed as

$$Y_U = (\mathbf{T}_d)^A_{UK}\, Y_K \quad \Leftrightarrow \quad \begin{Bmatrix} Y_{13} \\ Y_{17} \end{Bmatrix} = \begin{bmatrix} T_{13,3} & T_{13,7} \\ T_{17,3} & T_{17,7} \end{bmatrix} \begin{Bmatrix} Y_3 \\ Y_7 \end{Bmatrix} \tag{31}$$

These transmissibilities were obtained from eq. (8), using the FRFs calculated from the numerical model and measured experimentally. Two of them are plotted in Fig. 5, where one can observe that both ways are able to produce the transmissibility response of the structure, as they match reasonably well.

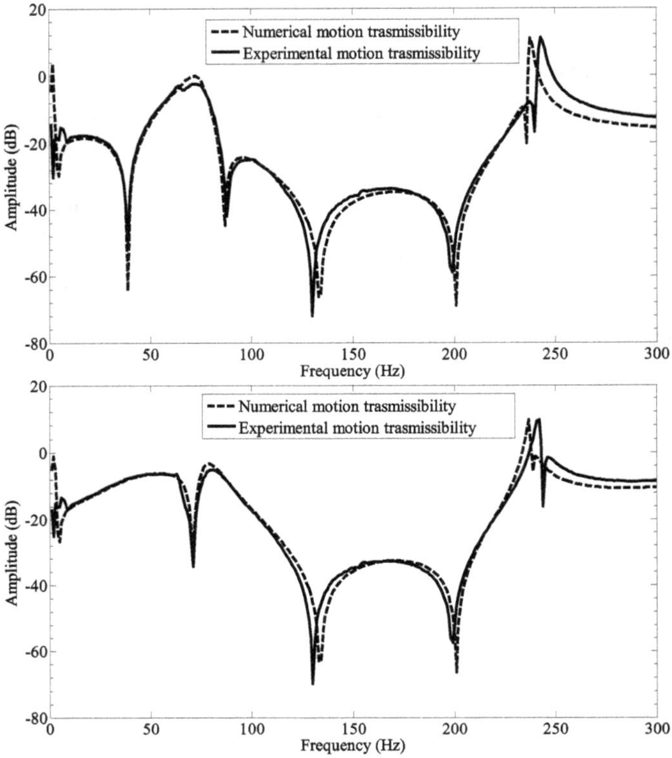

Figure 5. Numerical and experimental transmissibilities (upper plot, $T_{13,3}$; bottom plot, $T_{13,7}$)

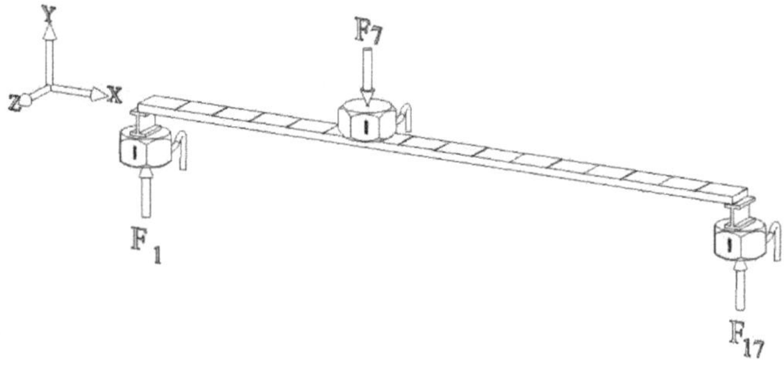

Figure 6. Schematic representation of the positions of the force transducers.

3.3.3. Example 2 — Transmissibility of forces

In this case, a simply supported beam is considered with one applied force at node 7 and reactions at nodes 1 and 17. Only the magnitude of the forces is measured, and the transmissibility is obtained directly from the measurements and compared with the numerical results. The experimental setup is illustrated in Fig. 6.

The force transmissibility relation between node 7 and the pair of nodes (1,17) may be expressed as

$$\begin{Bmatrix} F_1 \\ F_{17} \end{Bmatrix} = \begin{bmatrix} T_{1,7} \\ T_{17,7} \end{bmatrix} \{F_7\} \tag{32}$$

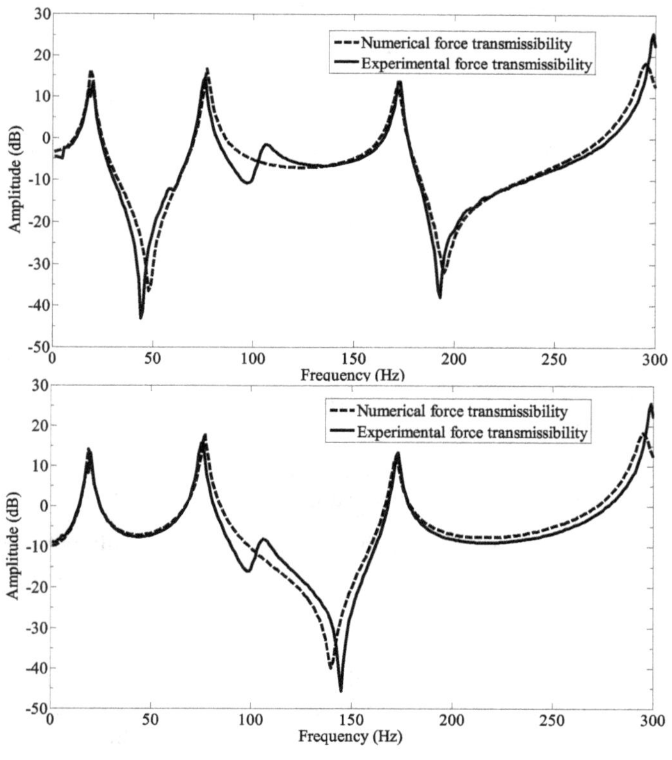

Figure 7. Numerical and experimental transmissibilities (upper plot, $T_{1,7}$; bottom plot, $T_{17,7}$)

The force transmissibilities were obtained using eq. (16) and are plotted in Fig. 7, where it is clear that both numerical and experimental FRFs are able to produce the transmissibility response of the structure. Note that around 100 Hz there is a "bump" in the experimental curve, due to the effect of the supports of the beam themselves; this effect has not been included in the numerical model because it was not important, as these results are only of an illustrative type.

4. Force identification

This section shall be divided into (i) part one for the force localization algorithm based on the transmissibility of motion and reconstruction using the measured responses and the updated numerical model, and (ii) part two for the force reconstruction using the transmissibility of forces.

4.1. Force localization based on the transmissibility of motion and force reconstruction

The force identification problem is a difficult matter, as one has a limited knowledge of the measured responses, due to the complexity of the structure, lack of access to some locations, etc. In other words, there are difficulties due to the incompleteness of the model.

Due to this difficulty in calculating the load vector directly, the authors propose to divide the process into two distinct steps:

1. the localization of the forces, i.e. the identification of the number and position of the applied forces using the concept of transmissibility of motion;
2. the load vector reconstruction.

For the first step, a search for the number and position of forces using the transmissibility of motion is performed. Essentially, this step consists of searching for the transmissibility matrix correspondent to the dynamics of the system and using the available measured data and the numerical model involved.

Once the corresponding transmissibility matrix is found, one has a solution for the number and position of the forces applied to the structure.

The second step consists of reconstructing the load vector with the results obtained in the first step. A more detailed description about this methodology is given in the following sections.

4.1.1. Force localization

In a first stage, to apply the method proposed in the previous section, one finds the transmissibility matrix that converts the dynamic responses $\boldsymbol{Y}_K$ into $\boldsymbol{Y}_U$. As one does not know the position of the applied forces, it was decided to cover all the possibilities until the calculated responses ($\boldsymbol{Y}_U$) match the measured ones $\tilde{\boldsymbol{Y}}$, over a range of frequencies. To calculate the vector $\boldsymbol{Y}_U$ one may use either eq. (7) or (11).

The maximum number of forces must be less or equal to the dimension of the known dynamic response vector $\boldsymbol{Y}$.

The successive combinations of the tested nodes are obtained according to the following scheme:

$$\text{For one force:} \quad \{(1),\dots\ (N);$$

$$\text{For two forces:} \quad \begin{cases} (1,2),\dots\ (1,N); \\ (2,3),\dots\ (2,N); \\ (3,4),\dots\ (3,N); \\ \quad \vdots \\ \quad (N-1,N); \end{cases}$$

$$\text{For three forces:} \quad \begin{cases} (1,2,3),\dots\ (1,2,N) \\ \quad \vdots \end{cases}$$

$$\vdots$$

The error in each combination is kept in a vector to identify the combination with the least associated error (in absolute value). Firstly, the algorithm scrolls through the possible combinations of position and number of forces. For each combination, the associated error between the calculated vector $\boldsymbol{Y}_U$ and the measured response vector $\tilde{\boldsymbol{Y}}$ is calculated; this is carried out over a frequency range defined by the user. The error between the predicted and the measured dynamic response at each co-ordinate i can be defined as:

$$\text{error}_i = \sum_{\omega}\left(\log\left(\text{abs}\left(\tilde{Y}_{U_i}(\omega)\right)\right) - \log\left(\text{abs}\left(Y_{U_i}(\omega)\right)\right)\right)^2 \tag{33}$$

For each combination, the calculated error is kept in an entry of the error vector and analyzed later on:

$$\boldsymbol{\varepsilon} = \{\text{error}_\text{i}\} \tag{34}$$

The accumulated error for a given combination of co-ordinates where $\boldsymbol{F}$ can be located is the norm of $\boldsymbol{\varepsilon}$. The calculations are repeated for sucessive combinations of number and position of forces. The combination of the force locations that gives the lowest error leads to the number and position of the forces applied to the structure. As already mentioned, the maximum number of forces that can be found is equal to the dimension of the known dynamic response vector.

As one does not know *a priori* how many forces exist, one has to follow a trial and error procedure that consists basically in assuming an increasing number of forces and the corresponding number of measurements; if the right number of forces is N_f, one has a minimum error ε for a certain set of co-ordinates. When one proceeds and assumes N_f *+1* forces and measurements, the error will be higher then ε, telling us that the right answer was effectively N_f at a certain set of co-ordinates.

It is clear that all the combinations of the N_f*+1* forces that contain the right combination of the N_f forces should exhibit a local minimum, though not the absolute one.

The method was implemented computationally (in MatLab®).

4.1.2. Force reconstruction

In a second step, the reconstruction of the force amplitudes consists of solving an inverse problem using the measured dynamic responses $\boldsymbol{Y}_K$:

$$\boldsymbol{F}_A = \left(\boldsymbol{H}_{KA}\right)^+ \boldsymbol{Y}_K \tag{35}$$

Note that for the given system to be invertible, the number of dynamic responses to be used (set K) must be higher or equal than the number of applied forces (set A). However, this is always verified, as in the first step one has already imposed it.

4.1.3. Example 3 — Localization of the applied forces

This is a numerical example, illustrated in Fig.8, where a set of uncorrelated forces is applied at co-ordinates 1 and 5 (set *A*), and one uses the three known responses (set *K*) to identify the number and location of forces.

A set of simulated results (to mimic the experimental measurements) are obtained at nodes 1, 3, 5, 11 and 17 (see Fig. 8); they define the following sets:

$$\mathbf{Y}_K = \{Y_3 \quad Y_5 \quad Y_{17}\}^T \quad \text{and} \quad \mathbf{Y}_U = \{Y_1 \quad Y_{11}\}^T \tag{36}$$

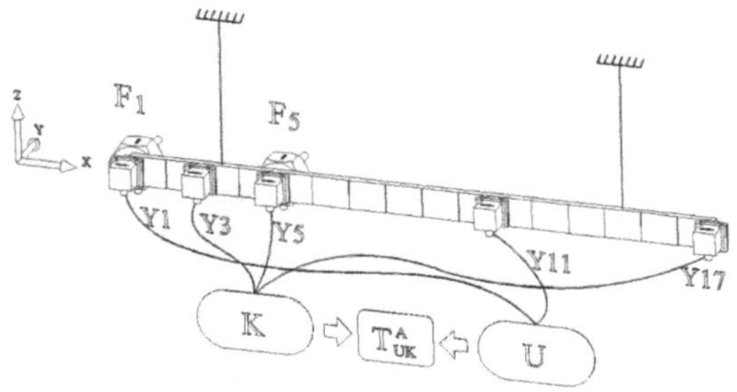

Figure 8. Illustration with the responses and applied force locations for example 3.

Considering these responses, the maximum number of identifiable applied forces is three, as explained before. The forces are uncorrelated and applied to the structure at co-ordinates 1 and 5. A series of force combinations have to be systematically generated as follows. In this case, all combinations up to three forces have been considered:

$$\text{one force}: \quad \{ \ (1),\ (2),\ (3),\ldots\ (17); \qquad \#\text{combination number } [1 \text{ to } 17]$$

$$\text{two forces}: \quad \begin{cases} (1,2),\ (1,3),\ (1,4),\ (1,5)\ \ldots(1,17); \\ \vdots \\ (16,17); \end{cases} \qquad \#\text{combination number } [18 \text{ to } 153]$$

$$\text{three forces}: \quad \begin{cases} (1,2,3)\ldots \\ \vdots \end{cases} \qquad \#\text{combination number } [154 \text{ to } 833]$$

Applying the localization method described in the previous subsection 4.1.1, one obtains the plot of the error defined in eq. (33), as shown in Fig.9.

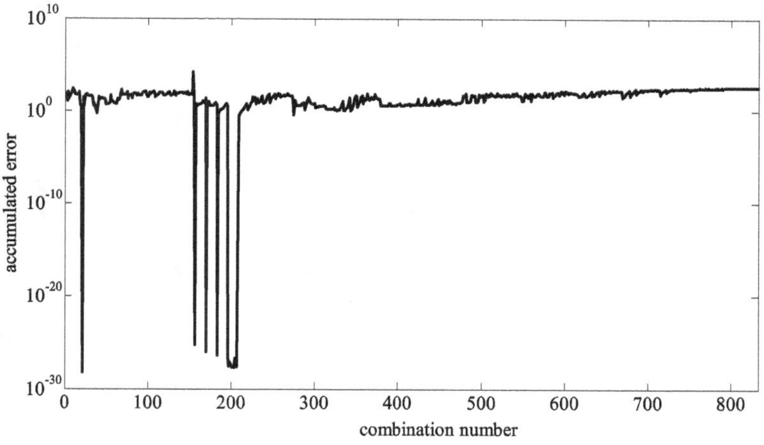

Figure 9. Accumulated error in frequency for each combination of forces.

It is clear that there are several situations (combinations) where the error is close to zero and other where is not.

The minimum error happens with the combination number 21, corresponding to two forces applied at co-ordinates 1 and 5, thus identifying the correct positions and number of forces (Table 2).

Combination	Number of forces	Position of the forces	Number of identified forces	Identified positions	Absolute error
21	2	1, 5	2	1, 5	2,69e-29

Table 2. Data of the combination with minimum error.

To better understand why there exist more combinations with small errors, Table 3 shows these combinations with its corresponding error value. All of them have a common group of co-ordinates, corresponding to the correct combination of number and positions of the forces. In this case the correct positions are obtained with success through the minimum error.

Combination	Real position of the forces	Absolute error
21	1,5	2,69e-29
156	1,2,5	7,99e-26
170	1,3,5	2,85e-27
183	1,4,5	6,99e-28
197	1,5,7	8,70e-29
198	1,5,8	1,86e-28
199	1,5,9	1,82e-28
200	1,5,10	5,54e-28

Table 3. Some combinations and their respective error.

This illustrates the localization step performed with two forces, whose number and location were not known at the beginning. From these results, it can be stated that the transmissibility of motion can be considered as adequate to perform this task. Note that, in spite of the high numbers of combinations that exist, the computations are relatively quick, as they involve only sub-matrices, and for the first permutations they are of a small order.

4.1.4. Example 4 — Localization and reconstruction 1

This is an experimental example, where a multisine signal is fed into the shaker, attached to the beam at co-ordinate 13. Later on, the applied force is compared with the reconstructed one.

The experimental measurements are obtained at nodes 5, 7, 11 and 15. The measured vectors are as follows:

$$\mathbf{Y}_K = \{Y_7 \quad Y_{15}\}^T \quad \text{and} \quad \mathbf{Y}_U = \{Y_5 \quad Y_{11}\}^T \tag{37}$$

Considering these responses, the maximum number of identifiable applied forces is two, as explained before.

A series of force combinations was systematically generated as described before. Applying the localization method, one obtains the graph of Fig. 10.

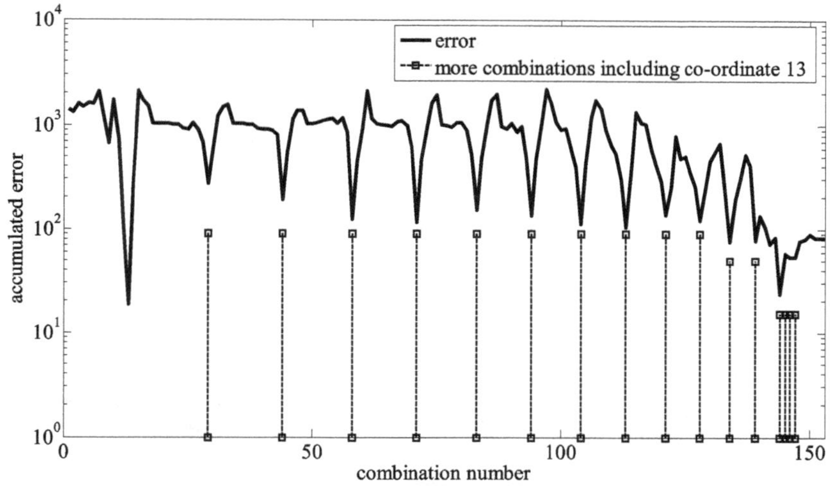

Figure 10. Accumulated error in frequency for each force combination.

As one can see, the absolute minimum corresponds to combination no. 13, which is right because this combination represents the force applied at co-ordinate 13. So, the method could localize correctly the position of the force at co-ordinate 13.

Once the localization of the force is accomplished, its reconstruction is a simple calculation, using the measured displacements relating those co-ordinates to the force location. As the force is located at node 13, taking the measurement at co-ordinates 5, 7, 11 and 15, for instance, it follows that:

$$\begin{Bmatrix} \tilde{Y}_5 \\ \tilde{Y}_7 \\ \tilde{Y}_{11} \\ \tilde{Y}_{15} \end{Bmatrix} = \begin{Bmatrix} H_{5,13} \\ H_{7,13} \\ H_{11,13} \\ H_{15,13} \end{Bmatrix} F_{13} \Leftrightarrow F_{13} = \begin{Bmatrix} H_{5,13} \\ H_{7,13} \\ H_{11,13} \\ H_{15,13} \end{Bmatrix}^{+} \begin{Bmatrix} \tilde{Y}_5 \\ \tilde{Y}_7 \\ \tilde{Y}_{11} \\ \tilde{Y}_{15} \end{Bmatrix} \tag{38}$$

Using the information from the measured responses, the reconstruction is now immediate. To validate this methodology, the result was ploted against its experimentally measured curve, as in Fig. 11. One may affirm that the method is able to predict the applied force. It is possible that a better matching of the curves may be obtained with a finer updated FE model.

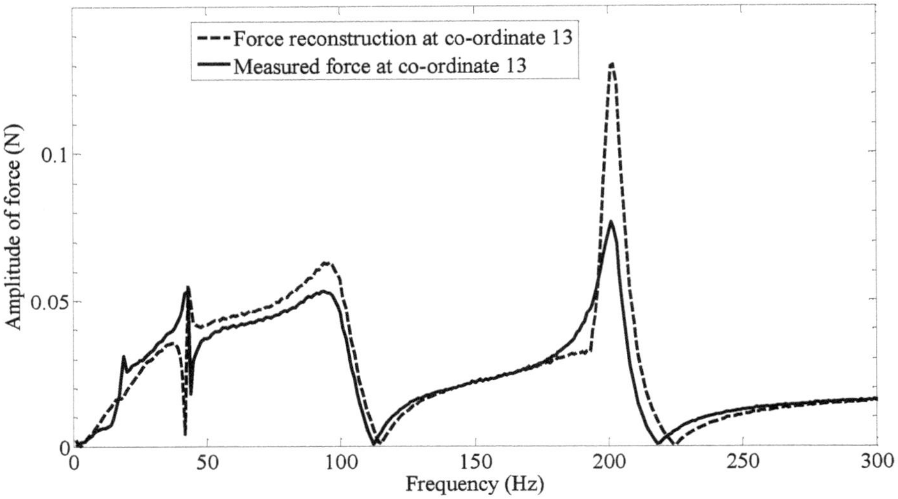

Figure 11. Comparison between the experimental and the reconstructed forces at co-ordinate 13.

4.1.5. Example 5 — Localization and reconstruction 2

Here, the same multisine signal is fed into two shakers, attached to the free-free beam at the co-ordinates 1 and 11. Later on, the applied forces are compared with the reconstructed ones.

The experimental measures are obtained at nodes 3, 7, 17 and 13. The measured vectors are as follows.

$$\mathbf{Y}_K = \{Y_3, Y_7, Y_{17}\}^T \quad \text{and} \quad \mathbf{Y}_U = \{Y_{13}\} \tag{39}$$

Considering these responses, the maximum number of identifiable applied forces is three. Applying the localization method, one obtains the plot shown in Fig.12.

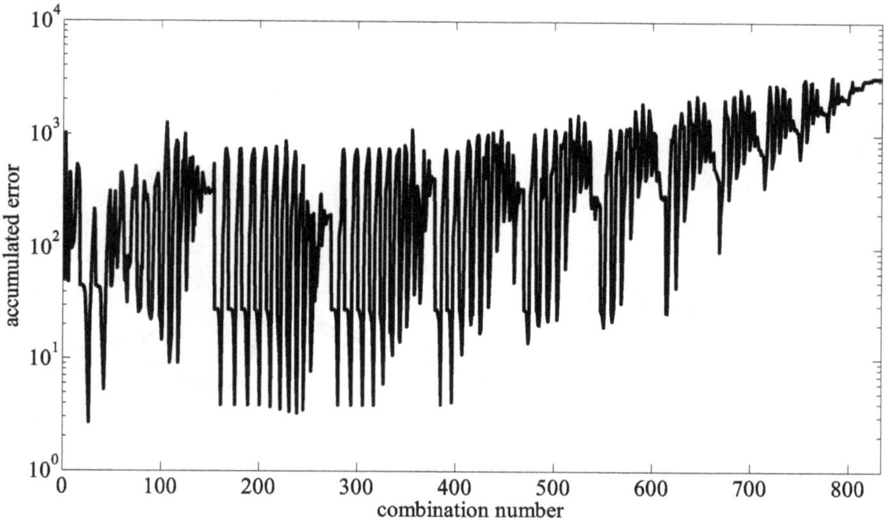

Figure 12. Accumulated error in frequency for each force combination.

As one can see, the absolute minimum corresponds to the right combination (no. 27), representing the forces applied at co-ordinates 1 and 11. So, the method located correctly the position of the forces. Note that, as there are two forces, a high number of combinations with small errors appear; observing the co-ordinates of those combinations, the best of them have in common the correct co-ordinates where the forces are applied and the others include co-ordinates physically close to them. Again, the force reconstruction is obtained using the measured displacements:

$$\begin{Bmatrix} \tilde{Y}_3 \\ \tilde{Y}_7 \\ \tilde{Y}_{13} \\ \tilde{Y}_{17} \end{Bmatrix} = \begin{bmatrix} H_{3,1} & H_{3,11} \\ H_{7,1} & H_{7,11} \\ H_{13,1} & H_{13,11} \\ H_{17,1} & H_{17,11} \end{bmatrix} \begin{Bmatrix} F_1 \\ F_{11} \end{Bmatrix} \Leftrightarrow \begin{Bmatrix} F_1 \\ F_{11} \end{Bmatrix} = \begin{bmatrix} H_{3,1} & H_{3,11} \\ H_{7,1} & H_{7,11} \\ H_{13,1} & H_{13,11} \\ H_{17,1} & H_{17,11} \end{bmatrix}^+ \begin{Bmatrix} \tilde{Y}_3 \\ \tilde{Y}_7 \\ \tilde{Y}_{13} \\ \tilde{Y}_{17} \end{Bmatrix} \quad (40)$$

Figs. 13 and 14 present the reconstructed forces versus the measured ones. Again, one can state that the method is able to predict the applied force.

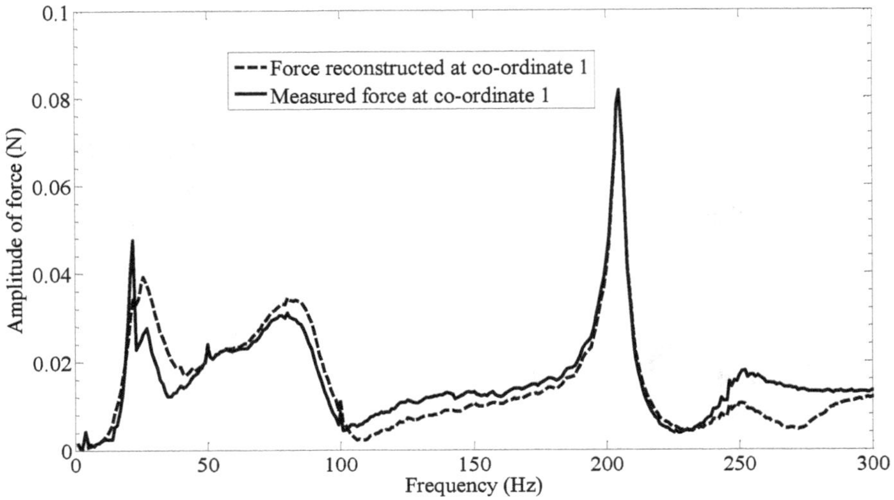

Figure 13. Comparison between the experimental and the reconstructed forces at co-ordinate1

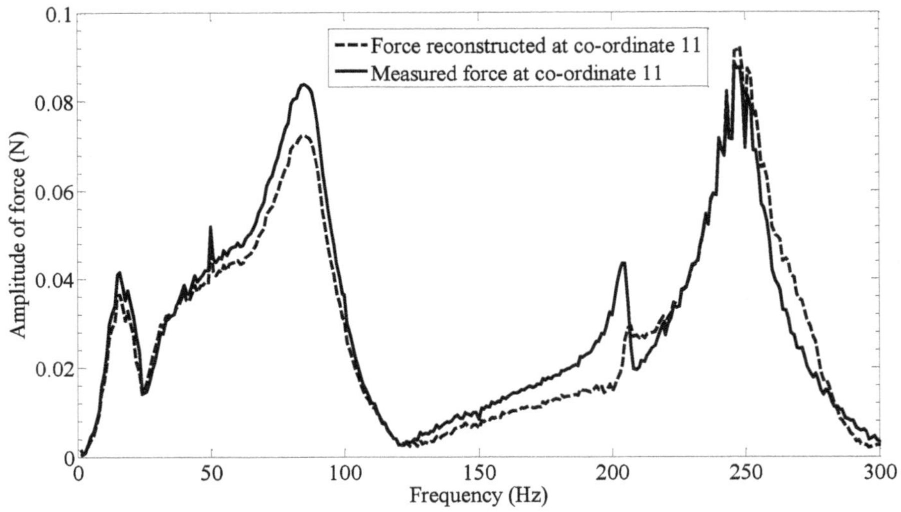

Figure 14. Comparison between the experimental and the reconstructed forces at co-ordinate11

4.2. Force reconstruction based on the transmissibility of forces

The main objective of this section is the estimation of the existing forces (reactions or applied forces) in the structure using the MDOF concept of transmissibility of forces. Two types of problems involving estimation of forces are here considered:

1. Reaction forces estimation, with the objective of calculating a set of unknown reactions from a set of known applied loads, as expressed by equation (15);

2. Applied forces estimation, with the objective of calculating a set of applied forces from a set of known reactions, as expressed by equation (17).

The method to estimate the applied forces is limited by the number of reactions, as it is not possible to perform the needed pseudo-inverse if the number of applied forces is greater than the number of reactions. So, it is a required condition that$n_K \leq n_U$.

4.2.1. Example 6 — Reaction forces estimation knowing the applied ones

The first experimental reconstruction case was carried out with the configuration presented in Fig. 6 (simple supported beam). One has a single applied force at node 7 (set *K*) and two reactions at nodes 1 and 17 (set *U*).

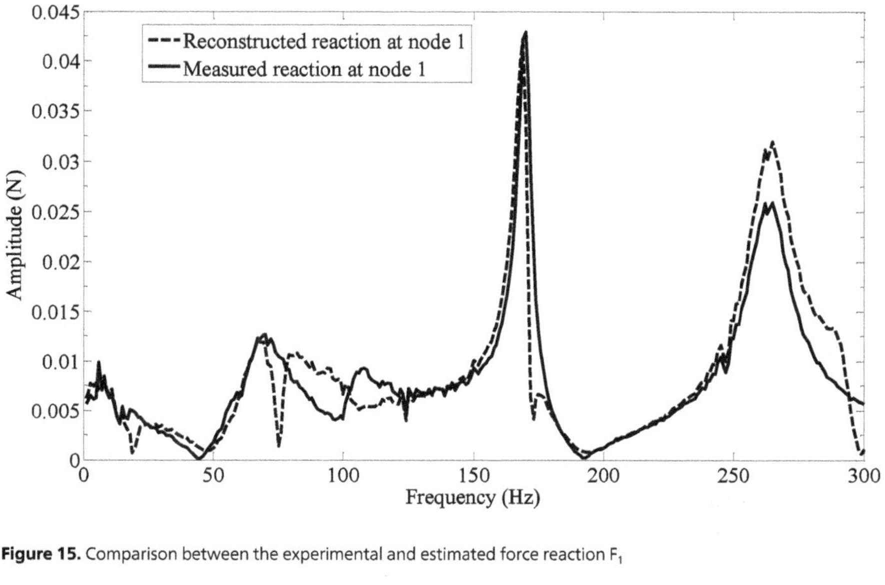

Figure 15. Comparison between the experimental and estimated force reaction F_1

In this case, with one applied force and two reactions, the transmissibility has a dimension of 2x1 and can be obtained either from the receptance matrix or from the dynamic stiffness matrix, as proposed in this work. The two different formulations are equivalent and are very close to the experimental results.

As the objective is to estimate the reaction forces, one needs the numerical model for the transmissibility matrix and to know the experimental vector of applied forces, which in this case has only one component. The calculation of the reactions is then reduced to the following form:

$$\begin{Bmatrix} F_1 \\ F_{17} \end{Bmatrix} = \begin{bmatrix} T_{1,7} \\ T_{17,7} \end{bmatrix} \{F_7\} \tag{41}$$

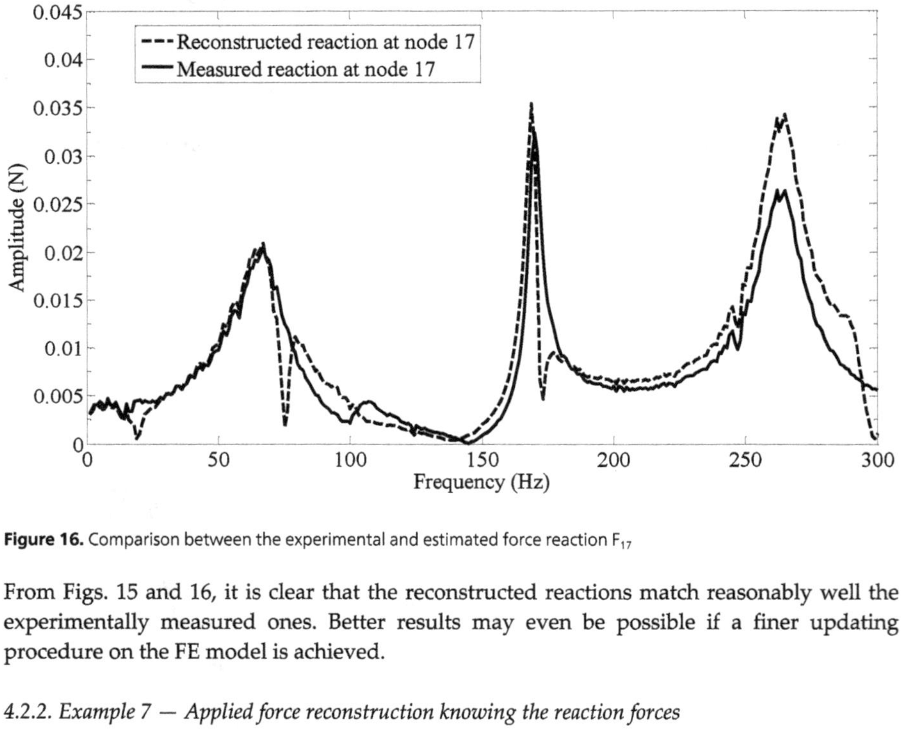

Figure 16. Comparison between the experimental and estimated force reaction F_{17}

From Figs. 15 and 16, it is clear that the reconstructed reactions match reasonably well the experimentally measured ones. Better results may even be possible if a finer updating procedure on the FE model is achieved.

4.2.2. Example 7 — Applied force reconstruction knowing the reaction forces

For the reconstruction of the applied forces (the inverse problem), one needs to known the vector of the reaction forces $\{F_U\}$ and the inverse transmissibility matrix that can be obtained from the numerical model.

In this case the same configuration presented in Fig. 6 was used (simple supported beam), with one applied force at node 7 (set K) and two reaction forces at nodes 1 and 17 (set U).

Knowing the reaction forces, the reconstruction of the applied force follows, in this case, the following expression:

$$\{F_7\} = \begin{bmatrix} T_{1,7} \\ T_{17,7} \end{bmatrix}^+ \begin{Bmatrix} F_1 \\ F_{17} \end{Bmatrix} \tag{42}$$

The reconstructed values are compared with the experimentally measured ones, in Fig. 17.

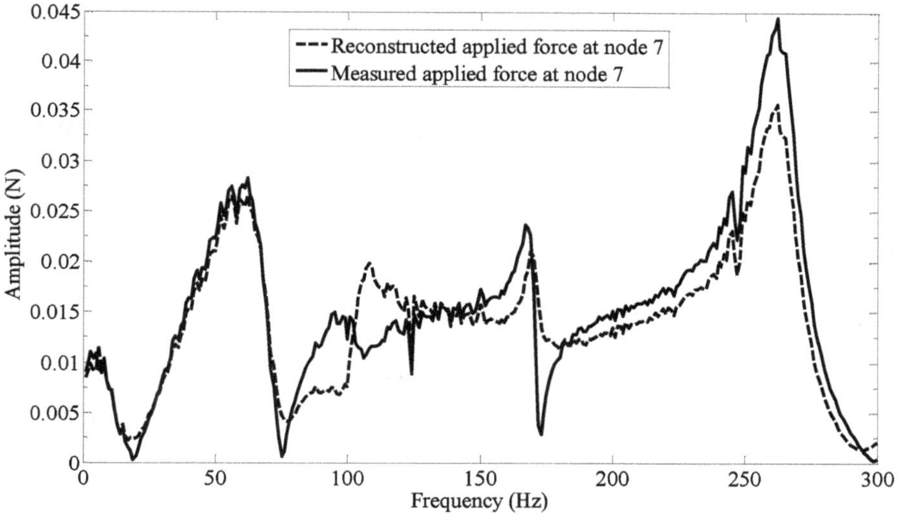

Figure 17. Comparison between the experimental and estimated applied load F7

In all the tested cases a good approach of the reconstructed forces was verified, as the values obtained by the direct and inverse problems are close enough to the experimentally measured ones.

5. Conclusions

In this work, the authors reviewed recent advances in the application of MDOF transmissibility-based methods for the identification of forces.

From these developments, one can draw the following main conclusions:

i. it is possible to localize forces acting on a structure, based on the motion transmissibility matrix, comparing the expected responses with the ones measured along the structure;

ii. finding where the forces are applied corresponds to finding the transmissibility matrix related to the smallest error between the expected responses and measured ones along the frequency range;

iii. in all the examples the identification of the number of forces and their localization have been accomplished;

iv. the magnitude of the estimated forces exhibits a very good correlation with the measured ones for most of the frequency range.

Acknowledgements

The current investigation had the support of IDMEC/IST and FCT, under the project PTDC/EME-PME/71488/2006.

Author details

N. M. M. Maia, Y. E. Lage and M. M. Neves

*Address all correspondence to: nmaia@dem.ist.utl.pt

IDMEC-IST, Technical University of Lisbon, Department Mechanical Engineering, Lisboa, Portugal

References

[1] Maia, N. M. M, Urgueira, A. P. V, & Almeida, R. A. B. Whys and Wherefores of Transmissibility. In: Dr. Francisco Beltran-Carbajal (ed.) Vibration Analysis and Control- New Trends and Developments, InTech; (2011). http://www.intechopen.com/books/vibration-analysis-and-control-new-trends-and-developments/whys-and-wherefores-of-transmissibility., 187-216.

[2] Hillary, B. Indirect Measurement of Vibration Excitation Forces. PhD Thesis. Imperial College of Science, Technology and Medicine, Dynamics Section, London, UK; (1983).

[3] Stevens, K. K. Force Identification Problems- an overview. Proceedings of the (1987). SEM Spring Conference on Experimental Mechanics. Houston, TX, USA; 1987.

[4] Mas, P, Sas, P, & Wyckaert, K. Indirect force identification based on impedance matrix inversion: a study on statistical and deterministic accuracy. Proceedings of 19th International Seminar on Modal Analysis, Leuven. Belgium, (1994)., 1049-1065.

[5] Dobson, B. J, & Rider, E. A review of the indirect calculation of excitation forces from measured structural response data. Proceedings of the Institution Mechanical Engineers, Part C, Journal of Mechanical Engineering Science,(1990)., 69-75.

[6] Ma, C. K, Chang, J. M, & Lin, D. C. Input Forces Estimation of Beam Structures by an Inverse Method. Journal of Sound and Vibration, (2003)., 387-407.

[7] Ma, C. K, & Ho, C. C. An Inverse Method for the Estimation of Input Forces Acting On Non-Linear Structural Systems. Journal of Sound and Vibration, (2004)., 953-971.

[8] Thite, A. N, & Thompson, D. J. The quantification of structure-borne transmission paths by inverse methods. Part 1: improved singular value rejection methods. Journal of Sound and Vibration, (2003)., 411-431.

[9] Thite, A. N, & Thompson, D. J. The quantification of structure-borne transmission paths by inverse methods. Part 2: use of regularization methods, Journal of Sound and Vibration, (2003)., 433-451.

[10] Choi, H, Thite, G, & Thompson, A. N. D. J. A threshold for the use of Tikhonov regularization in inverse force determination. Applied Acoustics, (2006)., 700-719.

[11] Michaels, J. E, & Pao, Y. H. The inverse source problem for an oblique force on an elastic plate. Journal of the Acoustic. Society of America, June (1985)., 2005-2011.

[12] Martin, M. T, & Doyle, J. F. Impact force identification from wave propagation responses. International Journal of Impact Engineering, January (1996)., 65-77.

[13] Huang, C. H. An inverse non-linear force vibration problem of estimating the external forces in a damped system with time-dependent system parameters. Journal of Sound and Vibration, (2001)., 749-765.

[14] Tadeusz UhlThe inverse identification problem and its technical application, Arch Appl Mech, (2007)., 325-337.

[15] Zhang, Q, Allemang, R. J, & Brown, D. L. Modal Filter: Concept and Applications, 8th International Modal Analysis Conference (IMAC VIII), Kissimmee, Florida, (1990)., 487-496.

[16] Kammer, D. C, & Steltzner, A. D. Structural identification of Mir using inverse system dynamics and Mir/shuttle docking data, Journal of Vibration and Acoustics, (2001)., 230-237.

[17] Allen, M. S, & Carne, T. G. Comparison of Inverse Structural Filter (ISF) and Sum of Weighted Accelerations (SWAT) Time Domain Force Identification Methods, Mechanical Systems and Signal Processing,(2008)., 1036-1054.

[18] Paulo, P. A Time-Domain Methodology For Rotor Dynamics: Analysis and Force Identification. MSc thesis. Instituto Superior Técnico Lisbon; (2011).

[19] Vakakis, A. F. Dynamic Analysis of a Unidirectional Periodic Isolator, Consisting of Identical Masses and Intermediate Distributed Resilient Blocks. Journal of Sound andVibration, (1985)., 25-33.

[20] Vakakis, A. F, & Paipetis, S. A. Transient Response of Unidirectional Vibration Isolatorswith Many Degrees of Freedom, Journal of Sound and Vibration, (1985). 0002-2460X., 99(4), 557-562.

[21] Vakakis, A. F, & Paipetis, S. A. The Effect of a Viscously Damped Dynamic Absorber ona Linear Multi-Degree-of-Freedom System, Journal of Sound and Vibration, (1986). (1), 49-60.

[22] Liu, W. Structural Dynamic Analysis and Testing of Coupled Structures. PhD thesis. Imperial College London; (2000).

[23] Liu, W, & Ewins, D. J. Transmissibility Properties of MDOF Systems, Proceedings of the16th International Operational Modal Analysis Conference (IMAC XVI), Santa Barbara, California, (1998)., 847-854.

[24] Varoto, P. S, & Mcconnell, K. G. Single Point vs Multi Point AccelerationTransmissibility Concepts in Vibration Testing, Proceedings of the 12th International Modal Analysis Conference (IMAC XVI), Santa Barbara, California, USA,(1998)., 83-90.

[25] Ribeiro, A. M. R. On the Generalization of the Transmissibility Concept, Proceedings of the NATO/ASI Conference on Modal Analysis and Testing, Sesimbra, Portugal, (1998)., 757-764.

[26] Fontul, M, Ribeiro, A. M. R, Silva, J. M. M, & Maia, N. M. M. Transmissibility Matrix in Harmonic and Random Processes, Shock and Vibration, (2004)., 563-571.

[27] Maia, N. M. M, Fontul, M, & Ribeiro, A. M. R. Transmissibility of Forces in Multiple-Degree-of-Freedom Systems, Proceedings of ISMA (2006). Noise and Vibration Engineering, Leuven, Belgium 2006.

[28] Rao, S. S. Mechanical Vibrations. Fourth International Edition. Prentice-Hall; (2004).

Chapter 7

Vibration and Optimization Analysis of Large-Scale Structures using Reduced-Order Models and Reanalysis Methods

Zissimos P. Mourelatos, Dimitris Angelis and John Skarakis

Additional information is available at the end of the chapter

1. Introduction

Finite element analysis (FEA) is a well-established numerical simulation method for structural dynamics. It serves as the main computational tool for Noise, Vibration and Harshness (NVH) analysis in the low-frequency range. Because of developments in numerical methods and advances in computer software and hardware, FEA can now handle much more complex models far more efficiently than even a few years ago. However, the demand for computational capabilities increases in step with or even beyond the pace of these improvements. For example, automotive companies are constructing more detailed models with millions of degrees of freedom (DOFs) to study vibro-acoustic problems in higher frequency ranges. Although these tasks can be performed with FEA, the computational cost can be prohibitive even for high-end workstations with the most advanced software.

For large finite element (FE) models, a modal reduction is commonly used to obtain the system response. An eigenanalysis is performed using the system stiffness and mass matrices and a smaller in size modal model is formed which is solved more efficiently for the response. The computational cost is also reduced using substructuring (superelement analysis). Modal reduction is applied to each substructure to obtain the component modes and the system level response is obtained using Component Mode Synthesis (CMS).

When design changes are involved, the FEA analysis must be repeated many times in order to obtain the optimum design. Furthermore in probabilistic analysis where parameter uncertainties are present, the FEA analysis must be repeated for a large number of sample points. In such cases, the computational cost is even higher, if not prohibitive. Reanalysis methods are intended to analyze efficiently structures that are modified due to various changes. They estimate the structural response after such changes without solving the complete set of modified analysis equations. Several reviews have been published on reanalysis methods [1-3] which are usually based on local and global approximations. Local approximations are very efficient but they are effective only for small structural changes. Global approximations are preferable for large changes, but they are usually computationally expensive especially for cases with many design parameters. The well-known Rayleigh-Ritz reanalysis procedure [4, 5] belongs to the category of local approximation methods. The mode shapes of a nominal design are used to form a Ritz basis for predicting the response in a small parametric zone around the nominal design point. However, it is incapable of capturing relatively large design changes.

A parametric reduced-order modeling (PROM) method, developed by Balmes [6, 7], expands on the Rayleigh-Ritz method by using the mode shapes from a few selected design points to predict the response throughout the design space. The PROM method belongs to the category of local approximation methods and can handle relatively larger parameter changes because it uses multiple design points. An improved component-based PROM method has been introduced by Zhang et al. [8, 9] for design changes at the component level. The PROM method using a 'parametric' approach has been successfully applied to design optimization and probabilistic analysis of vehicle structures. However, the 'parametric' approach is only applicable to problems where the mass and stiffness matrices can be approximated by a polynomial function of the design parameters and their powers. A new 'parametric' approach using Kriging interpolation [10] has been recently proposed [11]. It improves efficiency by interpolating the reduced system matrices without needing to assume a polynomial relationship of the system matrices with respect to the design parameters as in [6, 7].

The Combined Approximations (CA) method [12-14] combines the strengths of both local and global approximations and can be accurate even for large design changes. It uses a combination of binomial series (local) approximations, called Neumann expansion approximations, and reduced basis (global) approximations. The CA method is developed for linear static reanalysis and eigen-problem reanalysis [15-19]. Accurate results and significant computational savings have been reported. All reported studies on the CA method [12-19] use relatively simple frame or truss systems for static or dynamics analysis with a relatively small number of DOF and/or modes. For these problems, the computational efficiency was improved by a factor of 5 to 10. Such an improvement is beneficial for a single design change evaluation or even for gradient-based design optimization where only a limited number of reanalyses (e.g. less than 50) is performed. However, the computational efficiency of the CA method may not be adequate in simulation-based (e.g. Monte-Carlo) probabilistic dynamic analysis of large finite-element models where reanalysis must be performed hundreds or thousands of times in order to estimate accurately the reliability of a design.

A large number of modes must be calculated and used in a dynamic analysis of a large finite-element model with a high modal density, even if a reduced-order modeling approach (Section 2) is used. In such a case, the implicit assumption of the CA method that the cost of solving a linear system is dominated by the cost of matrix decomposition way no be longer valid (see Section 3.4) and the computational savings from using the CA method may not be substantial. For this reason, we developed a modified combined approximation (MCA) and integrated it with the PROM method to improve accuracy and computational efficiency. The computational savings can be substantial for problems with a large number of design parameters. Examples in this Chapter demonstrate the benefits of this reanalysis methodology.

The Chapter presents methodologies

1. for accurate and efficient vibration analysis methods of large-scale, finite-element models,
2. for efficient and yet accurate reanalysis methods for dynamic response and optimization, and
3. for efficient design optimization methods to optimize structures for best vibratory response.

The optimization is able to handle a large number of design variables and identify local and global optima. It is organized as follows. Section 2 presents an overview of reduced-order modeling and substructuring methods including modal reduction and component mode synthesis (CMS). Improvements to the CMS method are presented using interface modes and filtration of constraint modes. The section also overviews two Frequency Response Function (FRF) substructuring methods where two substructures, represented by FRFs or FE models, are assembled to form an efficient reduced-order model to calculate the dynamic response. Section 3 presents four reanalysis methods: the CDH/VAO method, the Parametric Reduced Order Modeling (PROM) method, the Combined Approximation (CA) method, and the Modified Combined Approximation (MCA) method. It also points out their strong and weak points in terms of efficiency and accuracy. Section 4 demonstrates how the reanalysis methods are used in vibration and optimization of large-scale structures. It also presents a new reanalysis method in Craig-Bampton substructuring with interface modes which is very useful for problems with many interface DOFs where the FRF substructuring methods cannot be used. Section 5 presents a vibration and optimization case study of a large-scale vehicle model demonstrating the value of reduced-order modeling and reanalysis methods in practice. Finally, Section 6 summarizes and concludes.

2. Reduced-Order Modeling for Dynamic Analysis

Computational efficiency is of paramount importance in vibration analysis of large-scale, finite-element models. Reduced-order modeling (or substructuring) is a common approach to reduce the computational effort. Substructuring methods can be classified in "mathematical" and "physical" methods. The "mathematical" substructuring methods include the Automatic Multi-level Substructuring (AMLS) and the Automatic Component Mode Synthesis (ACMS) in NASTRAN. The "physical" substructuring methods include the well known fixed-interface Craig-Bampton method. Both the AMLS and ACMS methods use graph theory to obtain an abstract (mathematical) substructuring using matrix partitioning of the entire finite-element model. The computational savings from the "mathematical" and "physical" methods can be comparable depending on the problem at hand.

2.1. Modal Reduction

For an undamped structure with stiffness and mass matrices K and M respectively, under the excitation force vector F, the equations of motion (EOM) for frequency response are

$$[\mathbf{K} - \omega^2\mathbf{M}]\mathbf{d} = \mathbf{F} \tag{1}$$

where the displacement **d** is calculated at the forcing frequency ω. If the response is required at multiple frequencies, the repeated direct solution of Equation (1) is computationally very expensive and therefore, impractical for large scale finite-element models.

A reduced order model (ROM) is a subspace projection method. Instead of solving the original response equations, it is assumed that the solution can be approximated in a subspace spanned by the dominant mode shapes. A modal response approach can be used to calculate the response more efficiently. A set of eigen-frequencies ω_i and corresponding eigenvectors (mode shapes) φ_i are first obtained. Then, the displacement **d** is approximated in the reduced space formed by the first n dominant modes as

$$\mathbf{d} = \hat{\mathbf{\Phi}}\mathbf{U} \tag{2}$$

where $\hat{\mathbf{\Phi}} = [\varphi_1 \quad \varphi_2 \quad \cdots \quad \varphi_n]$ is the modal basis and **U** is the vector of principal coordinates or modal degrees of freedom (DOF). Using the approximation of Equation (2), the EOM of Equation (1) can be transformed from the original physical to the modal degrees of freedom as

$$\left[\hat{\mathbf{\Phi}}^T\mathbf{K}\hat{\mathbf{\Phi}} - \omega^2\hat{\mathbf{\Phi}}^T\mathbf{M}\hat{\mathbf{\Phi}}\right]\mathbf{U} = \hat{\mathbf{\Phi}}^T\mathbf{F} \tag{3}$$

The response **d** can be recovered by solving Equation (3) for the modal response **U** and projecting it back to the physical coordinates using Equation (2). If $\omega_{\max}$ is the maximum excitation frequency, it is common practice to retain the mode shapes in the frequency range of $0 \div 2\omega_{\max}$. The system modes in the high frequency range can be safely truncated with minimal loss of accuracy.

Due to the modal truncation, the size of the ROM is reduced considerably, compared to the original model. However, the size increases with the maximum excitation frequency. An added benefit of the ROM is that Equations (3) are decoupled because of the orthogonality of the mode shapes and can be therefore, solved separately reducing further the overall computational effort.

Note that for a damped structure with a damping matrix C, Equation (1) becomes $[K + j\omega C - \omega^2 M]d = F$ and Equation (3) is modified as $\left[\hat{\Phi}^T K\hat{\Phi} + j\omega\hat{\Phi}^T C\hat{\Phi} - \omega^2\hat{\Phi}^T M\hat{\Phi}\right]U = \hat{\Phi}^T F$ by adding the modal damping term $j\omega\hat{\Phi}^T C\hat{\Phi}$. For proportional (structural or Rayleigh) damping, $[C]$ is a linear combination of $[M]$ and $[K]$; i.e. $[C]=\alpha[M]+\beta[K]$ where α and β are constants. In this case, the reduced Equations (3) of the modal model are decoupled. Otherwise, they are not. In this Chapter for simplicity, we present all theoretical concepts for undamped systems. However for forced vibrations of damped systems, the addition of damping is straightforward.

2.2. Substructuring with Component Mode Synthesis (CMS)

To model the dynamics of a complex structure, a finite-element analysis of the entire structure can be very expensive, and sometimes infeasible, due to computer hardware and/or software constraints. This is especially true in the mid-frequency range, where a fine finite element mesh must be used in order to capture the shorter wavelengths of vibration. Component mode synthesis (CMS) was developed as a practical and efficient approach to modeling and analyzing the dynamics of a structure in such circumstances [20–23]. The structure is partitioned in component structures and the dynamics are described by the normal modes of the individual components and a set of modes that couple all component. Besides the significant computational savings, this component-based approach also facilitates distributed design. Components may be modified or redesigned individually without re-doing the entire analysis.

One of the most accurate and widely-used CMS methods is the Craig-Bampton method [22] where the component normal modes are calculated with the interface between connected component structures held fixed. Attachment at the interface is achieved by a set of "constraint modes." A constraint mode shape is the static deflection induced in the structure by applying a unit displacement to one interface DOF while all other interface DOF are held fixed. The motion at the interface is thus completely described by the constraint modes. The Craig-Bampton reduced-order model (CBROM) results in great model size reduction by including only component normal modes within a certain frequency range. However, there is no size reduction for constraint modes because CBROM must have one DOF for each interface DOF. If the finite element mesh is sufficiently fine, the constraint-mode DOFs will dominate the CBROM mass and stiffness matrices, and increase the computational cost.

We address this problem by using interface modes (also called characteristic constraint – CC- modes) in order to reduce the number of retained interface DOFs of the Craig-Bampton approach. For that, a secondary eigenvalue analysis is performed using the constraint-mode partitions of the CMS mass and stiffness matrices. The CC modes are the resultant eigenvectors. The basic formulation is described in Sections 2.2.1 and 2.2.2. The interface modes represent more "natural" physical motion at the interface. Because they capture the characteristic motion of the interface, they may be truncated as if they were traditional modes of vibration, leading to a highly compact CC-mode-based reduced order model (CCROM). In addition, the CC modes provide a significant insight into the physical mechanisms of vibration transmission between the component structures. This information could be used, for example, to determine the design parameters that have a critical impact on power flow. Figure 1 compares a conventional constraint mode used in Craig-Bampton analysis with an interface mode, for a simple cantilever plate which is subdivided in two substructures.

It should be noted that the calculation of the CC modes is essentially a secondary modal analysis. Therefore, the benefits are the same as those of a traditional modal analysis. For instance, refining the finite-element mesh increases the accuracy of a CCROM without introducing any additional degrees of freedom. The ability of the CC mode approach to produce CCROM whose size does not depend on the original level of discretization makes it especially well suited for finite-element based analysis of mid-frequency vibration.

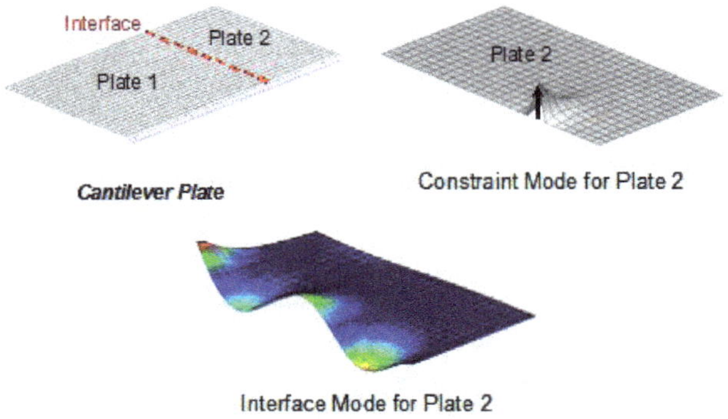

Figure 1. Illustration of interface modes.

2.2.1. Craig-Bampton Fixed Interface CMS

This section provides the basics of Craig-Bampton method using the fixed-interface assumption. The method is commonly used in CMS algorithms. The finite-element model of the entire structure is partitioned into a group of substructures. The DOFs in each substructure are divided into interface (Γ) DOF and interior (Ω) DOF. The equations of motion for the ith substructure are then expressed as

$$\begin{bmatrix} \mathbf{m}_i^{\Gamma\Gamma} & \mathbf{m}_i^{\Gamma\Omega} \\ \mathbf{m}_i^{\Omega\Gamma} & \mathbf{m}_i^{\Omega\Omega} \end{bmatrix} \begin{Bmatrix} \ddot{\mathbf{u}}_i^{\Gamma} \\ \ddot{\mathbf{u}}_i^{\Omega} \end{Bmatrix} + \begin{bmatrix} \mathbf{k}_i^{\Gamma\Gamma} & \mathbf{k}_i^{\Gamma\Omega} \\ \mathbf{k}_i^{\Omega\Gamma} & \mathbf{k}_i^{\Omega\Omega} \end{bmatrix} \begin{Bmatrix} \mathbf{u}_i^{\Gamma} \\ \mathbf{u}_i^{\Omega} \end{Bmatrix} = \begin{Bmatrix} \mathbf{f}_i^{\Gamma} \\ \mathbf{f}_i^{\Omega} \end{Bmatrix} \tag{4}$$

The fixed-interface Craig-Bampton CMS method utilizes two sets of modes to represent the substructure motion; substructure normal (N) modes$\boldsymbol{\Phi}_i^N$, and constraint (C) modes$\boldsymbol{\Phi}_i^C$, where i denotes the ith substructure. The size reduction of the Craig-Bampton method comes from the truncation of the normal modes $\boldsymbol{\Phi}_i^N = [\varphi_{i1} \;\; \varphi_{i2} \;\; \cdots \;\; \varphi_{in}]$ which are calculated by fixing all interface DOFs and solving the following eigenvalue problem

$$\begin{aligned} & k_i^{\Omega\Omega-1} \boldsymbol{\Phi}_i^C = \Lambda^N m_i^{\Omega\Omega} \boldsymbol{\Phi}_i^C \\ & [k_i^{\Omega\Omega}]\{\varphi_{in}\} = \lambda_n [m_i^{\Omega\Omega}]\{\varphi_{in}\} \quad for \quad n=1,\, 2,\, \ldots \end{aligned} \tag{5}$$

The static constraint modes$\boldsymbol{\Phi}_i^C$are calculated by enforcing a set of static unit constraints to the interface DOFs as

$$\left[\boldsymbol{\Phi}_i^C\right] = -\left[\mathbf{k}_i^{\Omega\Omega}\right]^{-1}\left[\mathbf{k}_i^{\Omega\Gamma}\right] \tag{6}$$

The original physical DOFs $\mathbf{u}_i^{\Gamma}$ and $\mathbf{u}_i^{\Omega}$can be thus represented by the constraint-mode DOFs $\mathbf{u}_i^C$ and the normal-mode DOFs $\mathbf{u}_i^N$ as

$$\begin{Bmatrix} \mathbf{u}_i^{\Gamma} \\ \mathbf{u}_i^{\Omega} \end{Bmatrix} = \begin{bmatrix} \mathbf{I} & \mathbf{0} \\ \boldsymbol{\Phi}_i^C & \boldsymbol{\Phi}_i^N \end{bmatrix} \begin{Bmatrix} \mathbf{u}_i^C \equiv \mathbf{u}_i^{\Gamma} \\ \mathbf{u}_i^N \end{Bmatrix} \tag{7}$$

Using the above Craig-Bampton transformation, the original EOM of Equation (7), can be expressed as

$$\begin{bmatrix} \mathbf{m}_i^{CC} & \mathbf{m}_i^{CN} \\ \mathbf{m}_i^{NC} & \mathbf{m}_i^{NN} \end{bmatrix} \begin{Bmatrix} \ddot{\mathbf{u}}_i^{C} \\ \ddot{\mathbf{u}}_i^{N} \end{Bmatrix} + \begin{bmatrix} \mathbf{k}_i^{CC} & \mathbf{k}_i^{CN} \\ \mathbf{k}_i^{NC} & \mathbf{k}_i^{NN} \end{bmatrix} \begin{Bmatrix} \mathbf{u}_i^{C} \\ \mathbf{u}_i^{N} \end{Bmatrix} = \begin{Bmatrix} \mathbf{f}_i^{C} \\ \mathbf{f}_i^{N} \end{Bmatrix} \tag{8}$$

where the superscripts C and N are used to indicate partition related to static constraint mode DOFs and fixed-interface normal mode DOFs, respectively. The matrix partitions of Equation (8) are

$$\mathbf{m}_i^{CC} = \mathbf{m}_i^{\Gamma\Gamma} + \mathbf{m}_i^{\Gamma\Omega}\boldsymbol{\Phi}_i^C + \left(\boldsymbol{\Phi}_i^C\right)^T \mathbf{m}_i^{\Omega\Gamma} + \left(\boldsymbol{\Phi}_i^C\right)^T \mathbf{m}_i^{\Omega\Omega}\boldsymbol{\Phi}_i^C \tag{9}$$

$$\mathbf{m}_i^{CN} = \left(\mathbf{m}_i^{NC}\right)^T = \mathbf{m}_i^{\Gamma\Omega}\mathbf{\Phi}_i^N + \left(\mathbf{\Phi}_i^C\right)^T \mathbf{m}_i^{\Omega\Omega}\mathbf{\Phi}_i^N \tag{10}$$

$$\mathbf{m}_i^{NN} = \left(\mathbf{\Phi}_i^N\right)^T \mathbf{m}_i^{\Omega\Omega}\mathbf{\Phi}_i^N \tag{11}$$

$$\mathbf{k}_i^{CC} = \mathbf{k}_i^{\Gamma\Gamma} + \mathbf{k}_i^{\Gamma\Omega}\mathbf{\Phi}_i^C = \mathbf{k}_i^{\Gamma\Gamma} - \mathbf{k}_i^{\Gamma\Omega}\left(\mathbf{k}_i^{\Omega\Omega}\right)^{-1}\mathbf{k}_i^{\Omega\Gamma} \tag{12}$$

$$\mathbf{k}_i^{CN} = \left(\mathbf{k}_i^{NC}\right)^T = \mathbf{0} \tag{13}$$

$$\mathbf{k}_i^{NN} = \left(\mathbf{\Phi}_i^N\right)^T \mathbf{k}_i^{\Omega\Omega}\mathbf{\Phi}_i^N = \mathbf{D}_i \equiv diag \tag{14}$$

$$\mathbf{f}_i^C = \mathbf{f}_i^\Gamma + \left(\mathbf{\Phi}_i^C\right)^T \mathbf{f}_i^\Omega \tag{15}$$

$$\mathbf{f}_i^N = \left(\mathbf{\Phi}_i^N\right)^T \mathbf{f}_i^\Omega \tag{16}$$

The matrices of each substructure are then assembled by applying displacement continuity and force balance along the interface to obtain the EOM of the reduced system. A secondary modal analysis is finally carried out using the mass and stiffness matrices of the reduced system to obtain the eigenvalues and eigenvectors.

Note that constraint mode matrix $\mathbf{\Phi}_i^C$ is usually a full matrix. Therefore Equation (9) can be computationally expensive due to the triple-product $\left(\mathbf{\Phi}_i^C\right)^T \mathbf{m}_i^{\Omega\Omega}\mathbf{\Phi}_i^C$ involving constraint modes. The computational cost of the Craig-Bamtpon method is mostly related to

1. solving for the normal modes,
2. solving for the constraint modes, and
3. the transformation calculation in Equation (9).

2.2.2. Craig-Bampton CMS with Interface Modes

In Craig-Bampton CMS, the matrices from all substructures are assembled into a global CBROM with substructures coupled at interfaces by enforcing displacement compatibility. This synthesis yields the modal displacement vector $\mathbf{d}^{CMS}$ of the synthesized system to be partitioned as

$$\mathbf{d}^{CMS} = \begin{bmatrix} \mathbf{d}^{CT} & \mathbf{d}_1^{NT} & \mathbf{d}_2^{NT} & \cdots & \mathbf{d}_{n^{ss}}^{N\,T} \end{bmatrix}^T \tag{17}$$

where n^{ss} is the number of substructures in the global structure. The corresponding synthesized CMS mass and stiffness matrices are as follows

$$\mathbf{M}^{CMS} = \begin{bmatrix} \bar{\mathbf{m}}^C & \mathbf{m}_1^{CN} & \mathbf{m}_2^{CN} & \cdots & \mathbf{m}_{n^{SS}}^{CN} \\ \mathbf{m}_1^{CNT} & \mathbf{m}_1^N & \mathbf{0} & \cdots & \mathbf{0} \\ \mathbf{m}_2^{CNT} & \mathbf{0} & \mathbf{m}_2^N & \cdots & \mathbf{0} \\ \vdots & \vdots & \vdots & \ddots & \vdots \\ \mathbf{m}_{n^{SS}}^{CNT} & \mathbf{0} & \mathbf{0} & \cdots & \mathbf{m}_{n^{SS}}^N \end{bmatrix}$$

$$\mathbf{K}^{CMS} = \begin{bmatrix} \bar{\mathbf{k}}^C & \mathbf{0} & \mathbf{0} & \cdots & \mathbf{0} \\ \mathbf{0} & \mathbf{k}_1^N & \mathbf{0} & \cdots & \mathbf{0} \\ \mathbf{0} & \mathbf{0} & \mathbf{k}_2^N & \cdots & \mathbf{0} \\ \vdots & \vdots & \vdots & \ddots & \vdots \\ \mathbf{0} & \mathbf{0} & \mathbf{0} & \cdots & \mathbf{k}_{n^{SS}}^N \end{bmatrix} \tag{18}$$

where the component modal matrices $\mathbf{m}_i^N$ and $\mathbf{k}_i^N$ are diagonalized and their sizes depend on the number of selected modes for the frequency range of interest. However, the number of constraint-mode DOFs, or the size of matrices $\bar{\mathbf{m}}^C$ and $\bar{\mathbf{k}}^C$, is equal to the number of DOFs of the interfaces between components and is therefore, determined by the finite-element mesh. If the mesh is fine in the interface regions, or if there are many substructures, the constraint-mode partitions of the CMS matrices may be relatively large. For this reason, we further reduce the CMS matrices by performing a modal analysis on the constraint-mode DOFs as follows

$$\bar{k}^C \psi_n = \Lambda_n \bar{m}^C \psi_n \quad for \quad n=1,\ 2,\ 3,\ \ldots \tag{19}$$

The eigenvectors ψ_n are transformed into the finite-element DOFs for the ith component structure using the following transformation

$$\mathbf{\Phi}_i^{CC} = \mathbf{\Phi}_i^C \boldsymbol{\beta}_i^C \mathbf{\Psi} \tag{20}$$

where

$$\boldsymbol{\Psi} = [\psi_1 \quad \psi_2 \quad \cdots \quad \psi_{n^{cc}}] \tag{21}$$

is a selected set of n^{CC} interface eigenvectors which are few compared to the number of the constraint-mode DOFs. The matrix $\boldsymbol{\beta}_i^C$ maps the global (system) interface DOFs $\mathbf{d}^C$ back to the local (subsystem i) DOF $\mathbf{d}_i^C$. The vectors in $\boldsymbol{\Phi}_i^{CC}$ are referred to as the *interface modes* or *characteristic constraint (CC) modes,* because they represent the characteristic physical motion associated with the constraint modes. Relatively few CC-mode DOFs are used compared to the number of interface DOFs.

Finally, the CMS matrices are transformed using the CC modes and the reduced-order CMS matrices are obtained similarly to Equations (18). Now, the unknown displacement vector $\mathbf{d}^{ROM}$ is partitioned as

$$\mathbf{d}^{ROM} = \begin{bmatrix} \mathbf{d}^{CCT} & \mathbf{d}_1^{NT} & \mathbf{d}_2^{NT} & \cdots & \mathbf{d}_{n^{ss}}^{N\,T} \end{bmatrix}^T \tag{22}$$

where superscript CC indicates the partition associated with the CC modes. The equations of motion of the reduced order CMS model (ROM) are expressed by

$$\left[-\omega^2 \mathbf{M}^{ROM} + \mathbf{K}^{ROM} \right] \mathbf{d}^{ROM} = \mathbf{f}^{ROM} \tag{23}$$

The mass matrix $\mathbf{M}^{ROM}$, the stiffness matrix $\mathbf{K}^{ROM}$, and the applied force vector $\mathbf{f}^{ROM}$, are explicitly written as

$$\mathbf{M}^{ROM} = \begin{bmatrix} \bar{\mathbf{m}}^{CC} & \bar{\mathbf{m}}_1^{CN} & \bar{\mathbf{m}}_2^{CN} & \cdots & \bar{\mathbf{m}}_{n^{ss}}^{CN} \\ \bar{\mathbf{m}}_1^{CNT} & \mathbf{m}_1^N & \mathbf{0} & \cdots & \mathbf{0} \\ \bar{\mathbf{m}}_2^{CNT} & \mathbf{0} & \mathbf{m}_2^N & \cdots & \mathbf{0} \\ \vdots & \vdots & \vdots & \ddots & \vdots \\ \bar{\mathbf{m}}_{n^{ss}}^{CNT} & \mathbf{0} & \mathbf{0} & \cdots & \mathbf{m}_{n^{ss}}^N \end{bmatrix} \tag{24}$$

$$\mathbf{K}^{ROM} = \begin{bmatrix} \bar{\mathbf{k}}^{CC} & \mathbf{0} & \mathbf{0} & \cdots & \mathbf{0} \\ \mathbf{0} & \mathbf{k}_1^N & \mathbf{0} & \cdots & \mathbf{0} \\ \mathbf{0} & \mathbf{0} & \mathbf{k}_2^N & \cdots & \mathbf{0} \\ \vdots & \vdots & \vdots & \ddots & \vdots \\ \mathbf{0} & \mathbf{0} & \mathbf{0} & \cdots & \mathbf{k}_{n^{ss}}^N \end{bmatrix} \tag{25}$$

$$\mathbf{f}^{ROM} = \begin{bmatrix} \mathbf{f}^{CCT} & \mathbf{f}_1^{NT} & \mathbf{f}_2^{NT} & \cdots & \mathbf{f}_{n^{ss}}^{N\,T} \end{bmatrix}^T \tag{26}$$

where

$$\bar{\mathbf{m}}^{CC} = \mathbf{\Psi}^T \bar{\mathbf{m}}^C \mathbf{\Psi},\ \bar{\mathbf{k}}^{CC} = \mathbf{\Psi}^T \bar{\mathbf{k}}^C \mathbf{\Psi} \tag{27}$$

and

$$\bar{\mathbf{m}}_i^{CN} = \mathbf{\Psi}^T \boldsymbol{\beta}_i^{C\,T} \mathbf{m}_i^{CN} \tag{28}$$

2.2.3. Filtration of Constraint Modes

Figure 2 shows a typical constraint mode for a plate substructure. The non-zero displacement field (indicated by red color) is usually limited to a very small region close to the perturbed interface DOF.

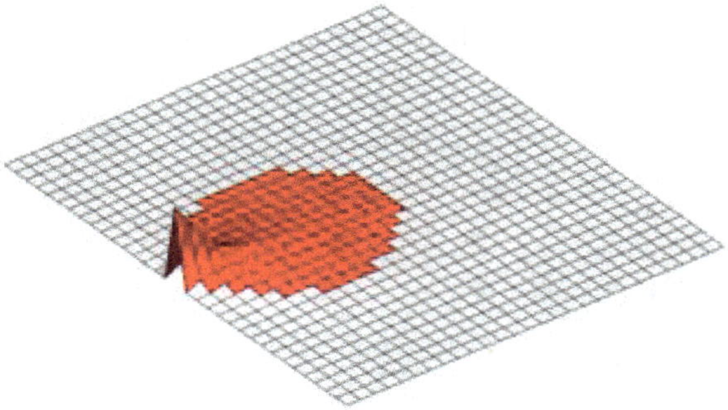

Figure 2. Illustration of a "filtered" constraint mode.

If the small-displacement part of the constraint mode shape is explicitly replaced by zero, the density of the resulting "filtered" constraint mode will be significantly reduced. Consequently, the computational cost in Equation (9) will be considerably reduced. To filter the constraint modes, the following criterion is used

$$\varphi_{pq}^C = 0 \quad if \quad |\varphi_{pq}^C| < \varepsilon * \max_p |\varphi_{pq}^C| \tag{29}$$

where φ_{pq}^C is the pth element of the qth constraint mode $\boldsymbol{\varphi}^C$. If the ratio of an element of the constraint mode vector to the maximum value in the vector is smaller than a defined small ε, the element of the constraint mode is truncated to zero. For the constraint mode of Figure 4, the constraint mode density reduces from 100% to 16% if $\varepsilon = 0.03$.

2.3. Frequency Response Function (FRF) Substructuring and Assembling

If the number of interface nodes (or DOFs) between connected substructures is small, a reduced-order model of small size can be developed using an FRF representation of each substructure. This is known as FRF substructuring. The FRF representation can be easily obtained from a finite element (FE) model or even experimentally. If the FE model of one substructure is very small (e.g. a vehicle suspension model), it can be easily coupled directly to another substructure which is represented using FRFs. This section provides the fundamentals of FRF substructuring for both FRF-FE and FRF-FRF coupling.

2.3.1. Algorithm for FRF/FE Coupling

The numerical algorithm is explained using the two-substructure example of Figure 3. Sub1 is an FRF type substructure, meaning that its dynamic behavior is described using FRF matrices which are denoted by **H** (see Equation 30 for notation). The elements of **H** are frequency dependent and complex if damping is present. A bold letter indicates a matrix or vector. According to Equation (30), $\mathbf{H}_{AC}$ for example, represents the displacement $\mathbf{X}_A$ of DOF A due to a unit force $\mathbf{F}_C$ on DOF C. Sub2 is a finite element (FE) type substructure. Its dynamic behavior is described using the stiffness **K**, mass **M** and damping **B** matrices.

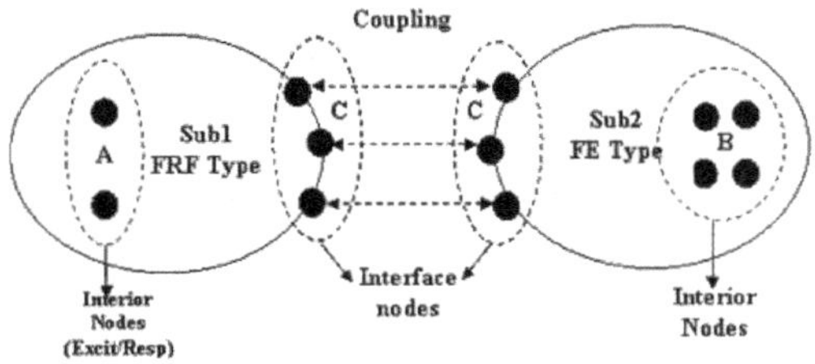

Figure 3. Two-substructure example and notation.

The FRF matrix of Sub1 can be calculated by either a direct frequency response method or a modal response method. In the former case, the original FE equations are used in the frequency domain while in the latter a modal model is first developed and then used to calculate the FRF matrix. The size of the FRF matrix is small and depends on the number of DOFs of the excitation, response and interface DOFs. Usually FRFs are calculated between excitation and response DOFs. However in order to assemble two substructures, FRFs are also calculated between interface DOFs and excitation/response DOFs. The Sub1 FRF matrix **H** in Equation (30) is thus partitioned into interior (A) DOFs and interface/coupling (C) DOFs. The interior DOFs include all excitation and all response DOFs (Figure 3).

The second substructure Sub2 is expressed in FE format. The system FE matrices **K**, **M**, and **B** form the frequency dependent dynamic matrix $\mathbf{Z}=\mathbf{K}+i\omega\mathbf{B}-\omega^2\mathbf{M}$ which is then partitioned according to the interior (B) and interface (C) DOFs. The interface DOFs for Sub1 and Sub2 have the same node IDs so that they can be assembled to obtain the system FRF matrix.

The procedure to assemble the **H** matrix of Sub1 with the **Z** matrix of Sub2 and calculate (solve for) the system matrix **H** is described below.

The equations of motion for Sub1 are expressed as

$$\begin{bmatrix} \mathbf{X}_A \\ \mathbf{X}_C \end{bmatrix} = \begin{bmatrix} \mathbf{H}_{AA} & \mathbf{H}_{AC} \\ \mathbf{H}_{CA} & \mathbf{H}_{CC} \end{bmatrix} \begin{bmatrix} \mathbf{F}_A \\ \mathbf{F}_C^1 \end{bmatrix} \tag{30}$$

where subscript A indicates the interior (excitation plus response) DOFs of Sub1, and subscript C indicates the connection/common/coupling DOFs between Sub1 and Sub2. The equations of motion for Sub2 are expressed as

$$\begin{bmatrix} \mathbf{F}_B \\ \mathbf{F}_C^2 \end{bmatrix} = \begin{bmatrix} \mathbf{Z}_{BB} & \mathbf{Z}_{BC} \\ \mathbf{Z}_{CB} & \mathbf{Z}_{CC} \end{bmatrix} \begin{bmatrix} \mathbf{X}_B \\ \mathbf{X}_C \end{bmatrix} \tag{31}$$

where subscript *B* indicates the interior DOFs of Sub2, and subscript *C* indicates the connection/common/coupling DOFs between Sub2 and Sub1. Because of displacement compatibility at the interface, $X_C$appears on the left-hand side of Equation (30) for Sub1 and on the right-hand side of Equation (31) for Sub2. Superscripts 1 and 2 are used to differentiate the interface forces $\mathbf{F}_C$ at Sub1 and Sub2.

To couple Sub1 and Sub2, compatibility of forces at the interface is applied as $\mathbf{F}_C = \mathbf{F}_C^1 + \mathbf{F}_C^2$ where the force vector $\mathbf{F}_C^1 = \mathbf{H}_{CC}^{-1}\mathbf{X}_C - \mathbf{H}_{CC}^{-1}\mathbf{H}_{CA}\mathbf{F}_A = \mathbf{H}_{CC}^{-1}\mathbf{X}_C - \mathbf{\Phi}\mathbf{F}_A$ with $\mathbf{\Phi} = \mathbf{H}_{CC}^{-1}\mathbf{H}_{CA}$ is obtained from the second row of Equation (30) and the second row of Equation (31) provides the force vector $\mathbf{F}_C^2 = \mathbf{Z}_{CC}\mathbf{X}_C + \mathbf{Z}_{CB}\mathbf{X}_B$. We thus have

$$\mathbf{F}_C = \left(\mathbf{H}_{CC}^{-1} + \mathbf{Z}_{CC}\right)\mathbf{X}_C - \mathbf{\Phi}\mathbf{F}_A + \mathbf{Z}_{CB}\mathbf{X}_B \tag{32}$$

From Equation (31),

$$\mathbf{X}_B = \mathbf{Z}_{BB}^{-1}\mathbf{F}_B - \mathbf{Z}_{BB}^{-1}\mathbf{Z}_{BC}\mathbf{X}_C = \mathbf{Z}_{BB}^{-1}\mathbf{F}_B + \mathbf{\Theta}^T\mathbf{X}_C \tag{33}$$

where $\mathbf{\Theta} = -\mathbf{Z}_{CB}\mathbf{Z}_{BB}^{-1}$. Substitution of Equation (33) in Equation (32) yields

$$\begin{aligned} \mathbf{F}_C &= \left(\mathbf{H}_{CC}^{-1} + \mathbf{Z}_{CC} + \mathbf{Z}_{CB}\mathbf{\Theta}^T\right)\mathbf{X}_C - \mathbf{\Phi}\mathbf{F}_A - \mathbf{\Theta}\mathbf{F}_B \\ &= \mathbf{R}\mathbf{X}_C - \mathbf{\Phi}\mathbf{F}_A - \mathbf{\Theta}\mathbf{F}_B \end{aligned} \tag{34}$$

where $\mathbf{R} = \mathbf{H}_{CC}^{-1} + \mathbf{Z}_{CC} + \mathbf{Z}_{CB}\mathbf{\Theta}^T$. From Equation (34), $\mathbf{X}_C$ can be expressed in terms of $\mathbf{F}_A$, $\mathbf{F}_B$ and $\mathbf{F}_C$ as

$$\mathbf{X}_C = \mathbf{R}^{-1}\mathbf{\Phi}\mathbf{F}_A + \mathbf{R}^{-1}\mathbf{\Theta}\mathbf{F}_B + \mathbf{R}^{-1}\mathbf{F}_C \tag{35}$$

Substitution of Equation (35) in Equation (33) gives $\mathbf{X}_B$ in terms of $\mathbf{F}_A$, $\mathbf{F}_B$ and $\mathbf{F}_C$ as

$$\begin{aligned}\mathbf{X}_B &= \mathbf{Z}_{BB}^{-1}\mathbf{F}_B + \mathbf{\Theta}^T\left(\mathbf{R}^{-1}\mathbf{\Phi}\mathbf{F}_A + \mathbf{R}^{-1}\mathbf{\Theta}\mathbf{F}_B + \mathbf{R}^{-1}\mathbf{F}_C\right) \\ &= \mathbf{\Theta}^T\mathbf{R}^{-1}\mathbf{\Phi}\mathbf{F}_A + \left(\mathbf{Z}_{BB}^{-1} + \mathbf{\Theta}^T\mathbf{R}^{-1}\mathbf{\Theta}\right)\mathbf{F}_B + \mathbf{\Theta}^T\mathbf{R}^{-1}\mathbf{F}_C\end{aligned} \tag{36}$$

Solving Equation (30) for $\mathbf{X}_A = \mathbf{H}_{AA}\mathbf{F}_A + \mathbf{H}_{AC}\mathbf{F}_C^1$ and substituting $\mathbf{F}_C^1 = \mathbf{H}_{CC}^{-1}\mathbf{X}_C - \mathbf{\Phi}\mathbf{F}_A$ yields

$$\begin{aligned}\mathbf{X}_A &= \mathbf{H}_{AA}\mathbf{F}_A + \mathbf{H}_{AC}\left(\mathbf{H}_{CC}^{-1}\mathbf{X}_C - \mathbf{\Phi}\mathbf{F}_A\right) \\ &= \left(\mathbf{H}_{AA} - \mathbf{H}_{AC}\mathbf{\Phi}\right)\mathbf{F}_A + \mathbf{\Phi}^T\mathbf{X}_C\end{aligned} \tag{37}$$

We can now express $\mathbf{X}_A$ in terms of $\mathbf{F}_A$, $\mathbf{F}_B$ and $\mathbf{F}_C$ by substituting Equation (35) in Equation (37), as

$$\begin{aligned}\mathbf{X}_A &= \left(\mathbf{H}_{AA} - \mathbf{H}_{AC}\mathbf{\Phi}\right)\mathbf{F}_A + \mathbf{\Phi}^T\mathbf{X}_C \\ &= \left(\mathbf{H}_{AA} - \mathbf{H}_{AC}\mathbf{\Phi}\right)\mathbf{F}_A + \mathbf{\Phi}^T\left(\mathbf{R}^{-1}\mathbf{\Phi}\mathbf{F}_A + \mathbf{R}^{-1}\mathbf{\Theta}\mathbf{F}_B + \mathbf{R}^{-1}\mathbf{F}_C\right) \\ &= \left(\mathbf{H}_{AA} - \mathbf{H}_{AC}\mathbf{\Phi} + \mathbf{\Phi}^T\mathbf{R}^{-1}\mathbf{\Phi}\right)\mathbf{F}_A + \mathbf{\Phi}^T\mathbf{R}^{-1}\mathbf{\Theta}\mathbf{F}_B + \mathbf{\Phi}^T\mathbf{R}^{-1}\mathbf{F}_C\end{aligned} \tag{38}$$

Based on Equations (38), (36) and (35), $\mathbf{X}_A$, $\mathbf{X}_B$ and $\mathbf{X}_C$ are expressed in terms of $\mathbf{F}_A$, $\mathbf{F}_B$ and $\mathbf{F}_A$ as follows

$$\begin{bmatrix}\mathbf{X}_A \\ \mathbf{X}_B \\ \mathbf{X}_C\end{bmatrix} = \left(\begin{bmatrix}\mathbf{H}_{AA} - \mathbf{H}_{AC}\mathbf{\Phi} & & \\ & \mathbf{Z}_{BB}^{-1} & \\ & & \mathbf{0}\end{bmatrix} + \begin{bmatrix}\mathbf{\Phi}^T \\ \mathbf{\Theta}^T \\ \mathbf{I}\end{bmatrix}\mathbf{R}^{-1}\begin{bmatrix}\mathbf{\Phi} & \mathbf{\Theta} & \mathbf{I}\end{bmatrix}\right)\begin{bmatrix}\mathbf{F}_A \\ \mathbf{F}_B \\ \mathbf{F}_C\end{bmatrix} \tag{39}$$

resulting in the following FRF of the assembled system

$$
\begin{aligned}
\mathbf{H}_{sys} &= \begin{bmatrix} \mathbf{H}_{AA}-\mathbf{H}_{AC}\boldsymbol{\Phi} & & \\ & \mathbf{Z}_{BB}^{-1} & \\ & & 0 \end{bmatrix} + \begin{bmatrix} \boldsymbol{\Phi}^T \\ \boldsymbol{\Theta}^T \\ \mathbf{I} \end{bmatrix} \mathbf{R}^{-1} \begin{bmatrix} \boldsymbol{\Phi} & \boldsymbol{\Theta} & \mathbf{I} \end{bmatrix} \\
&= diag\left(\mathbf{H}_{AA}-\mathbf{H}_{AC}\boldsymbol{\Phi},\ \ \mathbf{Z}_{BB}^{-1},\ \ 0\right)+\mathbf{S}^T\mathbf{R}^{-1}\mathbf{S} \\
&= \mathbf{HK}+\mathbf{S}^T\mathbf{R}^{-1}\mathbf{S}
\end{aligned}
\tag{40}
$$

where $\mathbf{S} = [\boldsymbol{\Phi}\ \boldsymbol{\Theta}\ \mathbf{I}]$ and $\mathbf{HK} = diag\left(\mathbf{H}_{AA}-\mathbf{H}_{AC}\boldsymbol{\Phi},\ \mathbf{Z}_{BB}^{-1},\ 0\right)$.

2.4.2. Algorithm for FRF/FRF Coupling

Figure 4 shows the coupling of two FRF type substructures. The equations of motion for Sub1 and Sub2 and are expressed as

$$
\begin{bmatrix} \mathbf{X}_A \\ \mathbf{X}_C \end{bmatrix} = \begin{bmatrix} \mathbf{H}_{AA} & \mathbf{H}_{AC} \\ \mathbf{H}_{CA} & \mathbf{H}_{CC}^1 \end{bmatrix} \begin{bmatrix} \mathbf{F}_A \\ \mathbf{F}_C^1 \end{bmatrix} \tag{41}
$$

and

$$
\begin{bmatrix} \mathbf{X}_B \\ \mathbf{X}_C \end{bmatrix} = \begin{bmatrix} \mathbf{H}_{BB} & \mathbf{H}_{BC} \\ \mathbf{H}_{CB} & \mathbf{H}_{CC}^2 \end{bmatrix} \begin{bmatrix} \mathbf{F}_B \\ \mathbf{F}_C^2 \end{bmatrix} \tag{42}
$$

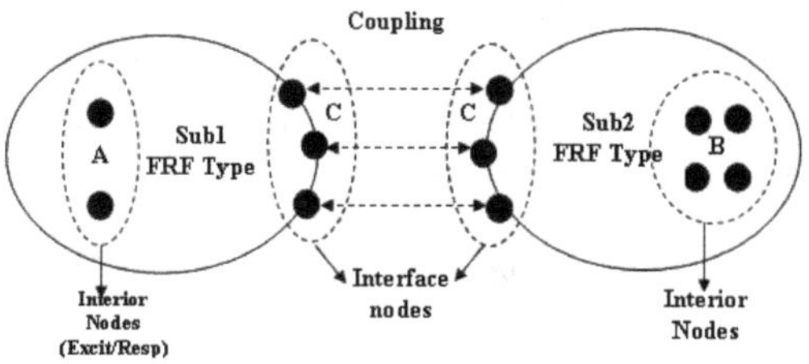

Figure 4. Two FRF type substructures example and notation

To couple Sub1 and Sub2, we enforce displacement compatibility at the interface and apply the interface force relationship $\mathbf{F}_C = \mathbf{F}_C^1 + \mathbf{F}_C^2$. In this case, the assembled system equations can be re-arranged in matrix form as

$$\begin{bmatrix} \mathbf{X}_A \\ \mathbf{X}_B \\ \mathbf{X}_C \end{bmatrix} = \left(\begin{bmatrix} \mathbf{H}_{AA} & \mathbf{H}_{AC} & \\ \mathbf{H}_{CA} & \mathbf{H}_{CC}^1 & \\ & & \mathbf{H}_{BB} \end{bmatrix} + \begin{bmatrix} \mathbf{H}_{AC} \\ \mathbf{H}_{CC}^1 \\ \mathbf{H}_{BC} \end{bmatrix} \begin{bmatrix} \mathbf{H}_{CC}^1 + \mathbf{H}_{CC}^2 \end{bmatrix}^{-1} \begin{bmatrix} \mathbf{H}_{AC} \\ \mathbf{H}_{CC}^1 \\ \mathbf{H}_{BC} \end{bmatrix}^T \right) \begin{bmatrix} \mathbf{F}_A \\ \mathbf{F}_C \\ \mathbf{F}_B \end{bmatrix} \tag{43}$$

The FRF/FRF coupling is a special case of the FRF/FE coupling.

3. Reanalysis Methods for Dynamic Analysis

3.1. CDH/VAO Method

The CDH/VAO method, developed by CDH AG for vibro-acoustic analysis, is a Rayleigh-Ritz type of approximation. If the stiffness and mass matrices of the baseline design structure are $\mathbf{K}_0$ and $\mathbf{M}_0$, the exact mode shapes in $\mathbf{\Phi}_0$ are obtained by solving the eigen-problem

$$\mathbf{K}_0\mathbf{\Phi}_0 = \mathbf{M}_0\mathbf{\Phi}_0\mathbf{\Lambda}_0 \tag{44}$$

where $\mathbf{\Lambda}_0$ is the diagonal matrix of the baseline eigenvalues. A new design (subscript p) has the following stiffness and mass matrices

$$\mathbf{K}_p = \mathbf{K}_0 + \Delta\mathbf{K} \qquad \mathbf{M}_p = \mathbf{M}_0 + \Delta\mathbf{M} \tag{45}$$

For a modest design change where $\Delta\mathbf{K}$ and $\Delta\mathbf{M}$ are small, it is assumed that the change in mode shapes is small and the new response can be therefore, captured in the sub-space spanned by the mode shapes $\mathbf{\Phi}_0$ of the initial design. The new stiffness and mass matrices are then condensed as$K_R = \Phi_0^T K_p \Phi_0$and $M_R = \Phi_0^T M_p \Phi_0$and the following reduced eigenvalue problem is solved to calculate the eigen-vector Θ

$$K_p\Theta = M_p\Theta\Lambda_p \tag{46}$$

The approximate eigenvalues of the new design are given by $\mathbf{\Lambda}_p$ and the approximate eigenvectors $\mathbf{\Phi}_p$ are

$$\Phi_p = R\Theta \tag{47}$$

where$R = \Phi_0$. Thus, the modal response of the modified structure can be easily obtained and the actual response can be recovered using the eigenvectorsΦ_p.

3.2. Parametric Reduced-Order Modeling (PROM) Method

The PROM method approximates the mode shapes of a new design in the subspace spanned by the dominant mode shapes of some representative designs, which are selected such that the formed basis captures the dynamic characteristics in each dimension of the parameter space. Balmes et al. [6, 7] suggested that these representative designs should correspond to the middle points on the faces of a box in the parameter space representing the range of design parameters. For a structure with m design variables, Zhang [9] suggested that the representative designs include a *baseline* design for which all parameters are at their lower limits plus m designs obtained by perturbing the design variables from their lower limits to their upper limits, one at a time. The points representing these designs in the space of the design variables are called *corner points* (see Figure 5). This selection of representative designs results in a more accurate PROM algorithm.

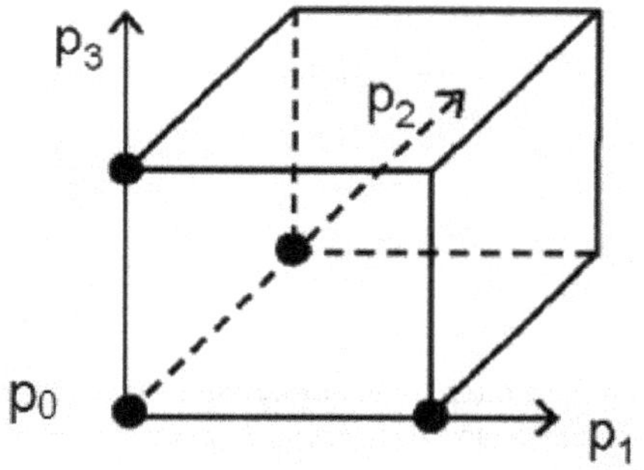

Figure 5. Design space of three parameters.

The mode shapes of a new design are approximated in the space of the mode shapes of the corner points as

$$\boldsymbol{\Phi} \approx \widetilde{\boldsymbol{\Phi}}_p = \boldsymbol{P}\boldsymbol{\Theta} \tag{48}$$

where the modal matrix **P** includes the basis vectors as in Equation (49) and $\boldsymbol{\Theta}$ represents the participation factors of these vectors. The columns of P are the dominant mode shapes of the above $(m+1)$ designs,

$$\boldsymbol{P} = [\boldsymbol{\Phi}_0 \quad \boldsymbol{\Phi}_1 \quad \dots \quad \boldsymbol{\Phi}_m] \tag{49}$$

where $\boldsymbol{\Phi}_0$ is the modal matrix composed of the dominant mode shapes of the baseline design, and $\boldsymbol{\Phi}_i$ is the modal matrix of the ith corner point. The mode shapes of the new design satisfy the following eigenvalue problem,

$$\boldsymbol{K}\widetilde{\boldsymbol{\Phi}}_p = \boldsymbol{M}\widetilde{\boldsymbol{\Phi}}_p\boldsymbol{\Lambda} \Leftrightarrow \boldsymbol{K}\boldsymbol{P}\boldsymbol{\Theta} = \boldsymbol{M}\boldsymbol{P}\boldsymbol{\Theta}\boldsymbol{\Lambda} \tag{50}$$

where $\boldsymbol{\Lambda}$ is a diagonal matrix of the first s eigenvalues.

A reduced eigenvalue problem is obtained by pre-multiplying both sides of Equation (50) by $\boldsymbol{P}^T$ as

$$\boldsymbol{K}_R\boldsymbol{\Theta} = \boldsymbol{M}_R\boldsymbol{\Theta}\boldsymbol{\Lambda} \tag{51}$$

where the reduced stiffness and mass matrices are

$$\boldsymbol{K}_R = \boldsymbol{P}^T\boldsymbol{K}\boldsymbol{P} \quad \text{and} \quad \boldsymbol{M}_R = \boldsymbol{P}^T\boldsymbol{M}\boldsymbol{P} \tag{52}$$

Thus, the matrix $\boldsymbol{\Theta}$ in Equation (48) consists of the eigenvectors of the reduced stiffness and mass matrices $\boldsymbol{K}_R$ and $\boldsymbol{M}_R$.

For m design variables, $(m+1)$ eigenvalue problems must be solved in order to form the basis $\boldsymbol{P}$ of Equation (49). Therefore, both the cost of obtaining the modal matrices $\boldsymbol{\Phi}_i$ and the size of matrix $\boldsymbol{P}$ increase linearly with m. The PROM approach uses the following algorithm to compute the mode shapes of a new design:

1. Calculate the mode shapes of the baseline design and the designs corresponding to the m corner points in the design space, and form subspace basis $\boldsymbol{P}$.

2. Calculate the reduced stiffness and mass matrices $\boldsymbol{K}_R$ and $\boldsymbol{M}_R$ from Equation (52).

3. Solve eigenproblem (51) for matrix $\boldsymbol{\Theta}$.

4. Reconstruct the approximated eigenvectors in $\widetilde{\boldsymbol{\Phi}}_p$ using Equation (48).

Step 1 is performed only once. A reanalysis requires only steps 2 to 4. For a small number of mode shapes and a small number of design variables, the cost of steps 2 to 4 is much smaller than the cost of a full analysis. The computational cost of PROM consists of

1. the cost of performing $(m+1)$ full eigen-analyses to form subspace basis $\boldsymbol{P}$ in Equation (49), and

2. the cost of reanalysis of each new design in steps 2 to 4.

The former is the fixed cost of PROM because it does not depend on the numbers of reanalyses and the latter is the variable cost of PROM because it is proportional to the number of reanalyses. The fixed cost is not attributed to the calculation of the response for a particular design. It is simply required to obtain the information needed to apply PROM. The variable cost (cost of reanalysis of a new design in part b) is small compared to the fixed cost. The fixed cost of PROM is proportional to the number of design variables m because it consists of the dominant eigenvectors $\boldsymbol{\Phi}_0$ of the baseline design and the dominant eigenvectors $\boldsymbol{\Phi}_i$, $i=1, ..., m$ of the m corner design points (see Equation 49). When the size of basis $\boldsymbol{P}$ increases so does the fixed cost because more eigenvalue problems and mode shapes must be calculated to form basis$\boldsymbol{P}$. The PROM method results in significant cost savings when applied to problems that involve few design variables (less than 10) and require many analyses (e.g. Monte Carlo simulation or gradient-free optimization using genetic algorithms).

3.3. Combined Approximations (CA) Method

The PROM method requires an eigenvalue analysis for multiple designs (corner points) to form a basis for approximating the eigenvectors at other designs. It is efficient only when the number of design parameters is relatively small. On the contrary, the CA method of this section does not have such a restriction because the reanalysis cost is not proportional to the number of design parameters. The CA method is thus more suitable than the PROM method, when the number of reanalyses is less than or comparable to the number of design parameters, such as in gradient-based design optimization.

The fundamentals of the combined approximations (CA) method [15-19] are given below. A subspace basis is formed through a recursive process for calculating the natural frequencies and mode shapes of a system. If $\boldsymbol{K}_0$ and $\boldsymbol{M}_0$ are the stiffness and mass matrices of the original (baseline) design, the exact mode shapes $\boldsymbol{\Phi}_0$ are obtained by solving the eigen-problem $\boldsymbol{K}_0\boldsymbol{\Phi}_0=\lambda_0\boldsymbol{M}_0\boldsymbol{\Phi}_0$. We want to approximate the mode shapes of a modified design (subscript p) with stiffness and mass matrices $\boldsymbol{K}_p=\boldsymbol{K}_0+\Delta\boldsymbol{K}$ and $\boldsymbol{M}_p=\boldsymbol{M}_0+\Delta\boldsymbol{M}$ where $\Delta\boldsymbol{K}$ and $\Delta\boldsymbol{M}$ represent large perturbations. The CA method estimates the new eigenvalues λ_p and eigenvectors $\boldsymbol{\Phi}_p$ without performing an exact eigenvalue analysis.

The eigen-problem for the modified design can be expressed as

$$\boldsymbol{\Phi}_p=\lambda_p\boldsymbol{K}_0^{-1}\boldsymbol{M}_p\boldsymbol{\Phi}_p-\boldsymbol{K}_0^{-1}\Delta\boldsymbol{K}\boldsymbol{\Phi}_p \tag{53}$$

leading to the following recursive equation

$$\boldsymbol{\Phi}_{p,j}=\lambda_p\boldsymbol{K}_0^{-1}\boldsymbol{M}_p\boldsymbol{\Phi}_{p,j-1}-\boldsymbol{K}_0^{-1}\Delta\boldsymbol{K}\boldsymbol{\Phi}_{p,j-1} \tag{54}$$

which produces a sequence of approximations of the mode shapes$\boldsymbol{\Phi}_{p,j}$, $j=1, 2, \ldots, s$. The CA method uses the changes $\boldsymbol{R}_j=\boldsymbol{\Phi}_{p,j}-\boldsymbol{\Phi}_{p,j-1}$ to form a subspace basis to approximate the modes of the new design. In order to simplify the calculations, $\lambda_p\boldsymbol{K}_0^{-1}\boldsymbol{M}_p\boldsymbol{\Phi}_{p,j-1}$in Equation (53) is replaced with $\lambda_p\boldsymbol{K}_0^{-1}\boldsymbol{M}_p\boldsymbol{\Phi}_0$ and Equation (54) becomes

$\boldsymbol{\Phi}_{p,j}=\lambda_p \boldsymbol{K}_0^{-1}\boldsymbol{M}_p\boldsymbol{\Phi}_0-\boldsymbol{K}_0^{-1}\Delta\boldsymbol{K}\boldsymbol{\Phi}_{p,j-1}$ showing that the basis vectors satisfy the following recursive equation

$$\mathbf{R}_j = -\mathbf{K}_0^{-1}\Delta\mathbf{K}\mathbf{R}_{j-1} \quad j=2,\ldots,s \tag{55}$$

where the first basis vector is assumed to be $\boldsymbol{R}_1=\boldsymbol{K}_0^{-1}\boldsymbol{M}_p\boldsymbol{\Phi}_0$.

The CA method forms a subspace basis

$$\mathbf{R}=\begin{bmatrix}\mathbf{R}_1 & \mathbf{R}_2 & \cdots & \mathbf{R}_s\end{bmatrix} \tag{56}$$

where s is usually between 3 and 6 [16-18, 23] and the mode shapes of the new $(\boldsymbol{K}_p, \boldsymbol{M}_p)$ design are then approximated in the subspace spanned by $\boldsymbol{R}$ using the following algorithm:

- Condense the stiffness and mass matrices as

$$\mathbf{K}_R = \mathbf{R}^T\mathbf{K}_p\mathbf{R} \qquad \mathbf{M}_R = \mathbf{R}^T\mathbf{M}_p\mathbf{R} \tag{57}$$

- Solve the reduced eigen-problem (using matrices $\mathbf{K}_R$ and $\mathbf{M}_R$) to calculate the eigenvector matrix $\boldsymbol{\Theta}$.
- Reconstruct the approximate eigenvectors of the new design $\widetilde{\boldsymbol{\Phi}}_p$ as

$$\widetilde{\boldsymbol{\Phi}}_p=\boldsymbol{R}\boldsymbol{\Theta} \tag{58}$$

The eigenvalues of the new design are approximated by the eigenvalues $\widetilde{\lambda}_p$ of the reduced eigen-problem.

The CA method has three main advantages:

1. it only requires a single matrix decomposition of stiffness matrix $\boldsymbol{K}_0$ in Equation (55) to calculate the subspace basis $\boldsymbol{R}$,
2. it is accurate because the basis is updated for every new design, and
3. the eigenvectors of a new design are efficiently approximated by Equation (58) where the eigenvectors $\boldsymbol{\Theta}$ correspond to a much smaller reduced eigen-problem.

However for a large number of reanalyses, the computational cost can increase substantially because a new basis and the condensed mass and stiffness matrices in Equation (57) must be calculated for every reanalysis. Examples where many analyses are needed are optimization problems in which a Genetic Algorithm is employed to search for the optimum, and probabilistic analysis problems using Monte-Carlo simulation. The PROM method can be more suitable for these problems because the subspace basis $\boldsymbol{R}$ does not change for every new design point. Note that steps 1 and 3 (Equations 57and 58) are similar to steps 2 and 4 of PROM. CA uses basis $\boldsymbol{R}$ and PROM uses basis$\boldsymbol{P}$.

The CA method is more efficient than PROM for design problems where few reanalyses are required for two reasons. First, it does not require calculation of the eigenvectors $\boldsymbol{\Phi}_i$, $i=1, \cdots, m$, of the m corner design points, and second the cost of matrix condensation of Equation (57) is much lower than that of Equation (52), because the size (number of columns) of basis $\boldsymbol{R}$ is not proportional to the number of parameters m as in basis $\mathbf{P}$. For problems with a large number of design parameters, the PROM approach is efficient only when a 'parametric' relationship is established [7] because a large overhead cost, proportional to the number of design parameters, is required. In contrast, the CA method does not require such an overhead cost because the reanalysis cost is not proportional to the number of design parameters. The CA method is therefore, more suitable than PROM, if the number of reanalyses is less than or comparable to the number of design parameters.

3.4. Modified Combined Approximations (MCA) Method

In the literature, the accuracy and efficiency of the CA method has been mostly tested on problems involving structures with up to few thousands of DOFs, such as frames or trusses [12-19]. We have tested the CA method using, among others, the structural dynamics response of a medium size (65,000 DOFs) finite-element model (Figure 7 of Section 4.3). Due to its high modal density, there were more than two hundred dominant modes in the frequency range of zero to 50 Hz. It was observed that the computational savings of the CA method, using the recursive process of Equation (55), were not substantial. For this reason, we developed a modified combined approximations method (MCA) by modifying the recursive process of Equation (55) which is much more efficient than the original CA method for large size models.

The cost of calculating the subspace basis in Equation (55) consists of one matrix decomposition (DCMP) and one forward-backward substitution (FBS). The DCMP cost is only related to the size and density of the symmetric stiffness matrix, while the FBS cost depends on both the size and density of the stiffness matrix and the number of columns of$\boldsymbol{\Phi}_0$. As the frequency range of interest increases, more modes are needed to predict the structural response accurately. In such a case, although a single DCMP is needed in Equation (55), the number of columns in $\boldsymbol{\Phi}_0$ may increase considerably, thereby increasing the cost of the repeated FBS. When the number of dominant modes becomes very large, the cost of performing the calculations in Equation (55) can be dominated by the FBS cost. For example, the vehicle model of Section 4.3 (Figure 7) has 65,000 degrees of freedom and 1050 modes in the frequency range of 0-300 Hz. The cost of one DCMP is 1.1 seconds (using a SUN ULTRA workstation and NASTRAN v2001) and the cost of one FBS is less than 0.1 seconds if $\boldsymbol{\Phi}_0$ has only one mode. In this case, the total cost is dominated by the DCMP, and the CA method reduces the cost of one reanalysis considerably. However, if $\boldsymbol{\Phi}_0$ has 1050 modes, the cost of FBS increases to 29 seconds dominating the cost of the DCMP. The CA method can therefore, improve the efficiency only when the number of retained modes is small. Otherwise, the computational savings do not compensate for the loss of accuracy from using K_0 (stiffness matrix of baseline design) instead of K_p (stiffness matrix of new design). The modified combined approximations (MCA) method of this section addresses this issue.

The MCA method uses a subspace basis **T** whose columns are constructed using the recursive process

$$\begin{aligned} T_1 &= K_p^{-1}(M_p\Phi_0) \\ T_i &= K_p^{-1}(M_pT_{i-1}) \qquad i=2,\ 3,\ \cdots,\ s \end{aligned} \tag{59}$$

instead of that of Equation (55). The selection of the appropriate number of basis vectors s is discussed later in this section. The only difference between Equations (55) and (59) is that matrix K_0 is inverted in the former while matrix K_p is inverted in the latter. The DCMP of K_p must be repeated for every new design. However, the cost of the repeated DCMP does not significantly increase the overall cost in Equation (59) because the latter is dominated by the FBS cost. The iterative process of Equation (59) provides a continuous mode shape updating of the new design. If the process converges in s iterations, the mode shapes T_s will be the exact mode shapesΦ_p. Equation (55) does not have the same property. The vectors T_i provide therefore, a more accurate approximation of the exact mode shapes Φ_p than the R_i vectors of the original CA method. This is an important advantage of MCA.

Because the mode shapes T_i in Equation (59) can quickly converge to the exact mode shapes Φ_p, for many practical problems only one iteration (i.e. $s = 1$) may be needed, resulting in

$$\mathbf{T} = \mathbf{T}_1 \tag{60}$$

If the convergence is slow, multiple sets of updated mode shapes can be used so that

$$T = [\Phi_0 \ \ T_1 \ \ T_2 \ \ \cdots \ \ T_s] \tag{61}$$

For better accuracy, the above basis also includes the mode shapes Φ_0 of the baseline design. Because the approximate modes T_i are more accurate than the CA vectors R_i in approximating the exact mode shapesΦ_p, MCA can achieve similar accuracy to the CA method using fewer modes. The example of Section 4.3 demonstrates that the MCA method achieves good accuracy with only 1 basis vector whereas the CA method requires 3 to 6 basis vectors [13-17].

In summary, the proposed MCA method involves four steps in calculating the approximate eigenvectors $\tilde{\Phi}_p$as follows

- Calculate basis T using Equation (60) or Equation (61).
- Calculate the condensed stiffness and mass matrices $\mathbf{K}_R$ and $\mathbf{M}_R$ as

$$\mathbf{K}_R = \mathbf{T}^T \mathbf{K}_p \mathbf{T} \qquad \mathbf{M}_R = \mathbf{T}^T \mathbf{M}_p \mathbf{T} \tag{62}$$

- Solve the following reduced eigen-problem to calculate the eigenvalues and the projections of the modes in the reduced space spanned by $\mathbf{T}$

$$(K_R - \lambda M_R)\Theta = 0 \tag{63}$$

- Reconstruct the approximate eigenvectors $\widetilde{\Phi}_p$ as

$$\widetilde{\Phi}_p = T\Theta \tag{64}$$

The slightly increased cost of using Equation (61) instead of Equation (60) is usually smaller than the realized savings in steps 2 through 4 of Equations (62) through (64) due to the smaller size of the reduced basis$T$. The bases of Equations (60) and (61) are smaller in size than the CA basis of Equation (56) for comparable accuracy. The MCA method requires therefore, less computational effort for steps 2 through 4. The computational savings compensate for the increased cost of DCMP for dynamic reanalysis of large finite-element models with a large number of dominant modes.

All mode shapes in Equation (63) must be calculated simultaneously in order to ensure that the approximate mode shapes $\widetilde{\Phi}_p$ are orthogonal with respect to the mass and stiffness matrices. However, the cost of estimating the mode shapes $\widetilde{\Phi}_p$using Equations (62) to (64) may increase quickly (quadratically) with the number of modes, and as a result, the MCA method may become more expensive than a direct eigen-solution when the number of dominant modes exceeds a certain limit. One way to circumvent this problem is to divide the frequency response into smaller frequency bands and calculate the frequency response in each band separately instead of solving for the frequency response in one step. The modal basis Tin Equation (61) is divided into k groups as

$$\mathbf{T} = \begin{bmatrix} \mathbf{T}^1 & \mathbf{T}^2 & \cdots & \mathbf{T}^k \end{bmatrix} \tag{65}$$

where

$$T^i = [\Phi_0^i \;\; T_1^i \;\; T_2^i \;\; \cdots \;\; T_s^i] \tag{66}$$

Each group T^icontains roughly n/k original modes Φ_0^i fromΦ_0, and their corresponding improved modes. The eigenvectors of the new design are calculated using T^iinstead of T in Equations (62) to (64). The process is repeated k times using a modal basis that is $1/k$ of the size of the original modal basis. All k groups of eigenvectors are then collected together to calculate the frequency response of the new design. As demonstrated in Section 4.3.2, this reduces the cost considerably with minimal loss of accuracy.

3.5. Comparison of CA/MCA and PROM Methods

As we have discussed, a large overhead cost which is proportional to the number of design parameters is required before the PROM reanalysis is carried out. However, the CA/MCA method does not require this overhead cost because it does not need the basis $\boldsymbol{P}$ of Equation (49) (see Section 3.2). Therefore, the CA/MCA method is more suitable, when the number of reanalyses is comparable to the number of design parameters. This is usually true in gradient-based design optimization. The CA and MCA methods can become expensive however, when many reanalyses are needed because, for each reanalysis, they require a new basis $\boldsymbol{R}$ or $\boldsymbol{T}$ (see Equations 56and 61, respectively) and new condensed mass and stiffness matrices in Equations (57) and (62). This is the case in gradient-free optimization problems employing a Genetic Algorithm for example, and in simulation-based probabilistic analysis problems employing the Monte-Carlo method. For these problems, the PROM method is more suitable because the subspace basis $\boldsymbol{P}$ does not change for every new design point. Table 1 summarizes the main characteristics, advantages and application areas of the CA/MCA and PROM methods.

	CA/MCA Method	PROM Method
Overhead Cost	None	Required cost to construct $\boldsymbol{P}$. Cost proportional to the number of design parameters m.
Basis Vector	Variable basis $\boldsymbol{R}/\boldsymbol{T}$. Size proportional to n and s.	Constant basis $\boldsymbol{P}$. Size proportional to n and m.
Reanalysis Cost	Moderate Relatively small size of $\boldsymbol{R}/\boldsymbol{T}$. Must recalculate $\boldsymbol{R}/\boldsymbol{T}$ at every new design.	High if no parametric relationship exists due to the condensation of large size and dense $\boldsymbol{P}$. Low if a parametric relationship exists.
Best Application	Small number of reanalyses compared to the number of design parameters. Evaluation of few design alternatives and gradient-based optimization.	Very large number of reanalyses. Gradient-free optimization (e.g. genetic algorithms) and probabilistic analysis.

Table 1. Comparison of the CA/MCA and PROM methods.

4. Reanalysis Methods in Dynamic Analysis and Optimization

The reanalysis methods of Section 3 can be used in different dynamic analyses such as modal or direct frequency response and free or forced vibration in time domain. Depending on the problem and the type of analysis, a particular reanalysis method may be preferred considering how many times it will be performed and how many design parameters will be allowed to change. This section demonstrates the computational efficiency and accuracy of reanalysis methods in dynamic analysis and optimization. It also introduces a new reanalysis method in Craig-Bampton substructuring with interface modes which is very useful for problems with many interface DOFs where FRF substructuring is not practically applicable.

4.1. Integration of MCA Method in Optimization

We have mentioned that the MCA method provides a good balance between accuracy and efficiency for problems that require a moderate number of reanalyses, as in gradient-based optimization. For problems where a large number of reanalyses is necessary, as in probabilistic analysis and gradient-free (e.g. genetic algorithms) optimization, a combination of the MCA and PROM methods is more suitable.

Figure 6 shows a flowchart of the optimization process for modal frequency response problems. The DMAP (Direct Matrix Abstraction Program) capabilities in NASTRAN have been used to integrate the MCA method and the NASTRAN modal dynamic response and optimization (SOL 200). The highlighted boxes indicate modifications to the NASTRAN optimizer. Starting from the original design, the code first calculates the design parameter sensitivities in order to establish a local search direction and determine an improved design along the local direction. At the improved design, an eigen-solution is obtained to calculate a modal model and the corresponding modal response. The dynamic response at certain physical DOFs is then recovered from the modal response. At this point, a convergence check is performed to decide if the optimal design is obtained. If not, further iterations are needed and the above procedure is repeated. Many iterations are usually needed for practical problems to obtain the final optimal design. Section 4.3.2 demonstrates how this process was used to optimize the vibro-acoustic behavior of a 65,000 DOF, finite-element model of a truck. Using the MCA method, the computational cost of the entire optimization process was reduced in half compared with the existing NASTRAN approach.

As for a stand alone modal frequency response, the eigen-solution accounts for a large part of the overall optimization cost for vibratory problems where a modal model is used. A reanalysis method can be inserted into the procedure as shown in Figure 6 to provide an approximate eigen-solution saving therefore, substantial computational cost. Other reanalysis methods such as the CDH/VAO, CA or PROM can also be used depending on the number of design variables and the number of expected iterations.

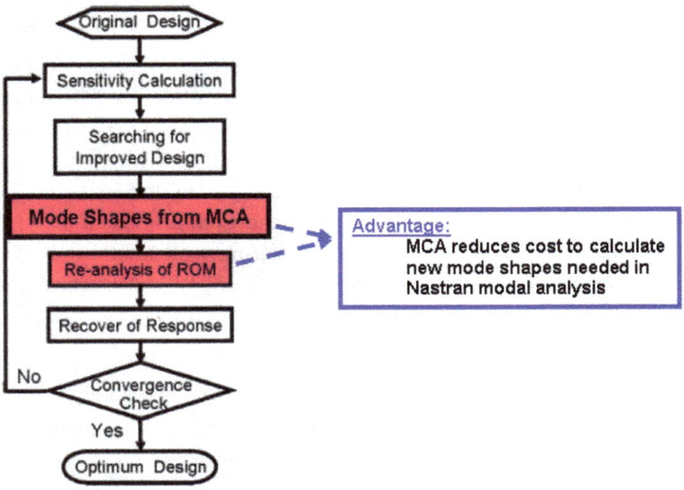

Figure 6. Flowchart for mca-enhanced nastran optimization.

4.2. Integration of MCA and PROM Methods

The PROM method requires exact calculation of the mode shapes for all designs corresponding to the corner points of the parameter space in order to calculate the subspace basis $\boldsymbol{P}$of Equation (49). The required computational effort can be prohibitive for a large number of parameters (optimization design variables). This effort can be reduced substantially if the modes of each corner point are approximated by the MCA method. In this case, an exact eigen-solution is required only for the baseline design. The following steps describe an algorithm to integrate the MCA and PROM methods:

- Perform an exact eigen-analysis at the baseline design point $\mathbf{p}_0$ all parameters are at their lower limit, to obtain the baseline mode shapes$\boldsymbol{\Phi}_0$.
- Use the MCA method at design point $\mathbf{p}_i$ with all parameters at their low limit except the ith parameter which is set at its upper limit. Obtain approximate mode shapes for the ith corner point using the following recursive process

$$\begin{aligned} &\boldsymbol{T}_{i,1} = \boldsymbol{K}_i^{-1}(\boldsymbol{M}_i \boldsymbol{\Phi}_o) \\ &\boldsymbol{T}_{i,j+1} = \boldsymbol{K}_i^{-1}(\boldsymbol{M}_i \boldsymbol{T}_{i,j}) \qquad j = 2,\ 3,\ \cdots,\ s \end{aligned} \tag{67}$$

- Form the subspace basis $\mathbf{T}$ as

$$T = [\Phi_o \quad T_{0,s} \quad T_{1,s} \quad \cdots \quad T_{m,s}] \tag{68}$$

where m is the total number of parameters.

- Obtain the approximate mode shapes $\tilde{\Phi}_p$ using the subspace projection procedure of Equations (50) through (52) where T is used instead of P.

The modal basis $\tilde{\Phi}_p$ can be subsequently used in a modal dynamic response solution. Only step 4 is repeated in reanalysis. The computational savings can be substantial especially for problems where many reanalyses are needed.

4.3. Combined MCA and PROM Methods: Vibro-Acoustic Response of a Vehicle

The pickup truck vehicle model with 65,000 DOFs of Figure 7 is used in this section to demonstrate the advantages of the combined MCA and PROM method in optimizing the vibro-acoustic response of a vehicle. The model has 78 components and roughly 11,000 nodes and elements. The example is performed on a SUN ULTRA workstation using NASTRAN v2001. The MCA and PROM methods have been implemented in NASTRAN DMAP.

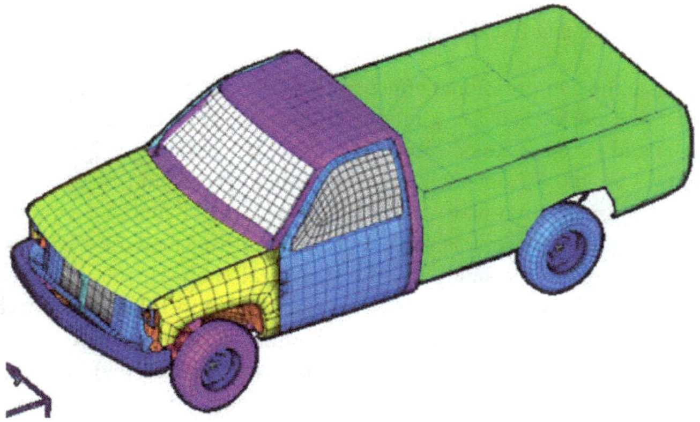

Figure 7. FE model of a pickup truck.

The sound pressure level at the driver's ear location is calculated using a vibro-acoustic analysis. The structural forced vibration response due to unit harmonic forces in x, y, and z directions at the engine mount locations, is coupled with an interior acoustic analysis. The first and second eigen-frequencies of the acoustic volume inside the cabin are 95.9 Hz and 128.3 Hz. The sound pressure level is calculated in the 80 to 140 Hz frequency range. The structure and fluid domains are coupled through boundary conditions ensuring continuity of vibratory displacement and acoustic pressure. A finite-element formulation of the coupled undamped problem yields the following system equations of motion [24].

$$\left(\begin{bmatrix} \mathbf{K}_S & -\mathbf{H}_{SF} \\ 0 & \mathbf{K}_F \end{bmatrix} - j\omega^2 \begin{bmatrix} \mathbf{M}_S & 0 \\ \rho_0 c_0^2 \mathbf{H}_{SF}^T & \mathbf{M}_F \end{bmatrix}\right) \begin{bmatrix} \mathbf{d}_S \\ \mathbf{p}_F \end{bmatrix} = \begin{bmatrix} \mathbf{f}_b \\ \mathbf{f}_q \end{bmatrix} \tag{69}$$

where the vibratory displacement $\boldsymbol{d}_S$ and the acoustic pressure $\boldsymbol{p}_F$ are the primary variables. The finite-element representation of the two domains consists of stiffness and mass matrix pairs $\langle \boldsymbol{K}_S, \boldsymbol{M}_S \rangle$ and$\langle \boldsymbol{K}_F, \boldsymbol{M}_F \rangle$, respectively. The air density and wave speed are ρ_0 andc_0. The right hand side of Equation (69) denotes the external forces.

The spatial coupling matrix $\boldsymbol{H}_{SF}$ indicates coupling between the two domains which is usually referred to as "two-way coupling." Due to this coupling, the combined structural-acoustic system of equations is not symmetric. If the acoustic effect on the structural response is small, the coupling term can be omitted, resulting in the so-called "one-way coupling," where the structural response is first calculated and then used as input ($\boldsymbol{f}_q$in Equation 69) to solve for the acoustic response. The coupled structure-acoustic system can be solved either by a direct method, or more efficiently by a modal response method which can be applied to both the structural and acoustic domains.

4.3.1. *Combined MCA and PROM Methods*

To demonstrate the computational effectiveness and accuracy of integrating MCA in PROM, a reanalysis was performed for a modified design where five plate thickness parameters vary; chassis and its cross links, cabin, truck bed, left door, and right door. All parameters were increased by 100% from their baseline values. The sound pressure at the driver's ear was calculated using "two-way" coupling. A structural modal frequency response was used. The acoustic response was calculated using a direct method because the size of the acoustic model is relatively small. For the structural analysis, 1050 modes were retained in the 0 to 300 Hz frequency range. The combined MCA and PROM approach was compared against the NASTRAN direct solution for a modified design where all five parameters were at their upper limits. Only one iteration was used in Equation (59) in order to get the set of once-updated mode shapes for each corner design point. The subspace basis, which includes information for all five design parameters, is therefore, represented by

$$\boldsymbol{T} = [\boldsymbol{\Phi}_o \quad \boldsymbol{T}_{0,1} \quad \boldsymbol{T}_{1,1} \quad \cdots \quad \boldsymbol{T}_{5,1}] \tag{70}$$

The maximum error in natural frequencies as predicted by the combined MCA and PROM method and NASTRAN, is less than 0.45% in the entire frequency range. Figure 8 indicates that the sound pressures calculated by both methods are almost identical. The computational effort for the MCA method to obtain approximate mode shapes at each corner design point is about 30 seconds. In contrast, it takes about 180 seconds for an exact eigen-solution using NASTRAN. The computational cost to construct the reduced basis (***P***in PROM and ***T*** in PROM+MCA) is compared in Table 2. The total cost was reduced from 1080 seconds to 330 seconds. The computational saving is more significant if the number of design parameter increases.

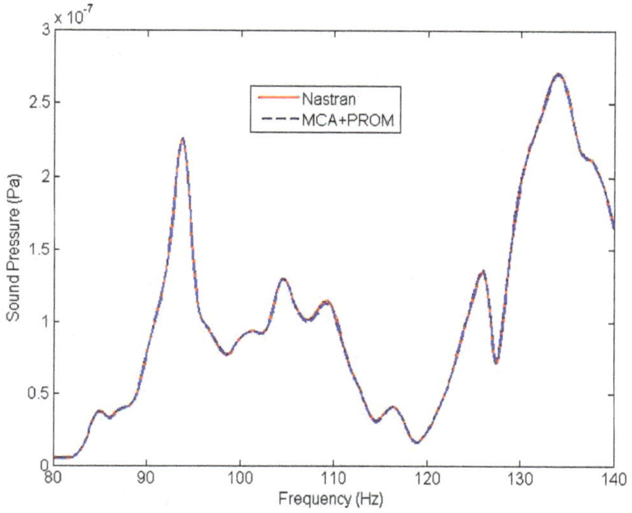

Figure 8. Comparison of sound pressure at driver's ear between combined MCA and PROM method and NASTRAN.

Method	Solving for mode shape Φ_0 at baseline design	Solving for mode shapes at 5 corner design points	Total Cost
PROM	180 sec	180*5=900 sec	1080 sec
PROM+MCA	180 sec	30*5=150 sec	330 sec

Table 2. CPU time to construct reduced basis.

4.3.2. Optimization using MCA Method

The goal here is to minimize the sound pressure at the driver's ear. A total of 41 design parameters are used representing the thickness of all vehicle components modeled with plate elements. All thicknesses are allowed to change by 100% from their baseline values. Table 3 describes all design parameters. At the initial point of the optimization process, all parameters are at their low bound.

Prm. #	Description (thickness of)	Prm. #	Description (thickness of)	Prm. #	Description (thickness of)
1	Bumper	15	Radiator mtg.	29	Tire, front right
2	Rails	16	Radiator mtg., mid.	30	Tire, rear left
3	A-arm, low left	17	Fan cover, low	31	Tire, rear right
4	A-arm, low right	18	Fan cover, up	32	Engine outer
5	A-arm, up left	19	Cabin	33	A-arm conn., up left
6	A-arm, up right	20	Cabin mtg. reinf.	34	A-arm conn., up right
7	Tire rim	21	Door, left	35	A-arm conn., low left
8	Engine Oil-box	22	Door, right	36	A-arm conn., low right
9	Fan	23	Bed	37	Glass, left
10	Hood	24	Brake, front left	38	Glass, right
11	Fender, left	25	Brake, front right	39	Glass, rear
12	Fender, right	26	Rail conn., rear	40	Glass, front
13	Wheel house, left	27	Rail mount	41	Rail conn., front
14	Wheel house, right	28	Tire, front left		

Table 3. Description of design parameters.

Because of the large number of design parameters, the combined MCA and PROM approach Section 4.3.1 is not computationally efficient because the size of the PROM basis is very large (see Equation 70). For this reason, we use the MCA reanalysis method and demonstrate its capability to handle a large number of parameters. It approximates the mode shapes at intermediate design points using only T_1 in Equation (59). The subspace basis at each optimization step is thus$T = [\Phi_o \quad T_1]$. Because 1050 modes exist in the frequency range of 0 to 300 Hz of the initial design, the size of the MCA modal basis is 2*1050 = 2100.

	k=1	k=21
Eq. (59)	31 sec	31 sec
Eq. (62)	258 sec	50 sec
Eq. (63)	48 sec	6 sec
Eq. (64)	67 sec	10 sec
Total Cost	404 sec	97 sec

Table 4. CPU time of the MCA method.

The cost of solving for 1050 modes directly from NASTRAN is 180 seconds (see Table 2). In the MCA method, the cost of solving the linear system of equations in Equation (59) is 31 seconds, and the additional combined cost of Equations (62) to (64) is 373 seconds, resulting in a total cost of 404 seconds (see Table 4). To reduce this cost, the 1050 modes are divided into 21 groups and the modes in each group are obtained separately as explained in the last paragraph of Section 3.4. This reduces the cost of Equations (62) to (64) to 66 seconds for a total cost of 97 seconds, which is about half the cost of the direct NASTRAN method.

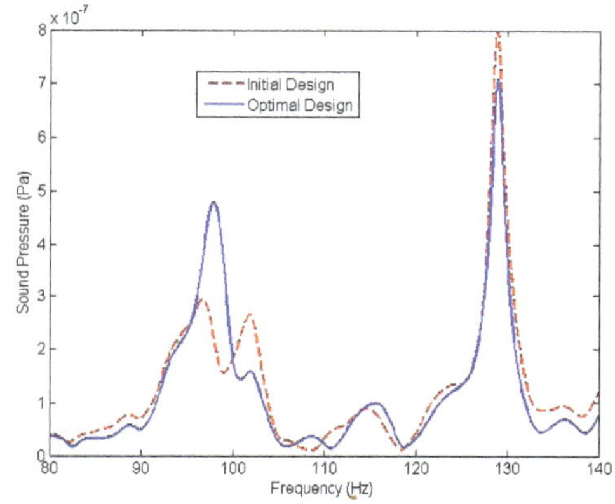

Figure 9. Comparison of sound pressure at driver's ear between initial and optimal designs.

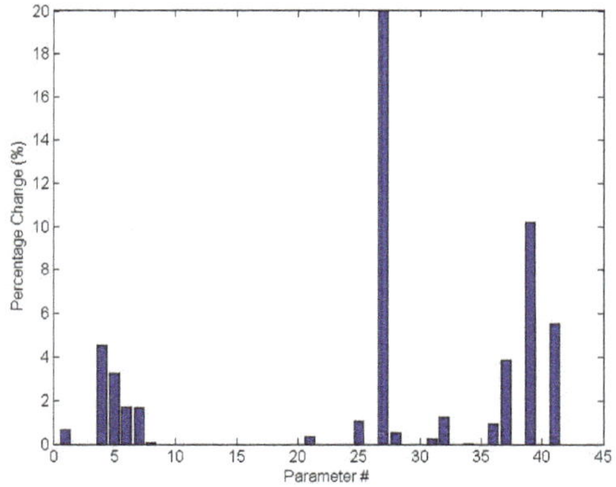

Figure 10. Percent increase of optimal design parameters relative to baseline design parameters.

The gradient-based optimizer in NASTRAN (SOL 200) using the optimization process of Figure 6 needed three iterations to calculate the optimal design. Figure 9 compares the sound pressure at the driver's ear between the optimal and initial designs. Figure 10 shows the percentage increase of optimal values relative to the initial values for all 41 design parameters. In the frequency range of 80-140Hz, the maximum sound pressure is slightly reduced from 7.9E-7 to 7.2E-7 Pascal. Most parameters are minimally changed. The largest increase is 20% for the rail mount thickness (parameter #27).

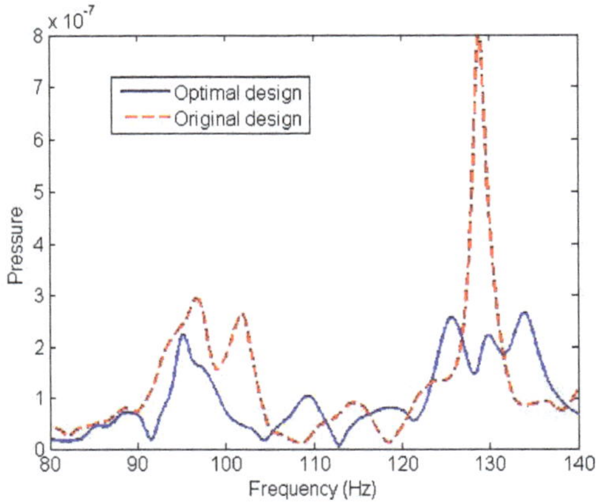

Figure 11. Comparison of sound pressure at driver's ear between baseline and optimal designs with 20 initial populations and 4 generations.

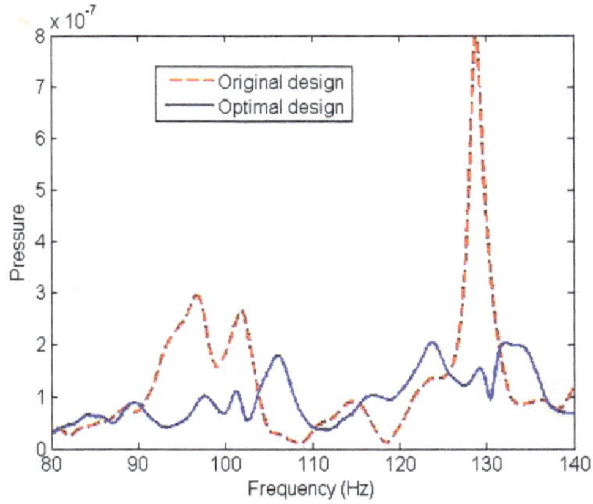

Figure 12. Comparison of sound pressure at driver's ear between baseline and optimal designs with 100 initial populations and 6 generations.

The Sequential Quadratic Programming (SQP) algorithm of NASTRAN can only find a local optimum. To obtain a more significant design improvement, two additional studies were performed using a Genetic Algorithm with the MCA method. The first study used 20 initial populations and 4 generations, and the second study used 100 initial populations and 6 generations. Figures 11and 12 show that the number of initial populations and the number of generations, affect the optimization results. While a higher number of initial populations and generations results in a slightly better result, both studies produced a much better optimum than the SQP algorithm. In the case of 100 initial populations and 6 generations, the sound pressure is reduced from 7.9E-7 Pascal to 2.0E-7 Pascal, which is equivalent to about 15 dB in sound pressure level (SPL).

To verify the accuracy of the MCA approximation, the sound pressure response at the optimal design from MCA+GA with 100 initial populations and 6 generations was evaluated by both direct NASTRAN and MCA. Figure 13 shows that the MCA method is very accurate. For a similar to MCA accuracy, the original CA method needed three sets of mode shapes to form the subspace basis, requiring 90 seconds to solve the linear equations. The much larger mode basis **R** in CA increases the computational cost to calculate the triple matrix products of Equation (57). Therefore for large scale, finite-element models with a high modal density, the proposed MCA method can be more efficient compared to either a complete NASTRAN analysis or the original CA method.

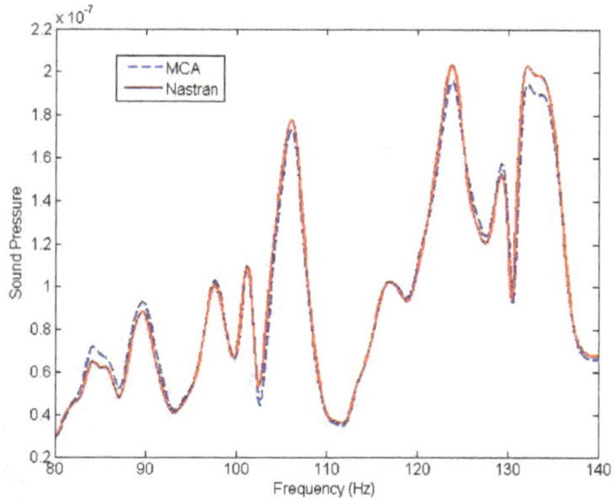

Figure 13. Comparison of sound pressure at driver's ear between direct nastran and mca.

4.4. Reanalysis in Craig-Bampton Substructuring with Interface Modes

The FRF substructuring of Section 2.3 couples two structures using FRF information between the coupling (interface) DOFs, and the excitation and/or response DOFs. Although this approach is very efficient, it is practical only if we have a few coupling DOFs; e.g. connection of a vehicle suspension to chassis or connection of the exhaust system to body through a few hangers. If the physical substructures have interfaces with many DOFs, a different reduced-order modeling (ROM) approach must be used such as the Craig-Bampton ROM of Section 2.2.1. The Craig-Bampton ROM can be large however, if the number of retained interface DOFs is large. We address this problem by performing a secondary eigenvalue analysis which yields the so-called *interface modes* (see Section 2.2.2). The following section describes a reanalysis methodology for physical substructuring with Craig-Bampton ROMs using interface modes. We show that its accuracy is very good and the computational savings are substantial.

4.4.1. Craig-Bampton with Interface Modes and Reanalysis

In the Craig-Bampton CMS method (Craig-Bampton reduced-order model or CBROM), the mass and stiffness matrices of each substructure are partitioned into interface sub-matrices, interior (omitted DOF) sub-matrices, and their coupling sub-matrices. The dynamics of a structure are then described by the normal modes of its individual components, plus a set of modes called *constraint modes* that couple the components. In CBROM, there is no size reduction for constraint modes since all of them are kept in the reduced equations. If the finite element mesh is sufficiently fine, the constraint-mode DOFs will dominate the size of CBROM mass and stiffness matrices and result in a large computational cost. This issue is addressed by using *interface modes* (also called *characteristic constraint –CC- modes*). For that, a secondary eigenvalue analysis is performed using the constraint-mode partitions of the CMS mass and stiffness matrices. The CC modes are the resultant eigenvectors. Details are provided in Sections 2.2.1 and 2.2.2.

The number of constraint modes n_c equals to the number of interface DOF. For many FE models of large structures, the number of interface DOF can be rather large. The calculation of constraint modes in Equation (6) involves a decomposition step and a FBS step. The cost of FBS is proportional ton_c. For any matrix multiplication that involves$\boldsymbol{\Phi}_i^C$, the cost is proportional ton_c. For any triple-product that involves $\boldsymbol{\Phi}_i^C$ the cost is proportional ton_c^2.

The matrices from all substructures are assembled into a global CBROM with substructures coupled at interfaces by enforcing displacement compatibility. If $\boldsymbol{k}_i^C$ and $\boldsymbol{m}_i^C$ are the component (substructure) matrices, the global matrices $\boldsymbol{K}^C$and $\boldsymbol{M}^C$are assembled as

$$\mathbf{K}^C = \sum \mathbf{k}_i^C, \quad \mathbf{M}^C = \sum \mathbf{m}_i^C \tag{71a}$$

and a secondary eigenvalue analysis is performed to calculate the *interface modes* $\boldsymbol{\Phi}^{CC}$ as

$$[\boldsymbol{K}^C - \lambda^{CC} \boldsymbol{M}^C] \boldsymbol{\Phi}^{CC} = 0 \tag{71b}$$

The matrices in Equations (9), (10) and (12) are then reduced as

$$
\begin{aligned}
m_i^{CC} &= \boldsymbol{\Phi}^{CCT^T} m_i^C \boldsymbol{\Phi}^{CC} \\
m_i^{CCN} &= \boldsymbol{\Phi}^{CCT^T} m_i^{CN} \\
k_i^{CC} &= \boldsymbol{\Phi}^{CCT^T} k_i^C \boldsymbol{\Phi}^{CC}
\end{aligned}
\tag{72}
$$

where the matrices $\mathbf{m}_i^{CC}$, $\mathbf{m}_i^{CCN}$ and $\mathbf{k}_i^{CC}$ are of much smaller size than matrices $\mathbf{m}_i^C$, $\mathbf{m}_i^{CN}$ and $\mathbf{k}_i^C$.

The interface modes reduce the interface size producing a smaller reduced order model (ROM) compared with the traditional Craig-Bampton ROM (CBROM). However, they are calculated from the assembled interface **K** and **M** matrices. Thus, the calculation of constraint modes and all matrix multiplications related to constraint modes are still necessary. The interface mode method reduces the size of ROM but it does not reduce the computational cost related to the constraint modes.

If the interface modes $\boldsymbol{\Phi}^{CC}$ were known before hand, the calculations in Equations (6), (9), (10) and (12) and Equation (72) could be performed much more efficiently as follows[55]

$$\hat{\boldsymbol{\Phi}}_i^C = \boldsymbol{\Phi}_i^C \boldsymbol{\Phi}^{CC} = -\mathbf{k}_i^{\Omega\Omega^{-1}} (\mathbf{k}_i^{\Omega\Gamma} \boldsymbol{\Phi}^{CC}) \tag{73}$$

$$\mathbf{m}_i^{CC} = (\boldsymbol{\Phi}^{CC})^T \mathbf{m}_i^{\Gamma\Gamma} \boldsymbol{\Phi}^{CC} + (\boldsymbol{\Phi}^{CC})^T \mathbf{m}_i^{\Gamma\Omega} \hat{\boldsymbol{\Phi}}_i^C + (\hat{\boldsymbol{\Phi}}_i^C)^T \mathbf{m}_i^{\Omega\Gamma} \boldsymbol{\Phi}^{CC} + (\hat{\boldsymbol{\Phi}}_i^C)^T \mathbf{m}_i^{\Omega\Omega} (\hat{\boldsymbol{\Phi}}_i^C)^T \tag{74}$$

$$\mathbf{m}_i^{CCN} = (\boldsymbol{\Phi}^{CC})^T \mathbf{m}_i^{\Gamma\Omega} \boldsymbol{\Phi}_i^N + (\hat{\boldsymbol{\Phi}}_i^C)^T \mathbf{m}_i^{\Omega\Omega} \boldsymbol{\Phi}_i^N \tag{75}$$

$$\mathbf{k}_i^{CC} = (\boldsymbol{\Phi}^{CC})^T \mathbf{k}_i^{\Gamma\Gamma} \boldsymbol{\Phi}^{CC} - (\boldsymbol{\Phi}^{CC})^T k_i^{\Gamma\Omega} \hat{\boldsymbol{\Phi}}_i^{CC} \tag{76}$$

The following observations can be made:

1. In Equations (74) to (76), the computation involves $\boldsymbol{\Phi}^{CC}$ and $\hat{\boldsymbol{\Phi}}_i^C$ and does not involve $\boldsymbol{\Phi}_i^C$. Therefore, the calculation of original constraint modes $\boldsymbol{\Phi}_i^C$ is no longer needed.
2. In Equation (73), the number of columns of matrix $(\mathbf{k}_i^{\Omega\Gamma} \boldsymbol{\Phi}^{CC})$ is equal to the number of interface modes n_{cc}which is usually smaller thann_c. Therefore, the FBS cost of solving for $\hat{\boldsymbol{\Phi}}_i^C$ is proportional to$n_{cc}$and it is much smaller than the FBS cost of solving for$\boldsymbol{\Phi}_i^C$.
3. Because both $\boldsymbol{\Phi}^{CC}$ and $\hat{\boldsymbol{\Phi}}_i^C$ are of size $n_{cc}$the cost of matrix multiplication and triple-product in Equations (74) to (76) are now proportional to n_{cc} andn_{cc}^2. Therefore, the cost is much smaller than the corresponding cost in Equations (9), (10) and (12).

In this **CCROM** method which is based on **CBROM**, the interface modes $\boldsymbol{\Phi}^{CC}$ are obtained using the assembled interface partitions of the CBROM formulation. Thus, it is impossible to know $\boldsymbol{\Phi}^{CC}$ before hand for a new design. For this reason, Equations (73) to (76) can not be theoretically implemented to improve efficiency. For this reason, we propose *a reanalysis approach where the calculated interface modes* Φ^{CC} *for original (baseline) design can be used as an approximation of the new interface modes at any modified design*. In this case, Equations (73) to (76) are applied to improve the computational efficiency.

4.4.2. A Car Door Example

The car door model of Figures 14 and 15 is used to demonstrate the proposed reanalysis method for substructuring with Craig-Bampton method using interface modes. It has 25,800 nodes and 25,300 elements and is divided into two substructures. The first substructure includes the outer door shell and a bar attached to it. The second substructure includes the rest of the door. There are 293 nodes (1758 DOFs) on the interface. Therefore, the **CBROM** or **CCROM** method must calculate 1758 constraint modes according to Equation (6) for both substructures. The 1758 constraint modes are involved in matrix multiplication or triple-products in Equations (9), (10) and (12). Figure 16 shows the interface nodes.

For the initial design using the CCROM method, 52 interface modes are calculated below 600 Hz. A modified design is created where the shell thicknesses for the outer door (substructure 1) and inner door (substructure 2) are doubled. To provide baseline numbers, the **CCROM** method is used on the new design to solve for the system natural frequencies. The new reanalysis approach is used on the new design to calculate approximate natural frequencies which are then compared with the baseline numbers. The interface modes calculated at the original design are used as an initial guess for the interface modes of the new design.

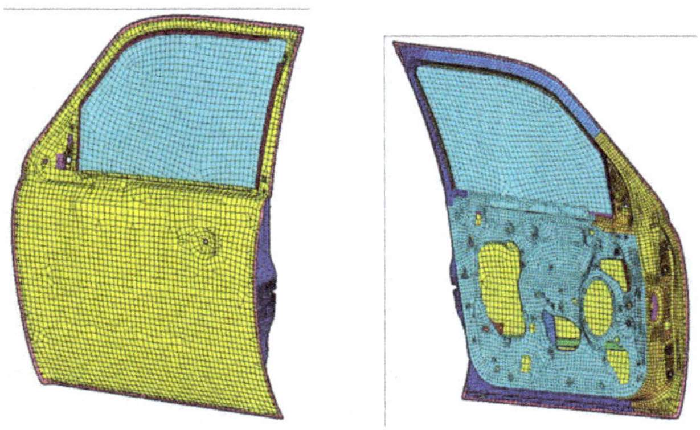

Figure 14. Outside and inside views of car door model.

Figure 15. Substructure 1 (outer door shell) and substructure 2 (rest of door).

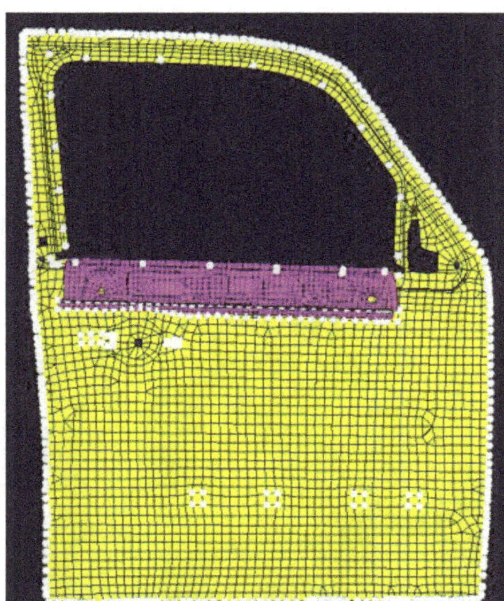

Figure 16. Interface nodes indicated by white dots.

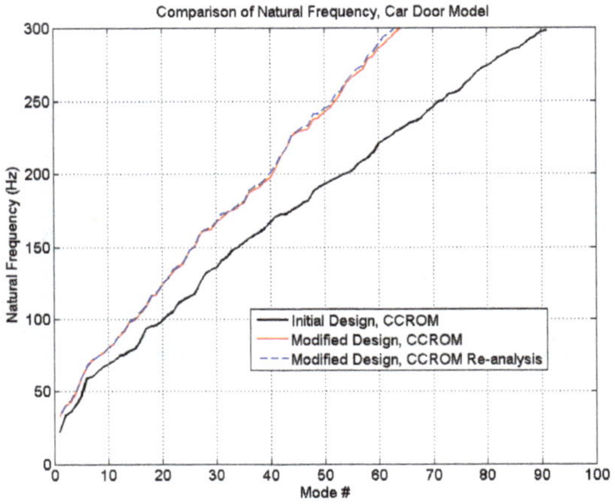

Figure 17. Comparison of natural frequencies between original CCROM method and CCROM with reanalysis for the car door example.

Figure 17 compares the natural frequencies of the new (modified) design between the original CCROM (Craig-Bampton with Interface modes) method and the new approach where reanalysis is used in CCROM to approximate the interface modes. We observe that the natural frequencies of the modified design are very different from those of the original design. Also, the accuracy of the proposed reanalysis method is excellent. The frequencies for the modified design calculated by the original CCROM and the proposed new approach are almost identical. The percentage error of the new approach versus the original CCROM approach is less than 1% on average. The computation cost is summarized in Table 5.

Substructure 1:					
CPU Time	**Normal Modes**	**Constraint Modes**	**Multiplication**	**Other Cost**	**Total Cost**
CCROM	8 sec	61 sec	65 sec	3 sec	137 sec
New Approach	7 sec	**2 sec**	**0.3 sec**	2 sec	11 sec
Substructure 2:					
CPU Time	**Normal Modes**	**Constraint Modes**	**Multiplication**	**Other Cost**	**Total Cost**
CCROM	108 sec	282 sec	927 sec	10 sec	1327 sec
New Approach	110 sec	**16 sec**	**3 sec**	10 sec	139 sec

Table 5. Summary of computational cost for the car door example.

In the new approach to reduce the cost related to constraint modes, the total remaining cost is dominated by the cost of calculating the normal modes for each substructure. For example, the calculation of the normal modes for Substructure 2 took 110 seconds out of a total of 139 seconds (see Table 5). It should be noted that the normal modes cost can be further reduced by applying another reanalysis method such as CDH/VAO, CA or MCA to approximate the normal modes. Therefore, the overall cost of substructuring based on Craig-Bampton with interface modes, can be drastically reduced by using the proposed reanalysis to approximate the constraint modes and a CDH/VAO or MCA reanalysis to approximate the normal modes at a new design.

5. Optimization of a Vehicle Model

A detailed optimization study is presented using a large-scale FE model of a vehicle. For simplicity, we call it "BETA" car model. It is composed of approximately 7.1 million DOFs and 1.1 million elements. Figure 18 shows all modeling details.

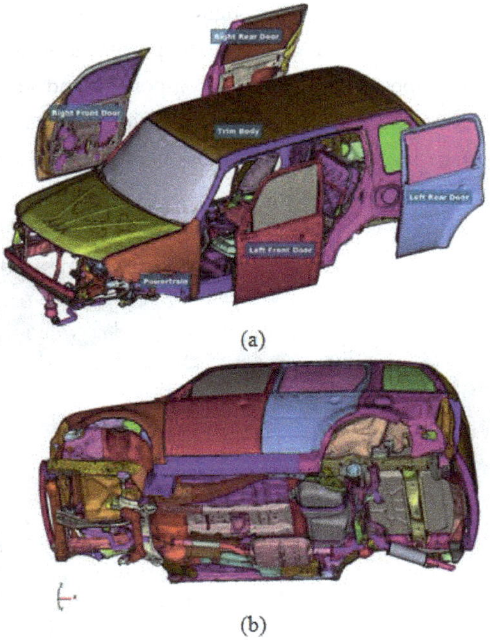

Figure 18. Details of "BETA" car model.

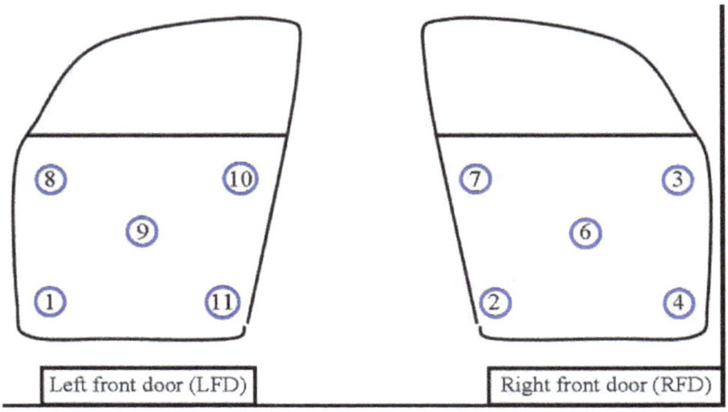

Figure 19. Ten response locations on two front doors.

We form an optimization problem in terms of the maximum vibratory displacement at any location of the outer shell of the two front doors by minimizing the maximum displacement among ten locations of the two front doors (Figure 19) due to a hypothetical engine excitation in the vertical (up-down) direction. The engine is represented by a lumped mass connected rigidly to the engine mounts (Figure 20). The powertrain-exhaust model has about 1.3 million DOFs and is composed of 29 PSHELL components and 12 PSOLID components. There are also some RBE2 and PBUSH elements which are used as connectors. The maximum displacement at each of the ten door locations is observed in the y direction (lateral direction – perpendicular to the door plane).

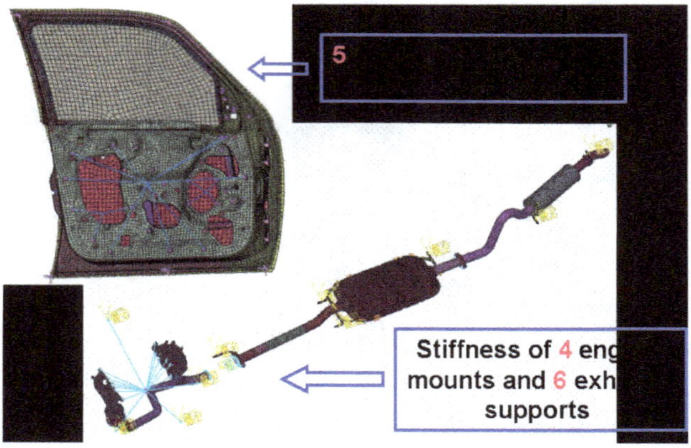

Figure 20. Description of the fifteen design variables.

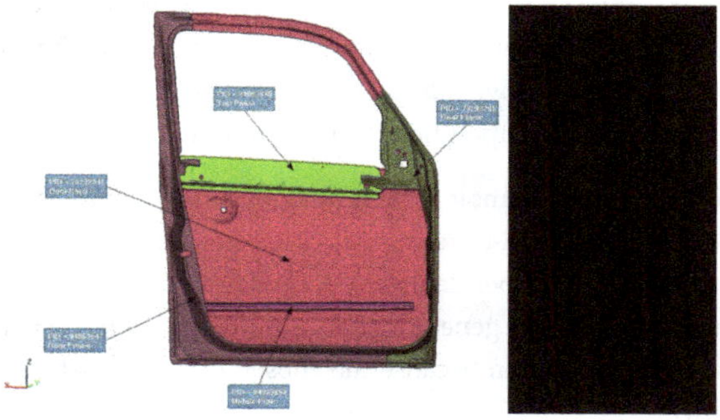

Figure 21. Description of the five design variables on the doors.

Fifteen design variables are chosen; five structural elements of each door (thickness of door shell, front frame, rear frame, top panel, middle pipe), vertical stiffness of each of the four engine mounts, and vertical stiffness of each of the six exhaust system supports. All design variables are schematically indicated in Figures 20 and 21.

The optimization problem is stated as follows:

$$\begin{aligned}
&Find\\
&X = [X_1 X_2 \cdots X_{15}]\\
&\min_{x}\left[\max_{i=1}^{10}(\mathrm{Re}\,sp_i)\right]\\
&such\ that: Mass \le Mass_{Nominal}\\
&where: \mathrm{Re}\,sp = \underbrace{|y(f)|}_{\uparrow\ Lateral\ door\ displacement}\ f \in \{100, 200\}\,Hz
\end{aligned}$$

The optimal value of each of the fifteen design variables is calculated in order to minimize the maximum response among the ten locations on the doors while the mass of the vehicle remains less or equal to the mass of the initial (nominal) vehicle. The response is calculated in the 100 Hz to 200 Hz frequency range and a 3% structural damping is used. The optimization problem is numerically very challenging because of

1. the many local optima and
2. the computational cost of each dynamic analysis.

The former was handled by using a hybrid optimization algorithm which first explores the entire design space using a Niching Genetic Algorithm (GA) [25] and then switches to a gradient-based optimizer (fmincon in MATLAB) using the best estimate of the optimal point from the GA as initial point. This ensures a rapid convergence to the final optimum because although all GA optimizers can move quickly to the vicinity of the final optimum, they have a very slow convergence rate in pinpointing the final optimum.

FRF substructuring is used to assemble all components of the vehicle (body, doors, and engine-exhaust) into a small reduced-order model. This keeps the computational cost of each dynamic analysis low (4 minutes per analysis). A modal model is created only once for the body subsystem and then used to generate an FRF representation. This modal model does not change during the optimization because the chosen design variables are not associated with the body. However, the modal models of the doors change during the optimization. The final model for the entire vehicle is created by assembling the FRF models of each component. The FRF assembly operation is repeated during optimization because the FRF models of the two doors keep changing.

The Niching GA optimizer maximizes a fitness function by modifying all design variables. A proper fitness function which minimizes the maximum response among the ten door locations while satisfying the vehicle mass constraint is chosen as follows

$$Fitness = \frac{\left[\max\limits_{i=1}^{10}(\text{Res}\, p_i)\right]_{Nominal}}{\left[\max\limits_{i=1}^{10}(\text{Res}\, p_i)\right]} * [1 + p * \min(c, 0)]$$

The ratio of the nominal maximum response over the actual maximum response is used so that the fitness value increases when the actual response is reduced. This ratio is multiplied by $1 + p * \min(c, 0)$ where p = 10 is a penalty value and $c = 1 - \frac{Mass}{Mass_{Nominal}}$. Thus, c is positive if *Mass* is less than $Mass_{Nominal}$ satisfying the constraint and the value of $1 + p * \min(c, 0)$ is equal to one.

Otherwise, c becomes negative if *Mass* is greater than $Mass_{Nominal}$ and the term $1 + p * \min(c, 0)$ assumes a large negative value which reduces the fitness value considerably. As a result, the GA optimizer always satisfies the mass constraint while maximizing the value of the fitness function.

Figure 22 summarizes the optimization results by comparing the maximum door response between the optimal and initial designs. The optimizer determined that the maximum response occurs at location 9 (center of left front door of Figure 19) at approximately 105 Hz. Figure 23 shows that this represents a vehicle local mode involving motion of the doors only. At the optimal design the maximum response was reduced from the initial 10^{-3} m to $0.47*10^{-3}$ m (Table 6).

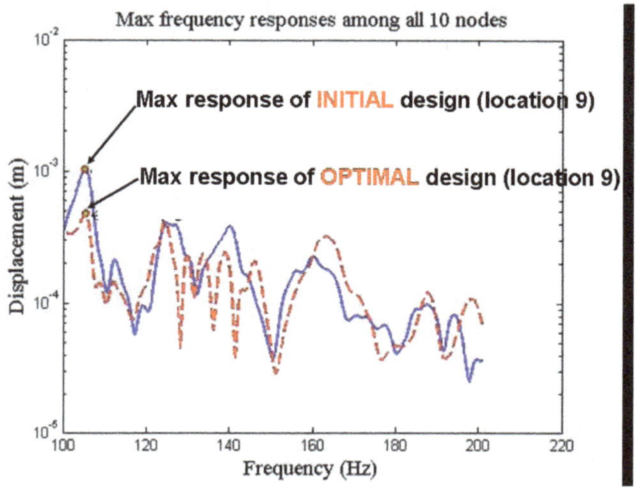

Figure 22. Comparison of optimal and initial designs.

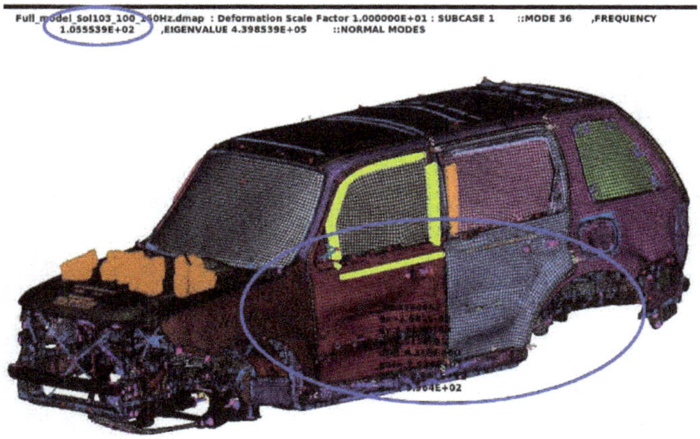

Figure 23. Vehicle local mode at 105 Hz indicating door deformation.

Table 6 compares the value of each design variable between the initial (nominal) and final optimal designs. It also indicates that all designed variables were allowed to vary within a lower and upper bound. The values of the five door design variables changed considerably between the initial and optimal designs. This is expected because the optimizer tried to suppress the local door mode. The stiffness of the four engine mounts and the six exhaust supports also changed. Although we intuitively expect the stiffness of the engine mounts to change but not the stiffness of the exhaust supports, this is not the case in this example. Table 6 also indicates that at the optimum we not only reduced the maximum response from 10^{-3} m to $0.47*10^{-3}$ m but the vehicle mass was also reduced from the initial 55.12 units to the final 51.92 units.

Design Variables	Thickness	Lower Bound	Upper Bound	Nominal Design	Optimal Design
X_1	Door Shell	0.1	1	0.7	0.6638
X_2	Front Frame	0.1	1	0.7	0.3084
X_3	Rear Frame	0.1	1	0.7	0.2019
X_4	Top Panel	0.1	2	0.7	0.2019
X_5	Middel Pipe	0.5	4	2.4	1.3722
X_6	Engine mount	19.5	370.5	195	110.6
X_7	Engine mount	19.5	370.5	195	161.8
X_8	Engine mount	19.5	370.5	195	108.8
X_9	Engine mount	19.5	370.5	195	132.1
X_{10}	Exhaust support	19.5	370.5	195	213.9
X_{11}	Exhaust support	19.5	370.5	195	218.1
X_{12}	Exhaust support	19.5	370.5	195	345.9
X_{13}	Exhaust support	19.5	370.5	195	286.1
X_{14}	Exhaust support	19.5	370.5	195	118.7
X_{15}	Exhaust support	19.5	370.5	195	233.4
Max Resp.				$1*10^{-3}$	$0.47*10^{-3}$
Door Mass				55.12	51.92

Table 6. Summary of optimal design.

Figure 24 shows the actual function evaluations (design points where the vehicle dynamic response was calculated) in the X_1-X_2-X_3 space and indicates the vicinity of the optimal design point. The GA optimizer needed only 359 function evaluations and used a population size of 5*(15+1) = 80 and a maximum of 10 generations. The population size and the number of allowed generations were kept at a minimum in order to locate the vicinity of the optimum quickly without "wasting" valuable computational effort.

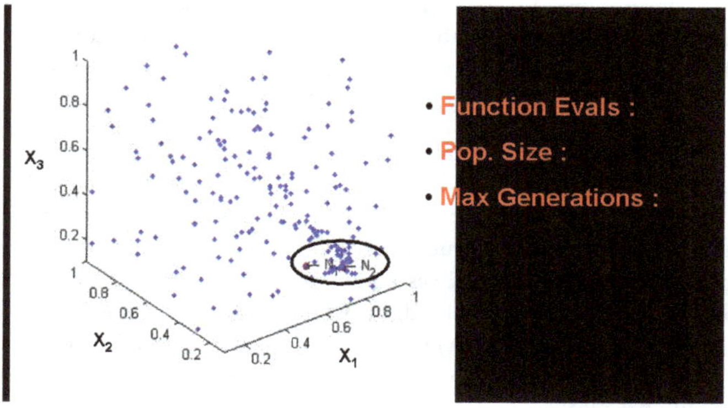

Figure 24. Function evaluations of the Niching GA in the X_1-X_2-X_3 space.

Considering that the computational cost for each function evaluation was 4 minutes, the total computational time was (359 function evaluations) * (4 minutes per evaluation) = 1436 minutes or 23.9 hours. This is acceptable considering the size and type of performed analysis. The computational cost, in terms of number of function evaluations, was kept low by coupling the Niching GA with a Lazy Learning metamodeling technique [26, 27]. The latter estimates the value of the fitness function from existing values at close by designs without calculating the actual response. It uses an error measure to figure out if the estimation is accurate. The error is small if enough previous designs, for which the fitness value was evaluated, are close to the new design. In this case, the metamodel estimates the current fitness value without running an actual dynamic response. If the error is large, an actual response is calculated and the fitness value of this new design is added to the "pool" of previous designs the Lazy Learning metamodeling technique will use downstream.

6. Conclusions and Future Work

Reduced-order models and reanalysis methodologies were presented for accurate and efficient vibration analysis of large-scale, finite element models, and for efficient design optimization of structures for best vibratory response. The optimization is able to handle a large number of design variables and identify local and global optima.

For large FE models, it is common to solve for the system response through modal reduction in order to improve computational efficiency. An eigenanalysis is performed using the system stiffness and mass matrices and a modal model is formed which is then solved for the response. The computational cost can be also reduced using substructuring (or reduced-order modeling) methods. A modal reduction is applied to each substructure to obtain the component modes and the system level response is then obtained using component mode synthesis. In optimization of dynamic systems involving design changes (e.g. thicknesses, material properties, etc) the FEA analysis must be repeated many times in order to obtain the optimum design. Also in probabilistic analysis where parameter uncertainties are present, the FEA analysis must be repeated for a large number of sample points. In such cases, the computational cost is very high, if not prohibitive.

To drastically reduce the computational cost without compromising accuracy beyond an acceptable level, we developed and used various reanalysis methods in conjunction with reduced-order modeling, in optimization of vibratory systems. Reanalysis methods are intended to efficiently calculate the structural response of a modified structure without solving the complete set of modified analysis equations. We presented a variety of reanalysis methods including the CDH/VAO method, the Combined Approximations (CA) and Modified Combined Approximations (MCA) method, and the Parametric Reduced-Order Modeling (PROM) method. Their advantages and limitations were fully described and demonstrated with practical examples.

Future work will concentrate on developing reanalysis methodologies for shape and topology optimization of vibratory systems and extend the presented work in optimization under uncertainty where efficient deterministic reanalysis methods will be combined with efficient probabilistic reanalysis methods.

Author details

Zissimos P. Mourelatos[1*], Dimitris Angelis[2] and John Skarakis[3]

*Address all correspondence to: mourelat@oakland.edu

1 Mechanical Engineering Department, Oakland University, U. S. A.

2 Beta CAE Systems S. A., Greece

3 Beta CAE Systems, U.S.A., Inc.

References

[1] Abu Kasim, A. M., & Topping, B. H. V. (1987). Topping, Static Reanalysis: A Review. *ASCE J. Str. Div*, 113, 1029-1045.

[2] Arora, J. S. (1976). Survey of Structural Reanalysis Techniques. *ASCE J. Str. Div*, 102, 783-802.

[3] Barthelemy, J. -F M., & Haftka, R. T. (1993). Approximation Concepts for Optimum Structural Design- A Review. *Structural Optimization*, 5, 129-144.

[4] Yasui, Y. (1998). Direct Coupled Load Verification of Modified Structural Component. *AIAA Journal*, 36(1), 94-101.

[5] Liu, J. K. (1999). A Universal Matrix Perturbation Technique for Structural Dynamic Modification using Singular Value Decomposition. *Journal of Sound and Vibration*, 228(2), 265-274.

[6] Balmes, E. (1996). Optimal Ritz Vectors for Component Mode Synthesis using the Singular Value Decomposition. *AIAA Journal*, 34(6), 1256-1260.

[7] Balmes, E., Ravary, F., & Langlais, D. (2004). Uncertainty Propagation in Modal Analysis. *Proceedings of IMAC-XXII: A Conference & Exposition on Structural Dynamics*, Dearborn, MI, January.

[8] Zhang, G., Castanier, M. P., & Pierre, C. (2005). Integration of Component-Based and Parametric Reduced-Order Modeling Methods for Probabilistic Vibration Analysis and Design. *Proceedings of the Sixth European Conference on Structural Dynamics*, Paris, France. Sep.

[9] Zhang, G. (2005). Component-Based and Parametric Reduced-Order Modeling Methods for Vibration Analysis of Complex Structures. Ph.D thesis, The University of Michigan, Ann Arbor, MI.

[10] Lophaven, S. N., et al. (2002). *DACE-A MATLAB Kriging Tool Box*, Technical Report IMM-TR-2002-12.

[11] Zhang, G., Mourelatos, Z. P., & Nikolaidis, E. (2007). An Efficient Reanalysis Methodology for Probabilistic Vibration of Complex Structures. *AIAA 8th Non-deterministic Analysis Conference*, Hawaii, April.

[12] Kirsch, U. (2002). *Design-Oriented Analysis of Structures*, Kluwer Academic Publishers, Dordrecht.

[13] Kirsch, U. (2003). Design-Oriented Analysis of Structures- A Unified Approach. *ASCE J. of Engrg Mech.*, 129, 264-272.

[14] Kirsch, U. (2003). A Unified Reanalysis Approach for Structural Analysis, Design and Optimization. *Structural and Multidisciplinary Optimization*, 25, 67-85.

[15] Chen, S. H., & Yang, X. W. (2000). Extended Kirsch Combined Method for Eigenvalue Reanalysis. *AIAA Journal*, 38, 927-930.

[16] Kirsch, U. (2003). Approximate Vibration Reanalysis of Structures. *AIAA Journal*, 41, 504-511.

[17] Kirsch, U., & Bogomolni, M. (2004). Procedures for Approximate Eigenproblem Reanalysis of Structures. *Intern. J. for Num. Meth. Engrg.*, 60, 1969-1986.

[18] Kirsch, U., & Bogomolni, M. (2004). Error Evaluation in Approximate Reanalysis of Structures. *Structural Optimization*, 28, 77-86.

[19] Rong, F., et al. (2003). Structural Modal Reanalysis for Topological Modificsatioms with Extended Kirsch Method. *Comp. Meth. in Appl. Mech. and Engrg.*, 192, 697-707.

[20] Hurty, W. C. (1965). Dynamic Analysis of Structural Systems Using Component Modes. *AIAA Journal*, 3(4), 678-685.

[21] Craig, R. R., & Bampton, M. C. C. (1968). Coupling of Substructures for Dynamics Analyses. *AIAA Journal*, 6(7), 1313-1319.

[22] Craig, R. R., & Chang, C. J. (1977). On the Use of Attachment Modes in Substructure Coupling for Dynamic Analysis. *Proceedings of the 18th AIAA/ASME/ASCE/AHS/ASC Structures, Structural Dynamics, and Materials Conference and Exhibit*, 89-99.

[23] Chen, S. H., et al. (2000). Comparison of Several Eigenvalue Reanalysis Methods for Modified Structures. *Structural Optimization*, 20, 253-259.

[24] Davidsson, P. (2004). Structure-Acoustic Analysis: Finite Element Modeling and Reduction Methods. Ph.D thesis, Lund University, Lund, Sweden.

[25] Li, J., & Mourelatos, Z. P. (2009). Time-Dependent Reliability Estimation for Dynamic Problems using a Niching Genetic Algorithm. *ASME Journal of Mechanical Design*, 131, 071009-1-13.

[26] Singh, A., Mourelatos, Z. P., & Li, J. (2010). Design for Lifecycle Cost using Time-Dependent Reliability. *ASME Journal of Mechanical Design*, 132, 091008-1-11.

[27] Aha, D. W. (1997). Editorial- Lazy Learning. *Artificial Intelligence Review*, 11(105), 1-6.

Chapter 8

Vibrations of Cylindrical Shells

Tiejun Yang, Wen L. Li and Lu Dai

Additional information is available at the end of the chapter

1. Introduction

Beams, plates and shells are the most commonly-used structural components in industrial applications. In comparison with beams and plates, shells usually exhibit more complicated dynamic behaviours because the curvature will effectively couple the flexural and in-plane deformations together as manifested in the fact that all three displacement components simultaneously appear in each of the governing differential equations and boundary conditions. Thus it is understandable that the axial constraints can have direct effects on a predominantly radial mode. For instance, it has been shown that the natural frequencies for the circumferential modes of a simply supported shell can be noticeably modified by the constraints applied in the axial direction [1]. Vibrations of shells have been extensively studied for several decades, resulting in numerous shell theories or formulations to account for the various effects associated with deformations or stress components.

Expressions for the natural frequencies and modes shapes can be derived for the classical homogeneous boundary conditions [2-9]. Wave propagation approach was employed by several researchers [10-13] to predict the natural frequencies for finite circular cylindrical shells with different boundary conditions. Because of the complexity and tediousness of the (exact) solution procedures, approximate procedures such as the Rayleigh-Ritz methods or equivalent energy methods have been widely used for solving shell problems [14-18]. In the Rayleigh-Ritz methods, the characteristic functions for a "similar" beam problem are typically used to represent all three displacement components, leading to a characteristic equation in the form of cubic polynomials. Assuming that the circumferential wave length is smaller than the axial wave length, Yu [6] derived a simple formula for calculating the natural frequencies directly from the shell parameters and the frequency parameters for the analogous beam case. Soedel [19] improved and generalized Yu's result by eliminating the short circumferential wave length restriction. However, since the wavenumbers for axial modal function are obtained from beam functions which do not exactly satisfy shell boundary conditions, it is mathematically difficult to access or ensure the accuracy and convergence of such a solution.

The free vibration of shells with elastic supports was studied by Loveday and Rogers [20] using a general analysis procedure originally presented by Warburton [3]. The effect of flexibility in boundary conditions on the natural frequencies of two (lower order) circumferential modes was investigated for a range of restraining stiffness values. The vibrations of circular cylindrical shells with non-uniform boundary constraints were studied by Amabili and Garziera [21] using the artificial spring method in which the modes for the corresponding less-restrained problem were used to expand the displacement solutions. The non-uniform spring stiffness distributions were systematically represented by cosine series and their presence was accounted for in terms of maximum potential energies stored in the springs.

A large number of studies are available in the literature for the vibrations of shells under different boundary conditions or with various complicating features. A comprehensive review of early investigations can be found in Leissa's book [22]. Some recent progresses have been reviewed by Qatu [23]. Regardless of whether an approximate or an exact solution procedure is employed, the corresponding formulations and implementations usually have to be modified or customized for different boundary conditions. This shall not be considered a trivial task in view that there exist 136 different combinations even considering the simplest (homogeneous) boundary conditions. Thus, it is useful to develop a solution method that can be generally applied to a wide range of boundary conditions with no need of modifying solution algorithms and procedures. Mathematically, elastic supports represent a general form of boundary conditions from which all the classical boundary conditions can be readily derived by simply setting each of the spring stiffnesses to either zero or infinity. This chapter will be devoted to developing a general analytical method for solving shell problems involving general elastically restrained ends.

2. Basic equations and solution procedures

Figure 1 shows an elastically restrained circular cylindrical shell of radius R, thickness h and length L. Each of the eight sets of elastic restraints shall be understood as a distributed spring along the circumference. Let u, v, and w denote the displacements in the axial x, circumferential θ and radial r directions, respectively. The equations of the motions for the shell can be written as

$$\begin{aligned}
&\frac{\partial N_1}{\partial x}+\frac{\partial N_{12}}{R\partial\theta}=\rho h\frac{\partial^2 u}{\partial t^2} && \text{(a)}\\
&\frac{\partial N_{12}}{\partial x}+\frac{\partial N_2}{R\partial\theta}=\rho h\frac{\partial^2 v}{\partial t^2} && \text{(b)}\\
&\frac{\partial Q_1}{\partial x}+\frac{\partial Q_1}{R\partial\theta}-\frac{N_2}{R}=\rho h\frac{\partial^2 w}{\partial t^2} && \text{(c)}
\end{aligned} \tag{1}$$

where ρ is the mass density of the shell material, and N_1, N_{12}, N_θ, Q_1 and Q_θ denote the resultant forces acting on the mid-surface.

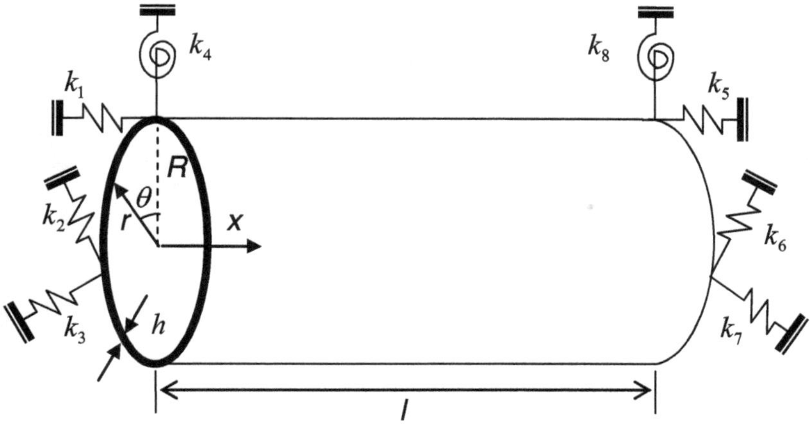

Figure 1. A circular cylindrical shell elastically restrained along all edges.

In terms of the shell displacements, the force and moment components can be expressed as

$$
\begin{aligned}
N_1 &= K\left(\frac{\partial u}{\partial x} + \sigma\frac{\partial v}{R\partial\theta} + \sigma\frac{w}{R}\right) && \text{(a)}\\
N_2 &= K\left(\frac{\partial v}{R\partial\theta} + \frac{w}{R} + \sigma\frac{\partial u}{\partial x}\right) && \text{(b)}\\
N_{12} &= K\frac{(1-\sigma)}{2}\left(\frac{\partial u}{R\partial\theta} + \frac{\partial v}{\partial x}\right) && \text{(c)}\\
M_1 &= -D\left(\frac{\partial^2 w}{\partial x^2} + \sigma\frac{\partial^2 w}{R^2\partial\theta^2}\right) && \text{(d)}\\
M_2 &= -D\left(\frac{\partial^2 w}{R^2\partial\theta^2} + \sigma\frac{\partial^2 w}{\partial x^2}\right) && \text{(e)}\\
M_{12} &= -D(1-\sigma)\frac{\partial^2 w}{R\partial x\partial\theta} && \text{(f)}\\
Q_1 &= \frac{\partial M_1}{\partial x} + \frac{\partial M_{12}}{R\partial\theta} = -D\left(\frac{\partial^3 w}{\partial x^3} + \frac{\partial^3 w}{R^2\partial x\partial\theta^2}\right) && \text{(g)}\\
Q_2 &= \frac{\partial M_2}{R\partial\theta} + \frac{\partial M_{12}}{\partial x} = -D\left(\frac{\partial^3 w}{R^3\partial\theta^3} + \frac{\partial^3 w}{R\partial x^2\partial\theta}\right) && \text{(h)}
\end{aligned}
\qquad (2)
$$

where

$$\begin{aligned} &K = Eh/(1-\sigma^2), && \text{(a)} \\ &D = Eh^3/12(1-\sigma^2) && \text{(b)} \\ &\kappa = D/K = h^2/12 && \text{(c)} \end{aligned} \tag{3}$$

E and σ are respectively the Young's modulus and Poisson ratio of the material; M_1, M_2 and M_{12} are the bending and twisting moments.

The boundary conditions for an elastically restrained shell can specified as:

at x=0,

$$\begin{aligned} &N_x - k_1 u = 0 && \text{(a)} \\ &N_{x\theta} - k_2 v = 0 && \text{(b)} \\ &Q_x + \frac{\partial M_{x\theta}}{R\partial\theta} - k_3 w = 0 && \text{(c)} \\ &M_x + k_4 \frac{\partial w}{\partial x} = 0 && \text{(d)} \end{aligned} \tag{4}$$

at x=L,

$$\begin{aligned} &N_x + k_5 u = 0 && \text{(a)} \\ &N_{x\theta} + k_6 v = 0 && \text{(b)} \\ &Q_x + \frac{\partial M_{x\theta}}{R\partial\theta} + k_7 w = 0 && \text{(c)} \\ &M_x - k_8 \frac{\partial w}{\partial x} = 0 && \text{(d)} \end{aligned} \tag{5}$$

where k_1, k_2, ..., k_8 are the stiffnesses for the restraining springs. The elastic supports represent a set of general boundary conditions, and all the classical boundary conditions can be considered as the special cases when the stiffness for each spring is equal to either zero or infinity.

The above equations are usually referred to as Donnell-Mushtari equations. Flügge's theory is also widely used to describe vibrations of shells. In terms of the shell displacements, the corresponding force and moment components are written as

$$
\begin{aligned}
&N_1 = K\left(\frac{\partial u}{\partial x} + \sigma \frac{\partial v}{R\partial \theta}\right) + K\sigma \frac{w}{R} - \frac{D}{R}\frac{\partial^2 w}{\partial x^2} && \text{(a)}\\
&N_{12} = \frac{1-\sigma}{2}K\left(\frac{\partial u}{R\partial \theta} + \frac{\partial v}{\partial x}\right) + \frac{D}{R}\frac{(1-\sigma)}{2}\left(\frac{\partial v}{R\partial x} - \frac{\partial^2 w}{R\partial \theta \partial x}\right) && \text{(b)}\\
&N_2 = K\left(\frac{\partial v}{R\partial \theta} + \frac{w}{R} + \sigma \frac{\partial u}{\partial x}\right) + \frac{D}{R}\left(\frac{\partial^2 w}{R^2 \partial \theta^2} + \frac{w}{R^2}\right) && \text{(c)}\\
&N_{21} = \frac{1-\sigma}{2}K\left(\frac{\partial u}{R\partial \theta} + \frac{\partial v}{\partial x}\right) + \frac{D}{R}\frac{(1-\sigma)}{2}\left(\frac{\partial u}{R^2\partial \theta} + \frac{\partial^2 w}{R\partial \theta \partial x}\right) && \text{(d)}\\
&M_1 = -D\left(\frac{\partial^2 w}{\partial x^2} + \sigma \frac{\partial^2 w}{R^2 \partial \theta^2} - \frac{\partial u}{R\partial x} - \sigma \frac{\partial v}{R^2 \partial \theta}\right) && \text{(e)} \qquad (6)\\
&M_{12} = -D(1-\sigma)\left(\frac{\partial^2 w}{R\partial \theta \partial x} - \frac{\partial v}{R\partial x}\right) && \text{(f)}\\
&M_2 = -D\left(\frac{w}{R^2} + \frac{\partial^2 w}{R^2 \partial \theta^2} + \sigma \frac{\partial^2 w}{\partial x^2}\right) && \text{(g)}\\
&M_{21} = -D(1-\sigma)\left(\frac{\partial^2 w}{R\partial \theta \partial x} + \frac{1}{2}\frac{\partial u}{R^2 \partial \theta} - \frac{1}{2}\frac{\partial v}{R\partial x}\right) && \text{(h)}\\
&Q_1 = -D\left(\frac{\partial^3 w}{\partial x^3} + \frac{1}{R^2}\frac{\partial^3 w}{\partial \theta^2 \partial x} - \frac{\partial^2 u}{R\partial x^2} + \frac{1-\sigma}{2R^3}\frac{\partial^2 u}{\partial \theta^2} - \frac{1+\sigma}{2R^2}\frac{\partial v}{\partial x \partial \theta}\right) && \text{(i)}
\end{aligned}
$$

A shell problem can be solved either exactly or approximately. An exact solution usually implies that both the governing equations and the boundary conditions are simultaneously satisfied exactly on a point-wise basis. Otherwise, a solution is considered approximate in which one or more of the governing equations and boundary conditions are enforced only in an approximate sense. Both solution strategies will be used below.

2.1. An approximate solution based on the Rayleigh-Ritz procedure

Approximate methods based on energy methods or the Rayleigh-Ritz procedures are widely used for the vibration analysis of shells with various boundary conditions and/or complicating factors. In such an approach, the displacement functions are usually expressed as

$$
\begin{aligned}
&u(x,\theta) = \sum_{m=0}^{\infty} a_m \varphi_u^m(x) \cos n\theta && \text{(a)}\\
&v(x,\theta) = \sum_{m=0}^{\infty} b_m \varphi_v^m(x) \sin n\theta && \text{(b)} \qquad (7)\\
&w(x,\theta) = \sum_{m=0}^{\infty} c_m \varphi_w^m(x) \cos n\theta && \text{(c)}
\end{aligned}
$$

where $\varphi_\alpha^m(x)$, $\alpha = u, v,$ and $w,$ are the characteristic functions for beams with *similar* boundary conditions. Although characteristic functions generally exist in the forms of trigonometric and hyperbolic functions, they also include some integration and frequency constants that have to be determined from boundary conditions. Consequently, each boundary condition basically calls for a special set of modal data. In the literature the modal parameters are well established for the simplest homogeneous boundary conditions. However, the determination of modal properties for the more complicated elastic boundary supports can become, at least, a tedious task since they have to be re-calculated each time when any of the stiffness values is changed. It should also be noted that calculating the modal properties will typically involve seeking the roots of a nonlinear transcendental equation, which mathematically requires an iterative root searching scheme and careful numerical implementations to ensure no missing data. To overcome this problem, a unified representation of the shell solutions will be adopted here in which the displacements, regardless of boundary conditions, will be invariably expressed as

$$u(x,\theta)=\left(\sum_{m=0}^{\infty} a_m \cos\lambda_m x + p_u(x)\right)\cos n\theta,\ (\lambda_m=\frac{m\pi}{L})\text{(a)}$$

$$v(x,\theta)=\left(\sum_{m=0}^{\infty} b_m \cos\lambda_m x + p_v(x)\right)\sin n\theta\ ,\text{(b)}$$

$$w(x,\theta)=\left(\sum_{m=0}^{\infty} c_m \cos\lambda_m x + p_w(x)\right)\cos n\theta\text{(c)}$$

(8)

where $p_\alpha(x)$, $\alpha=u, v,$ and w, denote three auxiliary polynomials which satisfy

$$\begin{aligned}
&\left.\frac{\partial p_u(x)}{\partial x}\right|_{x=0}=\left.\frac{\partial u(x,0)}{\partial x}\right|_{x=0}=\beta_1, &&\text{(a)}\\
&\left.\frac{\partial p_u(x)}{\partial x}\right|_{x=L}=\left.\frac{\partial u(x,0)}{\partial x}\right|_{x=L}=\beta_2 &&\text{(b)}\\
&\left.\frac{\partial p_v(x)}{\partial x}\right|_{x=0}=\left.\frac{\partial v(x,\pi/2)}{\partial x}\right|_{x=0}=\beta_3, &&\text{(c)}\\
&\left.\frac{\partial p_v(x)}{\partial x}\right|_{x=L}=\left.\frac{\partial v(x,\pi/2)}{\partial x}\right|_{x=L}=\beta_4, &&\text{(d)}\\
&\left.\frac{\partial p_w(x)}{\partial x}\right|_{x=0}=\left.\frac{\partial w(x,0)}{\partial x}\right|_{x=0}=\beta_5, &&\text{(e)}\\
&\left.\frac{\partial p_w(x)}{\partial x}\right|_{x=L}=\left.\frac{\partial w(x,0)}{\partial x}\right|_{x=L}=\beta_6, &&\text{(f)}\\
&\left.\frac{\partial^3 p_w(x)}{\partial x^3}\right|_{x=0}=\left.\frac{\partial^3 w(x,0)}{\partial x^3}\right|_{x=0}=\beta_7, &&\text{(g)}\\
&\left.\frac{\partial^3 p_w(x)}{\partial x^3}\right|_{x=L}=\left.\frac{\partial^3 w(x,0)}{\partial x^3}\right|_{x=L}=\beta_8 &&\text{(h)}
\end{aligned}\qquad(9)$$

It is clear from Eqs. (9) that these auxiliary polynomials are only dependent on the first and third derivatives β_i, (i=1,2,…,8) of the displacement solutions on the boundaries. In terms of boundary derivatives, the lowest-order polynomials can be explicitly expressed as [24, 25]

$$
\begin{aligned}
p_u(x) &= \zeta_1(x)\beta_1 + \zeta_2(x)\beta_2 && \text{(a)}\\
p_v(x) &= \zeta_1(x)\beta_3 + \zeta_2(x)\beta_4 && \text{(b)}\\
p_w(x) &= \zeta_1(x)\beta_5 + \zeta_2(x)\beta_6 + \zeta_3(x)\beta_7 + \zeta_4(x)\beta_8 && \text{(c)}
\end{aligned}
\qquad (10)
$$

where

$$
\zeta(x)^T = \begin{Bmatrix} \zeta_1(x) \\ \zeta_2(x) \\ \zeta_3(x) \\ \zeta_4(x) \end{Bmatrix} = \begin{Bmatrix} (6Lx - 2L^2 - 3x^2)/6L \\ (3x^2 - L^2)/6L \\ -(15x^4 - 60Lx^3 + 60L^2x^2 - 8L^4)/360L \\ (15x^4 - 30L^2x^2 + 7L^4)/360L \end{Bmatrix} \qquad (11)
$$

This alternative form of Fourier series recognizes the fact that the conventional Fourier series for a sufficiently smooth function $f(x)$ defined on a compact interval $[0, L]$ generally fails to converge at the end points. Introducing the auxiliary functions will ensure the cosine series in Eqs. (8) to converge uniformly and polynomially over the interval, including the end points. As a matter of fact, the polynomial subtraction techniques have been employed by mathematicians as a means to accelerate the convergence of the Fourier series expansion for an explicitly defined function [26-28].

The coefficients β_i represent the values of the first and third derivatives of the displacements at the boundaries, and are hence related to the unknown Fourier coefficients for the trigonometric terms. The relationships between the constants and the expansion coefficients can be derived either exactly or approximately.

In seeking an approximate solution based on an energy method, the solution is not required to explicitly satisfy the force or natural boundary conditions. Accordingly, the derivative parameters β_i in Eqs. (10) will be here determined from a simplified set of the boundary conditions, that is,

$$
\begin{aligned}
&\frac{\partial u}{\partial x} \mp \hat{k}_{1,5} u = 0 && \text{(a)}\\
&\frac{(1-\mu)}{2}\frac{\partial v}{\partial x} \mp \hat{k}_{2,6} v = 0 && \text{(b)}\\
&\kappa \frac{\partial^3 w}{\partial x^3} \pm \hat{k}_{3,7} w = 0 && \text{(c)}\\
&\kappa \frac{\partial^2 w}{\partial x^2} \mp \hat{k}_{4,8} \frac{\partial w}{\partial x} = 0 && \text{(d)}
\end{aligned}
\qquad (12)
$$

where

$$
\hat{k}_i = k_i / K. \qquad (13)
$$

By substituting Eqs. (8) and (10) into (12), one will obtain

$$\begin{aligned}
&\begin{Bmatrix}\beta_1\\ \beta_2\end{Bmatrix}=\sum_{m=0}^{\infty}\mathbf{H}_u^{-1}\mathbf{Q}_u^m a_m &&\text{(a)}\\
&\begin{Bmatrix}\beta_3\\ \beta_4\end{Bmatrix}=\sum_{m=0}^{\infty}\mathbf{H}_v^{-1}\mathbf{Q}_v^m b_m &&\text{(b)}\\
&\{\beta_5\quad \beta_6\quad \beta_7\quad \beta_8\}^T=\sum_{m=0}^{\infty}\mathbf{H}_w^{-1}\mathbf{Q}_w^m c_m &&\text{(c)}
\end{aligned}\tag{14}$$

where

$$\begin{aligned}
&\mathbf{H}_u=\begin{bmatrix}\dfrac{\hat{k}_1L}{3}+1 & \dfrac{\hat{k}_1L}{6}\\ \dfrac{\hat{k}_5L}{6} & \dfrac{\hat{k}_5L}{3}+1\end{bmatrix} &&\text{(a)}\\
&\mathbf{Q}_u^m=\left\{\hat{k}_1\quad (-1)^{m+1}\hat{k}_5\right\}^{\mathrm{T}} &&\text{(b)}\\
&\mathbf{H}_v=\begin{bmatrix}\dfrac{\hat{k}_2L}{3}+\dfrac{1-\mu}{2} & \dfrac{\hat{k}_2L}{6}\\ \dfrac{\hat{k}_6L}{6} & \dfrac{\hat{k}_6L}{3}+\dfrac{1-\mu}{2}\end{bmatrix} &&\text{(c)}\\
&\mathbf{Q}_v^m=\left\{\hat{k}_2\quad (-1)^{m+1}\hat{k}_6\right\}^{\mathrm{T}} &&\text{(d)}\\
&\mathbf{H}_w=\begin{bmatrix}-\dfrac{\hat{k}_3L}{3} & -\dfrac{\hat{k}_3L}{6} & \dfrac{\hat{k}_3L^3}{45}+\kappa & \dfrac{7\hat{k}_3L^3}{360}\\ -\dfrac{\hat{k}_7L}{6} & -\dfrac{\hat{k}_7L}{3} & \dfrac{7\hat{k}_7L^3}{360} & \dfrac{\hat{k}_7L^3}{45}+\kappa\\ \hat{k}_4+\dfrac{\kappa}{L} & -\dfrac{\kappa}{L} & \dfrac{\kappa L}{3} & \dfrac{\kappa L}{6}\\ -\dfrac{\kappa}{L} & \hat{k}_8+\dfrac{\kappa}{L} & \dfrac{\kappa L}{6} & \dfrac{\kappa L}{3}\end{bmatrix} &&\text{(e)}
\end{aligned}\tag{15}$$

and

$$\mathbf{Q}_w^m=\left\{-\hat{k}_3\quad (-1)^m\hat{k}_7\quad -\kappa\lambda_m^2\quad (-1)^m\kappa\lambda_m^2\right\}^T\tag{16}$$

In light of Eqs. (13), Eqs (8) can be reduced to Eqs. (7) with the axial functions being defined as

$$\varphi_\alpha^m(x) = \cos\lambda_m x + \zeta(x)^T \bar{\mathbf{Q}}_\alpha^m, \ (\alpha = u, v, w) \tag{17}$$

where

$$\begin{aligned} &\bar{\mathbf{Q}}_w^m = \tilde{\mathbf{Q}}_w^m && \text{(a)} \\ &\bar{\mathbf{Q}}_\alpha^m = \begin{Bmatrix} \tilde{\mathbf{Q}}_\alpha^m \\ 0 \\ 0 \end{Bmatrix} && \text{(b)} \\ &\tilde{\mathbf{Q}}_\alpha^m = (\mathbf{H}_\alpha)^{-1}\mathbf{Q}_\alpha^m && \text{(c)} \end{aligned} \tag{18}$$

Since the boundary conditions are not exactly satisfied by the displacements such constructed, the Rayleigh-Ritz procedure will be employed to find a weak form of solution. The current solution is noticeably different from the conventional Rayleigh-Ritz solutions in that: a) the shell displacements are expressed in terms of three independent sets of axial functions, rather than a single (set of) beam function(s), b) the basis functions in each displacement expansion constitutes a complete set so that the convergence of the Rayleigh-Ritz solution is guaranteed mathematically, and c) it does not suffer from the well-known numerical instability problem related to the higher order beam functions or polynomials. More importantly, the current method is that it provides a unified solution to a wide variety of boundary conditions.

The potential energy consistent with the Donnell-Mushtari theory can be expressed from

$$\begin{aligned} V = \frac{K}{2}\int_0^L\int_0^{2\pi} \Bigg\{ &\left(\frac{\partial u}{\partial x} + \frac{\partial v}{R\partial\theta} + \frac{w}{R}\right)^2 - 2(1-\nu)\frac{\partial u}{\partial x}\left(\frac{\partial v}{R\partial\theta} + \frac{w}{R}\right) + \frac{(1-\nu)}{2}\left(\frac{\partial v}{\partial x} + \frac{\partial u}{R\partial\theta}\right)^2 + \\ &\kappa\left[\left(\frac{\partial^2 w}{\partial x^2} + \frac{\partial^2 w}{R^2\partial\theta^2}\right)^2 - 2(1-\nu)\left(\frac{\partial^2 w}{\partial x^2}\frac{\partial^2 w}{R^2\partial\theta^2} - \left(\frac{\partial^2 w}{R\partial x\partial\theta}\right)^2\right)\right]\Bigg\} R\,dx\,d\theta && \text{(a)} \\ &+1/2\int_0^{2\pi} [k_1u^2 + k_2v^2 + k_3w^2 + k_4\left(\partial w/\partial x\right)^2]_{x=0} R\,d\theta \\ &+1/2\int_0^{2\pi} [k_5u^2 + k_6v^2 + k_7w^2 + k_8\left(\partial w/\partial x\right)^2]_{x=L} R\,d\theta \end{aligned} \tag{19}$$

and the total kinetic energy is calculated from

$$T = \frac{1}{2}\int_0^L\int_0^{2\pi} \rho h\left[\left(\partial u/\partial t\right)^2 + \left(\partial v/\partial t\right)^2 + \left(\partial w/\partial t\right)^2\right] R\,dx\,d\theta \qquad \text{(b)}$$

By minimizing the Lagrangian $L=V-T$ against all the unknown expansion coefficients, a final system of linear algebraic equations can be derived

$$\begin{bmatrix} \Lambda^{ss} & \Lambda^{s\theta} & \Lambda^{sr} \\ \Lambda^{s\theta T} & \Lambda^{\theta\theta} & \Lambda^{\theta r} \\ \Lambda^{srT} & \Lambda^{\theta rT} & \Lambda^{rr} \end{bmatrix} \begin{Bmatrix} \bar{\mathbf{a}} \\ \bar{\mathbf{b}} \\ \bar{\mathbf{c}} \end{Bmatrix} - \omega^2 \begin{bmatrix} \mathbf{M}^{ss} & \mathbf{0} & \mathbf{0} \\ \mathbf{0} & \mathbf{M}^{\theta\theta} & \mathbf{0} \\ \mathbf{0} & \mathbf{0} & \mathbf{M}^{rr} \end{bmatrix} \begin{Bmatrix} \bar{\mathbf{a}} \\ \bar{\mathbf{b}} \\ \bar{\mathbf{c}} \end{Bmatrix} = 0 \tag{20}$$

where

$$\begin{aligned} \bar{\mathbf{a}} &= \left\{ a_{00}, a_{01}, \dots, a_{mn}, \dots \right\}^{T}, & \text{(a)} \\ \bar{\mathbf{b}} &= \left\{ b_{01}, b_{02}, \dots, b_{mn}, \dots \right\}^{T}, & \text{(b)} \\ \bar{\mathbf{c}} &= \left\{ c_{00}, c_{01}, \dots, c_{mn}, \dots \right\}^{T}, & \text{(c)} \end{aligned} \tag{21}$$

$$\begin{aligned}
\Lambda^{ss}_{mn,m'n'} &= \delta_{nn'}[I^{mm'}_{uu,11} + \frac{(1-\mu)n^2}{2R^2} I^{mm'}_{uu,00} + \frac{2}{L}\hat{k}_1 \varphi^m_u(0)\varphi^{m'}_u(0) + \frac{2}{L}\hat{k}_5 \varphi^m_u(L)\varphi^{m'}_u(L)] & \text{(a)} \\
\Lambda^{s\theta}_{mn,m'n'} &= \delta_{nn'}[\frac{\mu n}{R} I^{mm'}_{uv,10} - \frac{(1-\mu)n}{2R} I^{mm'}_{uv,01}] & \text{(b)} \\
\Lambda^{sr}_{mn,m'n'} &= \delta_{nn'} \frac{\mu}{R} I^{mm'}_{uw,10} & \text{(c)} \\
\Lambda^{\theta\theta}_{mn,m'n'} &= \delta_{nn'}[\frac{n^2}{R^2} I^{mm'}_{vv,00} + \frac{(1-\mu)}{2} I^{mm'}_{vv,11} + \frac{2}{L}\hat{k}_2 \varphi^m_v(0)\varphi^{m'}_v(0) + \frac{2}{L}\hat{k}_6 \varphi^m_v(L)\varphi^{m'}_v(L)] & \text{(d)} \\
\Lambda^{\theta r}_{mn,m'n'} &= \delta_{nn'} \frac{n}{R^2} I^{mm'}_{vw,00} & \text{(e)} \\
\Lambda^{\theta\theta}_{mn,m'n'} &= \delta_{nn'}\{\frac{1}{R^2} I^{mm'}_{ww,00} + \kappa[I^{mm'}_{ww,22} + \frac{n^4}{R^4} I^{mm'}_{ww,00} + 2(1-\mu)\frac{n^2}{R^2} I^{mm'}_{ww,11} \\
&\quad - \frac{\mu n^2}{R^2}(I^{mm'}_{ww,02} + I^{mm'}_{ww,20})] + \frac{2}{L}\hat{k}_3 \varphi^m_w(0)\varphi^{m'}_w(0) + \frac{2}{L}\hat{k}_7 \varphi^m_w(L)\varphi^{m'}_w(L) + & \text{(f)} \\
&\quad + \frac{2}{L}\hat{k}_4 \frac{\partial \varphi^m_w(0)}{\partial x}\frac{\partial \varphi^{m'}_w(0)}{\partial x} + \frac{2}{L}\hat{k}_8 \frac{\partial \varphi^m_w(L)}{\partial x}\frac{\partial \varphi^{m'}_w(L)}{\partial x}\}
\end{aligned} \tag{22}$$

$$\begin{aligned} \mathbf{M}^{ss}_{mn,m'n'} &= \delta_{nn'} \rho h I^{mm'}_{uu,00} & \text{(a)} \\ \mathbf{M}^{\theta\theta}_{mn,m'n'} &= \delta_{nn'} \rho h I^{mm'}_{vv,00} & \text{(b)} \\ \mathbf{M}^{rr}_{mn,m'n'} &= \delta_{nn'} \rho h I^{mm'}_{ww,00} & \text{(c)} \end{aligned} \tag{23}$$

and

$$I^{mm'}_{\alpha\beta,pq} = 2/L \int_0^L \frac{\partial^p \phi^m_\alpha}{\partial x^p} \frac{\partial^q \phi^{m'}_\beta}{\partial x^q} dx \quad (\alpha,\beta = u,v,w) \tag{24}$$

The integrals in Eq. (23) can be calculated analytically; for instance

$$I^{mm'}_{\alpha\beta,00} = \frac{2}{L} \int_0^L \varphi^m_\alpha \varphi^{m'}_\beta \, dx$$

$$
\begin{aligned}
&= \frac{2}{L}\int_0^L (\cos\lambda_m x + \zeta(x)^T \bar{\mathbf{Q}}_\alpha^m)(\cos\lambda_{m'} x + \zeta(x)^T \bar{\mathbf{Q}}_\beta^{m'})\, dx \\
&= \frac{2}{L}\int_0^L (\cos\lambda_m x \cos\lambda_{m'} x + \cos\lambda_{m'} x\zeta(x)^T \bar{\mathbf{Q}}_\alpha^m + \cos\lambda_m x\zeta(x)^T \bar{\mathbf{Q}}_\beta^{m'} + \bar{\mathbf{Q}}_\alpha^{mT}\zeta(x)\zeta(x)^T \bar{\mathbf{Q}}_\beta^{m'})\, dx \\
&= \varepsilon_m \delta_{mm'} + S_\alpha^{m'm} + S_\beta^{mm'} + Z_{\alpha\beta}^{mm'}
\end{aligned}
\tag{25}
$$

where

$$
\begin{aligned}
\varepsilon_m &= (1+\delta_{m0}) && \text{(a)} \\
S_\alpha^{m'm} &= \mathbf{P}_c^{m'} \bar{\mathbf{Q}}_\alpha^m && \text{(b)} \\
Z_{\alpha\beta}^{mm'} &= \bar{\mathbf{Q}}_\alpha^{mT} \Xi \bar{\mathbf{Q}}_\beta^{m'} && \text{(c)}
\end{aligned}
\tag{26}
$$

and

$$
\mathbf{P}_c^m = \frac{2}{L}\int_0^L \zeta_w(x)^T \cos\lambda_m x\, dx \tag{27}
$$

$$
= \frac{2}{L}\begin{cases} \{0 \quad 0 \quad 0 \quad 0\}^T, & \text{for } m = 0 \\ \left\{\dfrac{-1}{\lambda_m^{\;2}} \quad \dfrac{(-1)^m}{\lambda_m^{\;2}} \quad \dfrac{1}{\lambda_m^{\;4}} \quad \dfrac{(-1)^{m+1}}{\lambda_m^{\;4}}\right\}^T, & \text{for } m \neq 0 \end{cases} \qquad \text{(a)}
$$

$$
\Xi = 2/L\int_0^L \zeta(x)^T \zeta(x) dx = \begin{bmatrix} \dfrac{2L^2}{45} & & & \\ \dfrac{7L^2}{180} & \dfrac{2L^2}{45} & & sym. \\ -\dfrac{4L^4}{945} & -\dfrac{31L^4}{7560} & \dfrac{2L^6}{4725} & \\ -\dfrac{31L^4}{7560} & -\dfrac{4L^4}{945} & \dfrac{127L^6}{302400} & \dfrac{2L^6}{4725} \end{bmatrix} \qquad \text{(b)} \tag{28}
$$

2.2. A strong form of solution based on Flügge's equations

As aforementioned, the displacement expressions in terms of beam functions cannot exactly satisfy the shell boundary conditions; instead they are made to satisfy the boundary conditions in a weak sense via the use of the Rayleigh-Ritz procedure. To overcome this problem, the displacement expressions, Eqs. (8), will now be generalized to

$$
\begin{aligned}
u(x,\theta) &= \sum_{n=0}^{\infty}\left(\sum_{m=0}^{\infty} A_{mn}\cos\lambda_m x + p_n^u(x)\right)\cos n\theta && \text{(a)} \\
v(x,\theta) &= \sum_{n=0}^{\infty}\left(\sum_{m=0}^{\infty} B_{mn}\cos\lambda_m x + p_n^v(x)\right)\sin n\theta && \text{(b)} \\
w(x,\theta) &= \sum_{n=0}^{\infty}\left(\sum_{m=0}^{\infty} C_{mn}\cos\lambda_m x + p_n^w(x)\right)\cos n\theta && \text{(c)}
\end{aligned}
\tag{29}
$$

which represent a 2-D version of the improved Fourier series expansions, Eqs. (8).

To demonstrate the flexibility in choosing the auxiliary functions $p_n^u(x)$, $p_n^v(x)$ and $p_n^w(x)$, an alternative set is used below:

$$\begin{aligned} p_n^u(x) &= \Lambda_n^u \alpha(x) \quad &(a) \\ p_n^v(x) &= \Lambda_n^v \alpha(x) \quad &(b) \\ p_n^w(x) &= \Lambda_n^w \beta(x) \quad &(c) \end{aligned} \tag{30}$$

here $\Lambda_n^u = [a_n \ b_n]$, $\Lambda_n^v = [c_n \ d_n]$, $\Lambda_n^w = [e_n \ f_n \ g_n \ h_n]$ with a_n, b_n,..., g_n and h_n being the unknown coefficients to be determined; $\boldsymbol{\alpha}(x) = \{\alpha_1(x) \ \alpha_2(x)\}^{\mathrm{T}}$ and $\boldsymbol{\beta}(x) = \{\beta_1(x) \ \beta_2(x) \ \beta_3(x) \ \beta_4(x)\}^{\mathrm{T}}$ and with their elements being defined as

$$\begin{aligned} \alpha_1(x) &= x(\frac{x}{l} - 1)^2 \quad &(a) \\ \alpha_2(x) &= \frac{x^2}{l}(\frac{x}{l} - 1) \quad &(b) \end{aligned} \tag{31}$$

and

$$\begin{aligned} \beta_1(x) &= \frac{9l}{4\pi}\sin(\frac{\pi x}{2l}) - \frac{l}{12\pi}\sin(\frac{3\pi x}{2l}) \quad &(a) \\ \beta_2(x) &= -\frac{9l}{4\pi}\cos(\frac{\pi x}{2l}) - \frac{l}{12\pi}\cos(\frac{3\pi x}{2l}) \quad &(b) \\ \beta_3(x) &= \frac{l^3}{\pi^3}\sin(\frac{\pi x}{2l}) - \frac{l^3}{3\pi^3}\sin(\frac{3\pi x}{2l}) \quad &(c) \\ \beta_4(x) &= -\frac{l^3}{\pi^3}\cos(\frac{\pi x}{2l}) - \frac{l^3}{3\pi^3}\cos(\frac{3\pi x}{2l}) \quad &(d) \end{aligned} \tag{32}$$

In Eqs. (27), the sums of x-related terms are here understood as the series expansions in x-direction, rather than characteristic functions for a beam with "similar" boundary condition. This distinction is important in that the boundary conditions and governing differential equations can now be exactly satisfied on a point-wise basis; that is, the solution can be found in strong form, as described below.

Substituting Eqs. (6) and (27) into (4) and (5) will lead to

$$
\begin{aligned}
&a_n-\left(\frac{7\sigma l}{3\pi R}+\frac{3\pi R\gamma}{4l}\right)f_n-\left(\frac{4\sigma l^3}{3\pi^3 R}+\frac{l\gamma R}{\pi}\right)h_n \\
&=\frac{k_1}{K}\sum_{m=0}^{\infty}A_{mn}-\frac{\sigma n}{R}\sum_{m=0}^{\infty}B_{mn}-\sum_{m=0}^{\infty}\left(\frac{\sigma}{R}+\lambda_m^2\gamma R\right)C_{mn}
\end{aligned}
\qquad \text{(a)}
$$

$$
\frac{1-\sigma}{2}(1+\gamma)c_n+\frac{(1-\sigma)\gamma n}{2}e_n=\frac{(1-\sigma)n}{2R}\sum_{m=0}^{\infty}A_{mn}+\frac{k_2}{K}\sum_{m=0}^{\infty}B_{mn} \qquad \text{(b)}
$$

$$
\begin{aligned}
&-\frac{4}{lR}a_n-\frac{2}{lR}b_n+\frac{(3-\sigma)n}{2R^2}c_n+\frac{(2-\sigma)n^2}{R^2}e_n+\frac{7lk_3}{3\pi D}f_n-g_n+\frac{4l^3k_3}{3\pi^3 D}h_n \\
&=\sum_{m=0}^{\infty}\left(\frac{\lambda_m^2}{R}-\frac{(1-\sigma)n^2}{2R^3}\right)A_{mn}+\frac{k_3}{D}\sum_{m=0}^{\infty}C_{mn}
\end{aligned}
\qquad \text{(c)}
$$

$$
\begin{aligned}
&-\frac{1}{R}a_n+\left(\frac{3\pi}{4l}+\frac{7vl\mathrm{n}^2}{3\pi R^2}\right)f_n+\left(\frac{l}{\pi}+\frac{4vl^3n^2}{3\pi^3R^2}\right)h_n-\frac{k_4}{D}e_n \\
&=\frac{\sigma n}{R^2}\sum_{m=0}^{\infty}B_{mn}+\sum_{m=0}^{\infty}\left(\lambda_m^2+\frac{\sigma n^2}{R^2}\right)C_{mn}
\end{aligned}
\qquad \text{(d)}
$$

$$
\begin{aligned}
&-b_n-\left(\frac{3\pi\gamma R}{4l}+\frac{7\sigma l}{3\pi R}\right)e_n-\left(\frac{l\gamma R}{\pi}+\frac{4\sigma l^3}{3\pi^3 R}\right)g_n \\
&=\frac{k_5}{K}\sum_{m=0}^{\infty}\cos(m\pi)A_{mn}+\frac{n\sigma}{R}\sum_{m=0}^{\infty}\cos(m\pi)B_{mn}+\sum_{m=0}^{\infty}\left(\frac{\sigma}{R}+\lambda_m^2\gamma R\right)\cos(m\pi)C_{mn}
\end{aligned}
\qquad \text{(e)} \qquad (33)
$$

$$
-\frac{1-\sigma}{2}(1+\gamma)d_n-\frac{(1-\sigma)n\gamma}{2}f_n=-\frac{(1-\sigma)n}{2R}\sum_{m=0}^{\infty}\cos(m\pi)A_{mn}+\frac{k_6}{K}\sum_{m=0}^{\infty}\cos(m\pi)B_{mn} \qquad \text{(f)}
$$

$$
\begin{aligned}
&-\frac{2}{lR}a_n-\frac{4}{lR}b_n-\frac{(3-\sigma)n}{2R^2}d_n-\frac{7lk_7}{3\pi D}e_n-\frac{(2-\sigma)n^2}{R^2}f_n-\frac{4l^3k_7}{3\pi^3 D}g_n+h_n \\
&=-\sum_{m=0}^{\infty}\left(\frac{\cos(m\pi)\lambda_m^2}{R}-\frac{(1-\sigma)\cos(m\pi)n^2}{2R^3}\right)A_{mn}+\frac{k_7}{D}\sum_{m=0}^{\infty}\cos(m\pi)C_{mn}
\end{aligned}
\qquad \text{(g)}
$$

$$
\begin{aligned}
&\frac{1}{R}b_n+\left(\frac{3\pi}{4l}+\frac{7\sigma l\mathrm{n}^2}{3\pi R^2}\right)e_n-\frac{k_8}{D}f_n+\left(\frac{l}{\pi}+\frac{4\sigma l^3\mathrm{n}^2}{3\pi^3R^2}\right)g_n \\
&=-\frac{\sigma n}{R^2}\sum_{m=0}^{\infty}\cos(m\pi)B_{mn}-\sum_{m=0}^{\infty}\left(\lambda_m^2+\frac{\sigma n^2}{R^2}\right)\cos(m\pi)C_{mn}
\end{aligned}
\qquad \text{(h)}
$$

Equations (31) represent a set of constraint conditions between the unknown (boundary) constants, a_n, b_n,..., g_n and h_n, and the Fourier expansion coefficients A_{mn}, B_{mn}, and C_{mn} (m, n = 0, 1, 2,...). The constraint equations (31a-h) can be rewritten more concisely, in matrix form, as

$$
\mathbf{Ly}=\mathbf{Sx} \qquad (34)
$$

The elements of the coefficient matrices can be readily derived from Eqs. (31); for example, Eq. (31a) implies

$$\begin{aligned}
&\{\mathbf{L}_{31}\}_{n,n'} = \delta_{nn'} && \text{(a)}\\
&\{\mathbf{L}_{36}\}_{n,n'} = -\left(\frac{7\sigma l}{3\pi R} + \frac{3\pi\gamma R}{4l}\right)\delta_{nn'} && \text{(b)}\\
&\{\mathbf{L}_{38}\}_{n,n'} = -\left(\frac{4\sigma l^3}{3\pi^3 R} + \frac{l\gamma R}{\pi}\right)\delta_{nn'} && \text{(c)}\\
&\{\mathbf{L}_{32}\}_{n,n'} = \{\mathbf{L}_{33}\}_{n,n'} = \{\mathbf{L}_{34}\}_{n,n'} = \{\mathbf{L}_{35}\}_{n,n'} = \{\mathbf{L}_{37}\}_{n,n'} = 0 && \text{(d)}\\
&\{\mathbf{S}_{11}\}_{mn,n'} = k_1\delta_{nn'} / K && \text{(e)}\\
&\{\mathbf{S}_{12}\}_{mn,n'} = -\sigma n\delta_{nn'} / R && \text{(f)}\\
&\{\mathbf{S}_{13}\}_{mn,n'} = -\delta_{nn'}\left(\frac{\sigma}{R} + \frac{m^2\pi^2\gamma R}{l^2}\right) && \text{(g)}
\end{aligned} \qquad (35)$$

Other sub-matrices can be similarly obtained from the remaining equations in Eqs. (31).

In actual numerical calculations, all the series expansions will have to be truncated to m=M and n=N. Thus there is a total number of (M+1)(3N+2)+8N+6 unknown expansion coefficients in the displacement functions. Since Eq. (33) represents a set of 8N+6 equations, additional (M+1)(3N+2) equations are needed to be able to solve for all the unknown coefficients. Accordingly, we will turn to the governing differential equations.

In Flügge's theory, the equations of motion are given as

$$\begin{aligned}
&\frac{\partial N_1}{\partial x} + \frac{\partial N_{21}}{R\partial\theta} = \rho h\frac{\partial^2 u}{\partial t^2} && \text{(a)}\\
&\frac{\partial N_{12}}{\partial x} + \frac{\partial N_2}{R\partial\theta} + \left(\frac{\partial M_2}{R^2\partial\theta} + \frac{\partial M_{12}}{R\partial x}\right) = \rho h\frac{\partial^2 v}{\partial t^2} && \text{(b)}\\
&\frac{\partial^2 M_1}{\partial x^2} + \frac{\partial^2 M_{12}}{R\partial x\partial\theta} + \frac{\partial^2 M_{21}}{R\partial x\partial\theta} + \frac{\partial^2 M_2}{R^2\partial\theta^2} - \frac{N_2}{R} = \rho h\frac{\partial^2 w}{\partial t^2} && \text{(c)}
\end{aligned} \qquad (36)$$

Substituting Eqs. (6) and (37) into Eqs. (34) results in

$$
\begin{aligned}
&\sum_{m=0}^{\infty}\sum_{n=0}^{\infty}\left(-\lambda_m{}^2-\frac{(1-\sigma)(1+\gamma)n^2}{2R^2}\right)\cos\lambda_m x A_{mn}+\sum_{n=0}^{\infty}\Lambda_n^u\left(\alpha''(x)-\frac{(1-\sigma)(1+\gamma)n^2}{2R^2}\right)\\
&\quad+\frac{(1+\sigma)}{2R}\left(-\sum_{m=0}^{\infty}\sum_{n=0}^{\infty}\lambda_m n\sin\lambda_m x B_{mn}+\sum_{n=0}^{\infty}n\Lambda_n^v\alpha'(x)\right)\\
&\quad-\sum_{m=0}^{\infty}\sum_{n=0}^{\infty}\left(\frac{\sigma}{R}\lambda_m+\gamma R\lambda_m{}^3-\frac{(1-\sigma)\gamma n^2}{2R}\lambda_m\right)\sin\lambda_m x C_{mn} &&\text{(a)}\\
&\qquad+\sum_{n=0}^{\infty}\Lambda_n^{w\mathrm{T}}\left(\frac{\sigma}{R}\beta'(x)-\gamma R\beta'''(x)-\frac{(1-\sigma)\gamma n^2}{2R}\beta'(x)\right)\\
&\qquad+\frac{\omega^2\rho h}{K}\left(\sum_{m=0}^{\infty}\sum_{n=0}^{\infty}\cos\lambda_m x A_{mn}+\sum_{n=0}^{\infty}\Lambda_n^u\alpha(x)\right)=0\\
&+\frac{(1+\sigma)}{2R}\left(\sum_{m=0}^{\infty}\sum_{n=0}^{\infty}\lambda_m n\sin\lambda_m x A_{mn}-\sum_{n=0}^{\infty}n\Lambda_n^u\alpha'(x)\right)\\
&-\sum_{m=0}^{\infty}\sum_{n=0}^{\infty}\left(\frac{n^2}{R^2}+\frac{(1-\sigma)(1+3\gamma)}{2}\lambda_m{}^2\right)\cos\lambda_m x B_{mn}\\
&-\sum_{n=0}^{\infty}\Lambda_n^{v\mathrm{T}}\left(\frac{n^2}{R^2}\alpha(x)-\frac{(1-\sigma)(1+3\gamma)}{2}\alpha''(x)\right) &&\text{(b)} \qquad (37)\\
&-\sum_{m=0}^{\infty}\sum_{n=0}^{\infty}\left(\frac{n}{R^2}+\frac{(3-\sigma)\gamma n}{2}\lambda_m{}^2\right)\cos\lambda_m x C_{mn}\\
&-\sum_{n=0}^{\infty}\Lambda_n^{w\mathrm{T}}\left(\frac{n}{R^2}\beta(x)-\frac{(3-\sigma)\gamma n}{2}\beta''(x)\right)\\
&+\frac{\omega^2\rho h}{K}\left(\sum_{m=0}^{\infty}\sum_{n=0}^{\infty}\cos\lambda_m x B_{mn}+\sum_{n=0}^{\infty}\Lambda_n^v\alpha(x)\right)=0\\
&-\sum_{m=0}^{\infty}\sum_{n=0}^{\infty}\left(\frac{\sigma}{R}-\frac{(1-\sigma)\gamma n^2}{2R}+\gamma R\lambda_m{}^2\right)\lambda_m\sin\lambda_m x A_{mn}\\
&+\sum_{n=0}^{\infty}\Lambda_n^{u\mathrm{T}}\left(\frac{\sigma}{R}\alpha'(x)-\frac{(1-\sigma)\gamma n^2}{2R}\alpha'(x)-\gamma R\alpha'''(x)\right)\\
&+\sum_{m=0}^{\infty}\sum_{n=0}^{\infty}\left(\frac{n}{R^2}+\frac{(3-\sigma)\gamma n}{2}\lambda_m{}^2\right)\cos\lambda_m x B_{mn}\\
&+\sum_{n=0}^{\infty}\Lambda_n^{u\mathrm{T}}\left(\frac{n}{R^2}\alpha(x)-\frac{(3-\sigma)\gamma n}{2}\alpha''(x)\right) &&\text{(c)}\\
&+\sum_{m=0}^{\infty}\sum_{n=0}^{\infty}\left(\frac{1+\gamma(n^2+1)^2}{R^2}+R^2\gamma\lambda_m{}^4+2\gamma n^2\lambda_m{}^2\right)\cos\lambda_m x C_{mn}\\
&+\sum_{n=0}^{\infty}\Lambda_n^{w\mathrm{T}}\left(\frac{1+\gamma(n^2+1)^2}{R^2}\beta(x)-2\gamma n^2\beta''(x)+\gamma R^2\beta''''(x)\right)\\
&-\frac{\omega^2\rho h}{K}\left(\sum_{m=0}^{\infty}\sum_{n=0}^{\infty}\cos\lambda_m x C_{mn}+\sum_{n=0}^{\infty}\Lambda_n^w\beta(x)\right)=0
\end{aligned}
$$

By expanding all non-cosine terms into Fourier cosine series and comparing the like terms, the following matrix equation can be obtained

$$\mathbf{Ex}+\mathbf{Fy}-\frac{\rho h\omega^2}{K}(\mathbf{Px}+\mathbf{Qy})=0. \tag{38}$$

where E, F, P and Q are coefficient matrices whose elements are given as:

$$\begin{aligned}
&\{\mathbf{E}_{11}\}_{mn,m'n'}=-\left(\lambda_m{}^2+\frac{(1-\sigma)(1+\gamma)n^2}{2R^2}\right)\delta_{mm'}\delta_{nn'} && \text{(a)}\\
&\{\mathbf{E}_{12}\}_{mn,m'n'}=-\frac{(1+\sigma)m'n\pi}{2lR}\chi_m^{m'}\delta_{mm'}\delta_{nn'} && \text{(b)}\\
&\{\mathbf{E}_{13}\}_{mn,m'n'}=-\left(\frac{\sigma m'\pi}{lR}+\gamma\left(R\lambda_{m'}{}^3-\frac{(1-\sigma)m'n^2\pi}{2lR}\right)\right)\chi_m^{m'}\delta_{mm'}\delta_{nn'} && \text{(c)}\\
&\{\mathbf{E}_{21}\}_{mn,m'n'}=\frac{(1+\sigma)m'n\pi}{2lR}\chi_m^{m'}\delta_{mm'}\delta_{nn'} && \text{(d)}\\
&\{\mathbf{E}_{22}\}_{mn,m'n'}=-\left(\frac{n^2}{R^2}+\frac{(1-\nu)(1+3\gamma)\lambda_m{}^2}{2}\right)\delta_{mm'}\delta_{nn'} && \text{(e)}\\
&\{\mathbf{E}_{23}\}_{mn,m'n'}=-\left(\frac{n}{R^2}+\frac{(3-\nu)\gamma n\lambda_m{}^2}{2}\right)\delta_{mm'}\delta_{nn'} && \text{(f)}\\
&\{\mathbf{E}_{31}\}_{mn,m'n'}=\left(-\frac{\nu\lambda_{m'}}{R}+\gamma\left(\frac{(1-\sigma)\lambda_{m'}n^2}{2R}-R\lambda_{m'}{}^3\right)\right)\chi_m^{m'}\delta_{mm'}\delta_{nn'} && \text{(g)}\\
&\{\mathbf{E}_{32}\}_{mn,m'n'}=\left(\frac{n}{R^2}+\frac{(3-\sigma)\lambda_m{}^2n\gamma}{2}\right)\delta_{mm'}\delta_{nn'} && \text{(h)}\\
&\{\mathbf{E}_{33}\}_{mn,m'n'}=\left(\frac{1}{R^2}+\gamma\left(\left(R\lambda_m+\frac{n^2-1}{R}\right)^2+2\lambda_m\right)\right)\delta_{mm'}\delta_{nn'} && \text{(i)}
\end{aligned} \tag{39}$$

$$\left\{\mathbf{F}_{11-a}\right\}_{mn,n'}=\left(\psi_{1m}-\frac{(1-\sigma)(1+\gamma)n^2}{2R^2}\phi_{1m}\right)\delta_{nn'} \qquad \text{(a)}$$

$$\left\{\mathbf{F}_{11-b}\right\}_{mn,n'}=\left(\psi_{2m}-\frac{(1-\sigma)(1+\gamma)n^2}{2R^2}\phi_{1m}\right)\delta_{nn'} \qquad \text{(b)}$$

$$\left\{\mathbf{F}_{12-c}\right\}_{mn,n'}=\frac{(1+\sigma)n}{2R}\varphi_{1m}\delta_{nn'} \qquad \text{(c)}$$

$$\left\{\mathbf{F}_{12-d}\right\}_{mn,n'}=\frac{(1+\sigma)n}{2R}\varphi_{2m}\delta_{nn'} \qquad \text{(d)}$$

$$\left\{\mathbf{F}_{13-e}\right\}_{mn,n'}=\left(\left(\frac{9\sigma}{8R}+\gamma\left(\frac{9R\pi^2}{32l^2}-\frac{9(1-\sigma)n^2}{16R}\right)\right)\kappa_{2m}-\left(\frac{\sigma}{8R}+\gamma\left(\frac{9R\pi^2}{32l^2}-\frac{(1-\sigma)n^2}{16R}\right)\right)\kappa_{4m}\right)\delta_{nn'} \qquad \text{(e)}$$

$$\left\{\mathbf{F}_{13-f}\right\}_{mn,n'}=\left(\left(\frac{9\sigma}{8R}+\gamma\left(\frac{9R\pi^2}{32l^2}-\frac{9(1-\sigma)n^2}{16R}\right)\right)\kappa_{1m}+\left(\frac{\sigma}{8R}+\gamma\left(\frac{9R\pi^2}{32l^2}-\frac{(1-\sigma)n^2}{16R}\right)\right)\kappa_{3m}\right)\delta_{nn'} \qquad \text{(f)}$$

$$\left\{\mathbf{F}_{13-g}\right\}_{mn,n'}=\left(\left(\frac{\sigma l^2}{2\pi^2R}+\gamma\left(\frac{R}{8}-\frac{(1-\sigma)l^2n^2}{4\pi^2R}\right)\right)\kappa_{2m}-\left(\frac{\sigma l^2}{2\pi^2R}+\gamma\left(\frac{9R}{8}-\frac{(1-\sigma)l^2n^2}{4\pi^2R}\right)\right)\kappa_{4m}\right)\delta_{nn'} \qquad \text{(g)}$$

$$\left\{\mathbf{F}_{13-h}\right\}_{mn,n'}=\left(\left(\frac{\sigma l^2}{2\pi^2R}+\gamma\left(\frac{R}{8}-\frac{(1-\sigma)l^2n^2}{4\pi^2R}\right)\right)\kappa_{1m}+\left(\frac{\sigma l^2}{2\pi^2R}+\gamma\left(\frac{9R}{8}-\frac{(1-\sigma)l^2n^2}{4\pi^2R}\right)\right)\kappa_{3m}\right)\delta_{nn'} \qquad \text{(h)}$$

$$\left\{\mathbf{F}_{21-a}\right\}_{mn,n'}=-\frac{(1+\sigma)n}{2R}\varphi_{1m}\delta_{nn'} \qquad \text{(i)}$$

$$\left\{\mathbf{F}_{21-b}\right\}_{mn,n'}=-\frac{(1+\sigma)n}{2R}\varphi_{2m}\delta_{nn'} \qquad \text{(j)}$$

$$\left\{\mathbf{F}_{22-c}\right\}_{mn,n'}=-\left(\frac{n^2}{R^2}\phi_{1m}-\frac{(1-\sigma)(1+3\gamma)}{2}\psi_{1m}\right)\delta_{nn'} \qquad \text{(k)}$$

$$\left\{\mathbf{F}_{22-d}\right\}_{mn,n'}=-\left(\frac{n^2}{R^2}\phi_{2m}-\frac{(1-\sigma)(1+3\gamma)}{2}\psi_{2m}\right)\delta_{nn'} \qquad \text{(l)}$$

$$\left\{\mathbf{F}_{23-e}\right\}_{mn,n'}=-\left(\left(\frac{9nl}{4\pi R^2}+\frac{9\pi(3-\sigma)n\gamma}{32l}\right)\kappa_{1m}-\left(\frac{nl}{12\pi R^2}+\frac{3\pi(3-\sigma)n\gamma}{32l}\right)\kappa_{3m}\right)\delta_{nn'} \qquad \text{(m)}$$

(40)

$$\left\{\mathbf{F}_{23-f}\right\}_{mn,n'}=\left(\left(\frac{9nl}{4\pi R^2}+\frac{9\pi(3-\sigma)n\gamma}{32l}\right)\kappa_{2m}+\left(\frac{nl}{12\pi R^2}+\frac{3\pi(3-\sigma)n\gamma}{32l}\right)\kappa_{4m}\right)\delta_{nn'} \qquad \text{(n)}$$

$$\left\{\mathbf{F}_{23-g}\right\}_{mn,n'}=-\left(\left(\frac{nl^3}{\pi^3R^2}+\frac{(3-\sigma)nl}{8\pi}\right)\kappa_{1m}-\left(\frac{nl^3}{3\pi^3R^2}+\frac{3(3-\sigma)nl\gamma}{8\pi}\right)\kappa_{3m}\right)\delta_{nn'} \qquad \text{(o)}$$

$$\left\{\mathbf{F}_{23-h}\right\}_{mn,n'}=\left(\left(\frac{nl^3}{\pi^3R^2}+\frac{(3-\sigma)nl}{8\pi}\right)\kappa_{2m}+\left(\frac{nl^3}{3\pi^3R^2}+\frac{3(3-\sigma)nl\gamma}{8\pi}\right)\kappa_{4m}\right)\delta_{nn'} \qquad \text{(p)}$$

$$\left\{\mathbf{F}_{31-a}\right\}_{mn,n'}=\left(\left(\frac{\sigma}{R}-\frac{(1-\sigma)\gamma n^2}{2R}\right)\varphi_{1m}-\frac{6R\gamma}{l^2}\delta_{m0}\right)\delta_{nn'} \qquad \text{(o)}$$

$$\left\{\mathbf{F}_{31-b}\right\}_{mn,n'}=\left(\left(\frac{\sigma}{R}-\frac{(1-\sigma)\gamma n^2}{2R}\right)\varphi_{2m}-\frac{6R\gamma}{l^2}\delta_{m0}\right)\delta_{nn'} \qquad \text{(r)}$$

$$\left\{\mathbf{F}_{32-c}\right\}_{mn,n'}=\left(\frac{n}{R^2}\phi_{1m}-\frac{(3-\sigma)n\gamma}{2}\psi_{1m}\right)\delta_{nn'} \qquad \text{(s)}$$

$$\left\{\mathbf{F}_{32-d}\right\}_{mn,n'}=\left(\frac{n}{R^2}\phi_{21m}-\frac{(3-\sigma)n\gamma}{2}\psi_{2m}\right)\delta_{nn'} \qquad \text{(t)}$$

$$\left\{\mathbf{F}_{33-e}\right\}_{mn,n'}=\left(\frac{9l}{4\pi}\left(\frac{1}{R^2}+\gamma\left(\left(\frac{R\pi^2}{4l^2}+\frac{n^2-1}{R}\right)^2+\frac{\pi^2}{2l^2}\right)\right)\kappa_{1m}-\frac{l}{3\pi}\left(\frac{1}{4R^2}+\gamma\left(\left(\frac{9R\pi^2}{8l^2}+\frac{n^2-1}{2R}\right)^2+\frac{9\pi^2}{8l^2}\right)\right)\kappa_{3m}\right)\delta_{nn'} \qquad \text{(u)}$$

$$\left\{\mathbf{F}_{33-f}\right\}_{mn,n'}=-\left(\frac{9l}{4\pi}\left(\frac{1}{R^2}+\gamma\left(\left(\frac{R\pi^2}{4l^2}+\frac{n^2-1}{R}\right)^2+\frac{\pi^2}{2l^2}\right)\right)\kappa_{2m}-\frac{l}{3\pi}\left(\frac{1}{4R^2}+\gamma\left(\left(\frac{9R\pi^2}{8l^2}+\frac{n^2-1}{2R}\right)^2+\frac{9\pi^2}{8l^2}\right)\right)\kappa_{4m}\right)\delta_{nn'} \qquad \text{(v)}$$

$$\left\{\mathbf{F}_{33-g}\right\}_{mn,n'}=\left(\frac{l}{\pi}\left(\frac{l^2}{\pi^2R^2}+\gamma\left(\left(\frac{R\pi}{4l}+\frac{\left(n^2-1\right)l}{\pi R}\right)^2-\frac{1}{2}\right)\right)\kappa_{1m}-\frac{l}{3\pi}\left(\frac{l^2}{\pi^2R^2}+\gamma\left(\left(\frac{9R\pi}{4l}+\frac{\left(n^2-1\right)l}{\pi R}\right)^2-\frac{9}{2}\right)\right)\kappa_{3m}\right)\delta_{nn'} \qquad \text{(w)}$$

$$\left\{\mathbf{F}_{33-h}\right\}_{mn,n'}=-\left(\frac{l}{\pi}\left(\frac{l^2}{\pi^2R^2}+\gamma\left(\left(\frac{R\pi}{4l}+\frac{\left(n^2-1\right)l}{\pi R}\right)^2-\frac{1}{2}\right)\right)\kappa_{2m}-\frac{l}{3\pi}\left(\frac{l^2}{\pi^2R^2}+\gamma\left(\left(\frac{9R\pi}{4l}+\frac{\left(n^2-1\right)l}{\pi R}\right)^2-\frac{9}{2}\right)\right)\kappa_{4m}\right)\delta_{nn'} \qquad \text{(x)}$$

$$
\begin{aligned}
&\{\mathbf{P}_{11}\}_{mn,m'n'} = \{\mathbf{P}_{22}\}_{mn,m'n'} = -\{\mathbf{P}_{33}\}_{mn,m'n'} = -\delta_{mm'}\delta_{nn'} && \text{(a)}\\
&\{\mathbf{Q}_{11-a}\}_{mn,n'} = \{\mathbf{Q}_{22-c}\}_{mn,n'} = -\phi_{1m}\delta_{nn'} && \text{(b)}\\
&\{\mathbf{Q}_{11-b}\}_{mn,n'} = \{\mathbf{Q}_{22-d}\}_{mn,n'} = -\phi_{2m}\delta_{nn'} && \text{(c)}\\
&\{\mathbf{Q}_{33-e}\}_{mn,n'} = \frac{l}{4\pi}\left(9\kappa_{1m} - \frac{\kappa_{3m}}{3}\right)\delta_{nn'} && \text{(d)}\\
&\{\mathbf{Q}_{33-f}\}_{mn,n'} = -\frac{l}{4\pi}\left(9\kappa_{2m} + \frac{\kappa_{4m}}{3}\right)\delta_{nn'} && \text{(e)}\\
&\{\mathbf{Q}_{33-g}\}_{mn,n'} = \frac{l^3}{\pi^3}\left(\kappa_{1m} - \frac{\kappa_{3m}}{3}\right)\delta_{nn'} && \text{(f)}\\
&\{\mathbf{Q}_{33-h}\}_{mn,n'} = -\frac{l^3}{\pi^3}\left(\kappa_{2m} + \frac{\kappa_{4m}}{3}\right)\delta_{nn'} && \text{(g)}
\end{aligned}
\qquad (41)
$$

The symbols κ_{1m}, κ_{2m}, κ_{3m}, κ_{4m}, ϕ_{1m}, ϕ_{2m}, φ_{1m}, φ_{2m}, ψ_{1m}, ψ_{2m}, and χ_m^i in the above equations are defined as

$$
\begin{aligned}
&\sin(\frac{\pi}{2l}x) = \sum_{m=0}^{\infty}\kappa_{1m}\cos\lambda_m x; && \kappa_{1m} = \begin{cases} \dfrac{2}{\pi} & m=0,\\ \dfrac{4}{(1-4m^2)\pi} & m\neq 0, \end{cases} && \text{(a,b)}\\
&\cos(\frac{\pi}{2l}x) = \sum_{m=0}^{\infty}\kappa_{2m}\cos\lambda_m x; && \kappa_{2m} = \begin{cases} \dfrac{2}{\pi} & m=0,\\ \dfrac{4(-1)^m}{(1-4m^2)\pi} & m\neq 0, \end{cases} && \text{(c,d)}\\
&\sin(\frac{3\pi}{2l}x) = \sum_{m=0}^{\infty}\kappa_{3m}\cos\lambda_m x; && \kappa_{3m} = \begin{cases} \dfrac{2}{3\pi} & m=0,\\ \dfrac{12}{(9-4m^2)\pi} & m\neq 0, \end{cases} && \text{(e,f)}\\
&\cos(\frac{3\pi}{2l}x) = \sum_{m=0}^{\infty}\kappa_{4m}\cos\lambda_m x; && \kappa_{4m} = \begin{cases} -\dfrac{2}{3\pi} & m=0,\\ \dfrac{12(-1)^{m+1}}{(9-4m^2)\pi} & m\neq 0, \end{cases} && \text{(g, h)}\\
&\alpha_1(x) = \sum_{m=0}^{\infty}\phi_{1m}\cos\lambda_m x; && \phi_{1m} = \begin{cases} \dfrac{l}{12} & m=0,\\ -\dfrac{2l\left(m^2\pi^2-6+6(-1)^m\right)}{m^4\pi^4} & m\neq 0, \end{cases} && \text{(i,j)}\\
&\alpha_2(x) = \sum_{m=0}^{\infty}\phi_{2m}\cos\lambda_m x; && \phi_{2m} = \begin{cases} -\dfrac{l}{12} & m=0,\\ \dfrac{2l\left(m^2\pi^2(-1)^m+6-6(-1)^m\right)}{m^4\pi^4} & m\neq 0, \end{cases} && \text{(k, l)}\\
&\alpha_1'(x) = \sum_{m=0}^{\infty}\varphi_{1m}\cos\lambda_m x; && \varphi_{1m} = \begin{cases} 0 & m=0,\\ \dfrac{4\left(2+(-1)^m\right)}{m^2\pi^2} & m\neq 0, \end{cases} && \text{(m,n)}\\
&\alpha_2'(x) = \sum_{m=0}^{\infty}\varphi_{2m}\cos\lambda_m x; && \varphi_{2m} = \begin{cases} 0 & m=0,\\ \dfrac{4\left(1+2(-1)^m\right)}{m^2\pi^2} & m\neq 0, \end{cases} && \text{(o,p)}\\
&\alpha_1''(x) = \sum_{m=0}^{\infty}\psi_{1m}\cos\lambda_m x; && \psi_{1m} = \begin{cases} -\dfrac{1}{l} & m=0,\\ \dfrac{12\left(-1+(-1)^m\right)}{lm^2\pi^2} & m\neq 0, \end{cases} && \text{(q,r)}\\
&\alpha_2''(x) = \sum_{m=0}^{\infty}\psi_{2m}\cos\lambda_m x; && \psi_{2m} = \begin{cases} \dfrac{1}{l} & m=0,\\ \dfrac{12\left(-1+(-1)^m\right)}{lm^2\pi^2} & m\neq 0, \end{cases} && \text{(s, t)}\\
&\sin\lambda_m x = \sum_{i}^{\infty}\chi_i^m\cos\lambda_i x = \sin\lambda_i x = \sum_{m}^{\infty}\chi_m^i\cos\lambda_m x; && \chi_m^i = \begin{cases} 0 & i=0,\\ \begin{cases} \dfrac{1-(-1)^i}{i\pi} & m=0,\\ \dfrac{2i\left[(-1)^{m+i}-1\right]}{(m^2-i^2)\pi} & m\neq 0, \end{cases} & i\neq 0. \end{cases} && \text{(u,v)}
\end{aligned}
\qquad (42)
$$

All the unmentioned elements in matrices P and Q are identically equal to zero.

Equations (32) and (36) can be combined into

$$\left(\mathbf{K} - \frac{\rho h \omega^2}{K}\mathbf{M}\right)\mathbf{x} = 0, \tag{43}$$

where $K = E + FL^{-1}S$ and$M = P + QL^{-1}S$.

The final system of equations, Eq. (19) or (41), represents a standard characteristic equation for a matrix eigen-problem from which all the eigenvalues and eigenvectors can be readily calculated. It should be mentioned that the elements in each eigenvector are actually the expansion coefficients for the corresponding mode; its "physical" mode shape can be directly obtained from Eqs. (7) or (27).

In the above discussions, the stiffness distribution for each restraining spring is assumed to be axisymmetric or uniform along the circumference. However, this restriction is not necessary. For non-uniform elastic boundary restraints, the displacement expansions, Eq. (27), shall be used, and any and all of stiffness constants can be simply understood as varying with spatial angle θ. For simplicity, we can universally expand these functions into standard cosine series and modify Eq. (31) accordingly to reflect this complicating factor.

3. Results and discussion

Several numerical examples will be given below to verify the two solution strategies described earlier.

3.1. Results about the approximate Rayleigh-Ritz solution

We first consider a familiar simply-supported cylindrical shell. The simply supported boundary condition, $N_x = M_x = v = w = 0$at each end, can be considered as a special case when $k_{2,6} = k_{3,7} = \infty$ and $k_{1,5} = k_{4,8} = 0$ (in actual calculations, infinity is represented by a sufficiently large number). To examine the convergence of the solution, Table 1 shows the frequency parameters, $\Omega = \omega R\sqrt{\rho(1-\sigma^2)/E}$, calculated using different numbers of terms in the series expansions. It is seen that the solution converges nicely with only a small number of terms. In the following calculations, the expansions in axial direction will be simply truncated to M=15. Given in Table 2 are the frequencies parameters for some lower-order modes. Exact solution is available for the simply supported case and the results are also shown there for comparison. An excellent agreement is observed between these two sets of results. Although the simply supported boundary condition represents the simplest case in shell analysis, this problem is not trivial in testing the reliability and sophistication of the current solution method. From numerical analysis standpoint, it may actually represent a quite challenging case because of the extreme stiffness values involved. The non-trivialness can also been seen mathematically from the fact that the simple sine function (in the axial direction) in the exact solution is actually expanded as a cosine series expansion in the current solution.

Number of terms used in the series	$\Omega = \omega R\sqrt{\rho(1-\sigma^2)/E}$				
	n=0	*n=1*	*n=2*	*n=3*	*n=4*
M=5	0.464652	0.257389	0.127132	0.143329	0.234823
M=7	0.464649	0.257386	0.127129	0.143327	0.234822
M=9	0.464648	0.257385	0.127128	0.143327	0.234822
M=10	0.464648	0.257385	0.127128	0.143327	0.234822

Table 1. Frequency parameters, $\Omega = \omega R\sqrt{\rho(1-\sigma^2)/E}$, obtained using different numbers of terms in the displacement expansions.

Mode	$\Omega = \omega R\sqrt{\rho(1-\sigma^2)/E}$				
	n=0	*n=1*	*n=2*	*n=3*	*n=4*
m=1, Current	0.464648	0.257385	0.127128	0.143327	0.234822
Exact	0.464648	0.257385	0.127128	0.143327	0.234822
m=2, Current	0.928907	0.574179	0.337652	0.248813	0.285620
Exact	0.928907	0.574176	0.337649	0.248810	0.285619
m=3, Current	0.948172	0.764375	0.532951	0.399893	0.383688
Exact	0.948172	0.764355	0.532923	0.399865	0.383667

Table 2. Frequency parameters, $\Omega = \omega R\sqrt{\rho(1-\sigma^2)/E}$, for a simply-supported shell; $L=4R$, $h/R=0.05$ and $\mu=0.3$.

Next, consider a cylindrical shell clamped at each end, that is, $u = v = w = \partial w/\partial x = 0$. The clamped-clamped boundary condition is a case when the stiffnesses of the restraining springs all become infinitely large. The related shell and material parameters are as follows: L=0.502 m, R=0.0635 m, h=0.00163 m, $E=2.1\times10^{11}$, μ=0.28, and ρ=7800. Listed in Table 3 are some of the lowest natural frequencies for this clamped-clamped shell. The reference results given there are calculated from

$$\Omega^6 - A_2\Omega^4 + A_1\Omega^2 - A_0 = 0 \tag{44}$$

where $\Omega = \omega R\sqrt{\rho(1-\sigma^2)/E}$, and the coefficients A_0, A_1 and A_2 are the functions of the modal indices, shell parameters, and the boundary conditions [27]. Equation (42) can be derived from the Rayleigh-Ritz procedure by adopting the beam characteristic functions as the axial functions for all three displacement components. A noticeable difference between these two sets of results may be attributed to the fact that: a) Eq. (42) given in ref. [29] is based on the Flügge shell theory, rather than the Donnell-Mushtari theory, and b) Eq. (42) uses only a single beam characteristic function in contrast to the three complete sets (of basis functions) in the current method.

Mode	Current m=1	**Eq. (42)**	Current m=2	**Eq. (42)**
n=1	1886.74	2035.05	3854.75	4302.05
2	934.220	971.531	2039.66	2189.59
3	982.265	990.339	1454.80	1500.07
4	1598.55	1600.90	1769.54	1782.28
5	2484.78	2486.49	2572.31	2578.07

Table 3. The natural frequencies in Hz for a clamped-clamped shell; L=0.502 m, R=0.0635 m, h=0.00163 m, E=2.1E+11, μ=0.28, ρ=7800 kg/m³.

Mode	***n*=2**	***n*=3**	***n*=4**	***n*=5**	***n*=6**	***n*=7**
Translation:						
current	0.00413	0.00986	0.01792	0.02830	0.04099	0.05599
(2.130) [22]	0.00310	0.00876	0.01680	0.02717	0.03986	0.05487
FEA	0.00310	0.00876	0.01679	0.02714	0.03980	0.05475
Rotation:						
Current	0.01907	0.01676	0.02068	0.02995	0.04220	0.05713
(2.132) [22]	0.00343	0.00924	0.01734	0.02774	0.04045	0.05546
FEA	0.00343	0.00923	0.01731	0.02769	0.04037	0.05533
m=1:						
Current	0.24075	0.13190	0.08343	0.06292	0.05906	0.06606
FEA	0.23810	0.12836	0.07938	0.05893	0.05555	0.06332

Table 4. Frequency parameters, $\Omega = \omega R\sqrt{\rho(1-\sigma^2)/E}$, for a free-free shell; R=0.5 m, L=4R, h=0.002R, and μ=0.28.

Another classical example involves a completely free shell. Vibration of a free-free shell is of particular interest as manifested in the debate between two legendary figures, Rayleigh and Love, about the validity of the inextensional theory of shells. The lower-order modes are typically related to the rigid-body motions in the axial direction. Theoretically, the H_w matrix given in Eqs. (14) will become non-invertible for a completely free shell. However, this numerical irregularity can be easily avoided by letting one of the bending-related springs have a very small stiffness, such as,$\hat{k}_4=10^{-6}$. Table 4 shows a comparison of the frequency parameters calculated using different techniques. While the results obtained from the current technique agree reasonably well with the other two reference sets, perhaps within the variance of different shell theories, the frequency parameters for the two lower order modes with rigid-body rotation (n=2 and 3) are clearly inaccurate which probably indicates that the inability of exactly satisfying the shell boundary conditions by the "beam functions" tends to have more serious consequence in such a case. Amazingly enough, the inextensional theory works very well in predicting the frequency parameters for the "rigid-body" modes (those with rigid-body motions in the axial direction). It is also seen that the frequency parameters of the rigid-body modes increases monotonically with the circumferential modal index n.

After it has been adequately illustrated how the classical boundary conditions can be easily and universally dealt with by simply changing the stiffness values of the restraining springs, we will direct our attention to shells with elastic end restraints. For the purpose of comparison, the problems previously studied in ref. [20] will be considered here. It was observed in that study that the tangential stiffness had the greatest effect on the natural frequency of the cylinder supported at both ends while the axial boundary stiffness had the greatest influence on the natural frequency of the cylinder supported at one end. It was also determined that natural frequencies varied rapidly with the boundary flexibility when the non-dimensionalized stiffness is between 10^{-2} and 10^{2}.

The frequency parameters for the "clamped"-free shell are shown in Table 5 for the reduced axial stiffness $\hat{k}_1 L\ (1-\mu^2)=1$ (corresponding to $k_u^*=1$ in ref. [20]). It is seen that the current results are slightly larger than those taken from ref. [20]. The possible reasons include: 1) the difference in shell theories (the Flügge theory, rather than the Donnell-Mushtari, was used there), and 2) different Poisson ratios may have been used in the calculations.

Mode	*n=0*	*n=1*	*n=2*	*n=3*	*n=4*	*n=5*
m=1	0.9752	0.514686	0.32866 (0.315*)	0.361036	0.532604 (0.498)	0.782661
m=2	1.22044	1.12788	1.08573	1.10467	1.16021	1.432

Note: the numbers in parentheses are taken from ref. [20]

Table 5. Frequency parameters, $\Omega=\omega R\sqrt{\rho(1-\sigma^2)/E}$, for a "clamped"-free shell; R=0.00625 m, $L=R$, $h=0.1R$, μ=0.28, and$\hat{k}_1 L\ (1-\mu^2)=1$.

Although all eight sets of springs can be independently specified here, for simplicity we will only consider a simple configuration: a cantilevered shell with an elastic support being attached to its free (right) end in the radial direction. Listed in Table 6 are the four lowest natural frequencies for several different stiffness values. Obviously, the cases for $\hat{k}_7=0$ and ∞ represent the clamped-free and clamped-simply supported boundary conditions, respectively.

The mode shapes for the three intermediate stiffness values are plotted in Figs. 2-4. It is seen that the modal parameters can be significantly modified by the stiffness of the restraining springs. The four modes in Fig. 2 for $\hat{k}_7=0.01\ m^{-1}$ closely resemble their counterparts in the clamped-free case, even though the natural frequencies have been modified noticeably. While all the first four natural frequencies happen to increase, more or less, with the spring stiffness, the modal sequences are not necessarily the same. For example, when the spring stiffness $\hat{k}_7$ is increased from 0.01 to 0.1 m^{-1}, the third natural frequency goes from 886.66 to 926.17 Hz. However, this frequency drift may not necessarily reflect the direct effect of the stiffness change on the (original) third mode. It is evident from Figs. 2 and 3 that the third and fourth modes are actually switched in these two cases: the original third mode now becomes the fourth at 1200.88 Hz. It is also interesting to note that while stiffening the elastic support $\hat{k}_7$ (from 0.01 to 0.1 m^{-1}) has significantly raised the natural frequencies for the first two modes, the fourth mode is adversely affected: its frequency has actually dropped from 1023.61 to 926.17 Hz (see Figs. 2 and 3). A similar trend is also observed between the fourth mode for $\hat{k}_7$=0.1 m^{-1} and the second mode for $\hat{k}_7$=1 m^{-1}, as shown in Figs. 3 and 4.

Mode	$\hat{k}_7=0$	$\hat{k}_7=0.01$	$\hat{k}_7=0.1$	$\hat{k}_7=1$	$\hat{k}_7=10^{10}$
1	404.108	451.242	627.345	729.593	742.920
2	487.598	513.222	679.082	935.745	936.719
3	865.603	886.656	926.173	1084.91,	1269.58
4	1003.38	1023.61	1200.88	1319.99, ,	1333.37

Table 6. Natural frequencies in Hz for a clamped-elastically supported shell; L=0.502 m, R=0.0635 m, h=0.00163 m, E=2.1E+11, μ=0.28, ρ=7800 kg/m^3;$\hat{k}_5=\hat{k}_6=\hat{k}_8=0$.

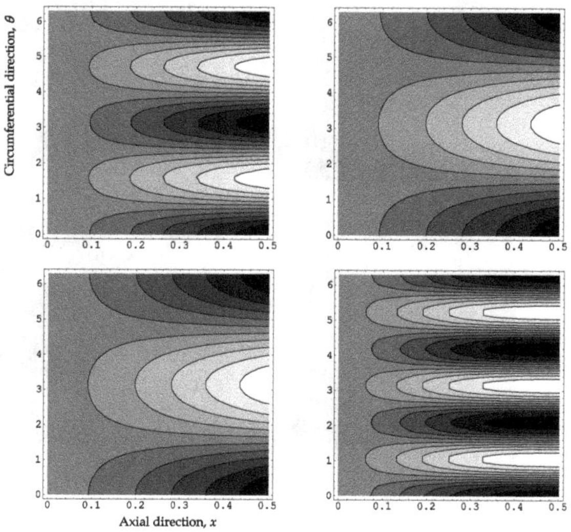

Figure 2. First four modes for the clamped-elastically supported shell; $\hat{k}_7$=0.01 m^{-1}and$\hat{k}_5=\hat{k}_6=\hat{k}_8=0$.

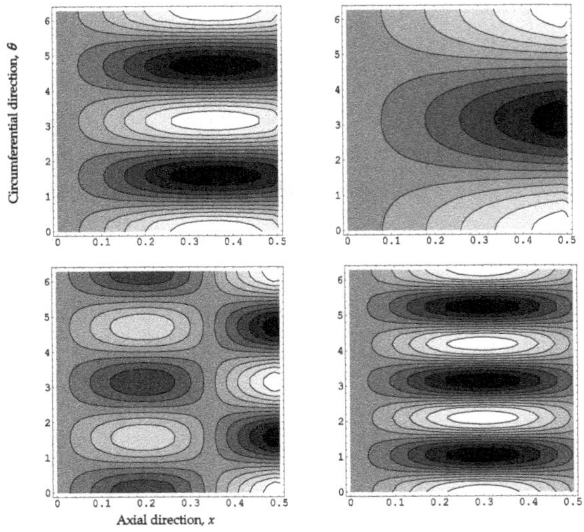

Figure 3. First four modes for the clamped-elastically supported shell; $\hat{k}_7=0.1\ \text{m}^{-1}$ and $\hat{k}_5=\hat{k}_6=\hat{k}_8=0$.

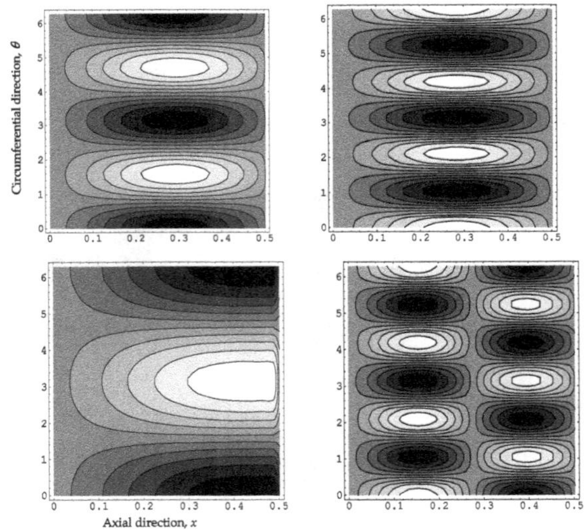

Figure 4. First four modes for the clamped-elastically supported shell; $\hat{k}_7=1\ \text{m}^{-1}$ and $\hat{k}_5=\hat{k}_6=\hat{k}_8=0$.

3.2. An exact solution based on the Flügge's equations

To validate the exact solution method, the simply supported shell is considered again. Given in Table 7 are the calculated natural frequency parameters $\Omega = \omega R \sqrt{\rho(1-\sigma^2)/E}$. The current results agree well with the exact solutions based on Flügge's theory [30], solutions based on beam functions [31] and three-dimensional linear elasticity solutions [30].

h/R	n	$\Omega = \omega R \sqrt{\rho(1-\sigma^2)/E}$			
		Ref. [30][a]	Ref. [31]	Ref. [30][b]	Present
0.05	0	0.0929586	0.0929682	0.0929296	0.0929590
	1	0.0161065	0.0161029	0.0161063	0.0161064
	2	0.0393038	0.0392710	0.0392332	0.0393035
	3	0.1098527	0.1098113	0.1094770	0.1098468
	4	0.2103446	0.2102770	0.2090080	0.2103419
0.002	0	0.0929296	0.0929298	0.0929296	0.0929299
	1	0.0161011	0.0161011	0.0161011	0.0161023
	2	0.0054532	0.0054530	0.0054524	0.0054547
	3	0.0050419	0.0050415	0.0050372	0.0050427
	4	0.0085341	0.0085338	0.0085341	0.0085344

[a] Exact solutions based on Flügge's theory.

[b] Three-dimensional linear elasticity solutions.

Table 7. Comparison of values of the natural frequency parameter $\Omega = \omega R \sqrt{\rho(1-\sigma^2)/E}$ for a circular cylindrical shell with simply supported boundary conditions, $m = 1$, $R/l = 0.05$, $\sigma = 0.3$.

n	*m = 1*			*m = 2*		
	FEM	present	difference (%)	FEM	present	difference (%)
0	3229.8	3230.3	0.015%	5131.4	5131.1	0.006%
1	2478.6	2479.3	0.028%	4830.4	4830.6	0.004%
2	269.20	269.30	0.037%	276.62	278.58	0.704%
3	761.25	761.01	0.032%	770.99	771.62	0.082%
4	1459.2	1458.6	0.041%	1469.6	1469.3	0.020%
5	2359.4	2358.6	0.034%	2369.9	2369.0	0.038%

Table 8. Comparison of values of the natural frequency for a circular cylindrical shell with free-free boundary conditions, L=0.502 m, R=0.0635 m, h=0.00163 m, σ=0.28, E=2.1E+11 N/m^3, ρ=7800 kg/m^3.

The current solution method is also compared with the finite element model (ANSYS) for shells under free-free boundary condition. In the FEM model, the shell surface is divided into 8000 elements with 8080 nodes. The calculated natural frequencies are compared in Tables 8. An excellent agreement is observed between these two solution methods.

In most techniques, such as the wave approach, the beam functions for the analogous boundary conditions are often used to determine the axial modal wavenumbers. While such an approach is exact for a simply supported shell, and perhaps acceptable for slender thin shells, it may become problematic for shorter shells due to the increased coupling of the radial and two in-plane displacements. To illustrate this point, we consider relatively shorter and thicker shell (l=8R and R =39h). The calculated natural frequencies are compared in Table 9 for a clamped-clamped shell. It is seen that while the current and FEM results are in good agreement, the frequencies obtained from the wave approach (based on the use of beam functions) are significantly higher, especially for the lower order modes.

n	m = 1			m = 2		
	FEM	Ref. [32]	present	FEM	Ref. [32]	Present
0	3229.8	4845.5	3230.3	5146.0	8075.8	5139.8
1	1882.8	2350.2	1880.9	3850.7	4775.6	3848.9
2	899.59	985.48	898.18	2017.8	2303.4	2014.1
3	896.97	919.01	896.56	1390.9	1479.2	1388.9
4	1501.9	1517.45	1501.6	1676.4	1714.0	1676.0
5	2386.1	2402.05	2386.0	2472.5	2501.8	2472.6

Table 9. Comparison of the natural frequencies for a circular cylindrical shell with clamped-clamped boundary conditions, L=0.502 m, R=0.0635 m, h=0.00163 m, σ=0.28, E=2.1E+11 N/m^3, ρ=7800 kg/m^3.

The exact solution method can be readily applied to shells with elastic boundary supports. Since the above examples are considered adequate in illustrating the reliability and accuracy of the current method, we will not elaborate further by presenting the results for elastically restrained shells. Instead, we will simply point out that the solution method based on Eqs. (27) is also valid for non-uniform or varying boundary restraint along the circumferential direction, which represents a significant advancement over many existing techniques.

4. Conclusion

An improved Fourier series solution method is described for vibration analysis of cylindrical shells with general elastic supports. This method can be easily and universally applied to a wide variety of boundary conditions including all the 136 classical homogeneous boundary conditions. The displacement functions are invariantly expressed as series expansions in terms of the complete set of trigonometric functions, which can mathematically ensure the accuracy and convergence of the present solution. From practical point of view, the change of boundary conditions here is as simple as varying a typical shell or material parameter (e.g., thickness or mass density), and does not involve any solution algorithm and procedure modifications to adapt to different boundary conditions. In addition, the proposed method does not require pre-determining any secondary data such as modal parameters for an "analogous" beam, or modifying the implementation algorithms to avoid the numerical instabilities resulting from computer round-off errors. It should be mentioned that the current method can be readily extended to shells with arbitrary non-uniform elastic restraints. The accuracy and reliability of the current solutions have been demonstrated through numerical examples involving various boundary conditions.

Acknowledgments

The authors gratefully acknowledge the financial support from the National Natural Science Foundation of China (No. 50979018).

Author details

Tiejun Yang[1], Wen L. Li[2] and Lu Dai[1]

1 College of Power and Energy Engineering, Harbin Engineering University, Harbin, PR China

2 Department of Mechanical Engineering, Wayne State University, Detroit, USA

References

[1] Influence of boundary conditions on the modal characteristics of thin cylindrical shells,. AIAA Journal ; , 2-2150.

[2] Forsberg, K. Axisymmetric and beam-type vibrations of thin cylindrical shells,. AIAA Journal (1969). , 7-221.

[3] Warburton, G. B. Vibrations of thin circular cylindrical shell, J. Mech. Eng. Sci. (1965). , 7-399.

[4] Warburton, G. B., & Higgs, J. Natural frequencies of thin cantilever cylindrical shells. J. Sound Vib. (1970). , 11-335.

[5] Goldman, R. L. (1974). Mode shapes and frequencies of clamped-clamped cylindrical shells. *AIAA Journal*, 12-1755.

[6] Yu, Y. Y. Free vibrations of thin cylindrical shells having finite lengths with freely supported and clamped edges, J. Appl. Mech. (1955). , 22-547.

[7] Berglund, J. W., & Klosner, J. M. Interaction of a ring-reinforced shell and a fluid medium. J. Appl. Mech., (1968). , 35-139.

[8] De Silva, C. N., & Tersteeg, C. E. Axisymmetric vibrations of thin elastic shells. J. Acoust. Soc. Am. (1964). , 4-666.

[9] Smith, B. L., & Haft, E. E. (1968). Natural frequencies of clamped cylindrical shells,. *AIAA Journal*, 6-720.

[10] Li, X. A., new, approach., for, free., vibration, analysis., of, thin., circular, cylindrical., & shell, J. Sound Vib. (2006). , 296, 91-98.

[11] Wang, C., & Lai, J. C. S. (2000). Prediction of natural frequencies of finite length circular cylindrical shells. *Applied Acoustics*, 59-385.

[12] Zhang, X. M., Liu, G. R., & Lam, K. Y. Vibration analysis of thin cylindrical shells using wave propagation approach,. Journal of Sound and Vibration (2001). , 239(3), 397-403.

[13] Li, X. B. (2008). Study on free vibration analysis of circular cylindrical shells using wave propagation. *Journal of Sound and Vibration*, 311-667.

[14] Arnold, R. N., & Warburton, G. B. Flexural vibrations of the walls of thin cylindrical shells having freely supported ends. Proc. Roy. Soc., A (1949). , 197-238.

[15] Arnold, R. N., & Warburton, G. B. The flexural vibrations of thin cylinders. Proc. Inst. Mech. Engineers, A (1953). , 167, 62-80.

[16] Sharma, C. B., & Johns, D. J. Vibration characteristics of a clamped-free and clamped-ring-stiffened circular cylindrical shell, J. Sound Vib. 1971;. 14-459.

[17] Sharma, C. B., & Johns, D. J. Free vibration of cantilever circular cylindrical shells-a comparative study, J. Sound Vib. (1972). , 25-433.

[18] Sharma, C. B. Calculation of natural frequencies of fixed-free circular cylindrical shells. J. Sound Vib. (1974). , 35-55.

[19] Soedel, W. A., new, frequency., formula, for., closed, circular., cylindrical, shells., for, a., large, variety., of, boundary., & conditions, . J. Sound Vib. (1980). , 70-309.

[20] Loveday, P. W., & Rogers, C. A. Free vibration of elastically supported thin cylinders including gyroscopic effects,. J. Sound Vib, (1998). , 217, 547-562.

[21] Amabili, M., & Garziera, R. Vibrations of circular cylindrical shells with nonuniform constraints, elastic bed and added mass: part I: empty and fluid-filled shells. J. Fluids Struct, (2000). , 14, 669-690.

[22] Leissa A. W. Vibration of Shells, Acoustical Society of America; 1993.

[23] Qatu M.S. (2002). Recent research advances in the dynamic behavior of shells. part 2: homogeneous shells,. *Applied Mechanics Reviews*, 55, 415-434.

[24] Li, W. L. Free vibrations of beams with general boundary conditions. J. Sound Vib. (2000). , 237-709.

[25] Li, W. L. Vibration analysis of rectangular plates with general elastic boundary supports. J. Sound Vib. (2004). , 273-619.

[26] Lanczos, C. Discourse on Fourier series. Hafner, New York; (1966).

[27] Jones, W. B., & Hardy, G. Accelerating convergence of trigonometric approximations. Math. Comp. (1970). , 24-547.

[28] Baszenski, G., Delvos, J., Tasche, M. A., united, approach., to, accelerating., trigonometric, expansions., & Comput, . Math. Appl. (1995). , 30-33.

[29] Blevins, R. D. Formulas for Natural Frequency and Mode Shape. New York: Van Nostrand Reinhold Company; (1979).

[30] Khdeir, A. A., & Reddy, J. N. Influence of edge conditions on the modal characteristics of cross-ply laminted shells, Computers and Structures (1990). , 34-817.

[31] Lam, K. Y., & Loy, C. T. (1995). Effects of boundary conditions on frequencies of a multi-layered cylindrical shell. *Journal of Sound and Vibration*, 363-384.

[32] Zhang, X. M., Liu, G. R., & Lam, K. Y. Vibration analysis of thin cylindrical shells using wave propagation approach,. Journal of Sound and Vibration (2001). , 239(3), 397-403.

Chapter 9

Vibration Analysis of Cracked Beams Using the Finite Element Method

A. S. Bouboulas, S. K. Georgantzinos and
N. K. Anifantis

Additional information is available at the end of the chapter

1. Introduction

Most of the members of engineering structures operate under loading conditions, which may cause damages or cracks in overstressed zones. The presence of cracks in a structural member, such as a beam, causes local variations in stiffness, the magnitude of which mainly depends on the location and depth of the cracks. These variations, in turn, have a significant effect on the vibrational behavior of the entire structure. To ensure the safe operation of structures, it is extremely important to know whether their members are free of cracks, and should any be present, to assess their extent. The procedures often used for detection are direct procedures such as ultrasound, X-rays, etc. However, these methods have proven to be inoperative and unsuitable in certain cases, since they require expensive and minutely detailed inspections [1]. To avoid these disadvantages, in recent decades, researchers have focused on more efficient procedures in crack detection using vibration-based methods [2]. Modelling of a crack is an important aspect of these methods.

The majority of published studies assume that the crack in a structural member always remains open during vibration [3-7]. However, this assumption may not be valid when dynamic loadings are dominant. In this case, the crack breathes (opens and closes) regularly during vibration, inducing variations in the structural stiffness. These variations cause the structure to exhibit non-linear dynamic behavior [8]. The main distinctive feature of this behavior is the presence of higher harmonic components. In particular, a beam with a breathing crack shows natural frequencies between those of a non-cracked beam and those of a faulty beam with an open crack. Therefore, in these cases, vibration-based methods should employ breathing crack models to provide accurate conclusions regarding the state of damage. Several researchers [9-11] have developed breathing crack models considering only the fully open and fully closed crack states. However, experiments have indicated that the transition between these two crack states does not occur instantaneously [12]. In reference [13] represented the interaction forces between two segments of a beam, separated by a crack, using time-varying connection matrices. These matrices were expanded in Fourier series to simulate the alternation of a crack opening and closing. However, the implementation of this study requires excessive computer time. In references [14, 15] considered a simple periodic function to model the time-varying stiffness of a beam. However, this model is limited to the fundamental mode, and thus, the equation of motion for the beam must be solved.

A realistic model of a breathing crack is difficult to create due to the lack of fundamental understanding about certain aspects of the breathing mechanism. This involves not only the identification of variables affecting the breathing crack behavior, but also issues for evaluating the structural dynamic response of the fractured material. It is also not yet entirely clear how partial closure interacts with key variables of the problem. The actual physical situation requires a model that accounts for the breathing mechanism and for the interaction between external loading and dynamic crack behavior. When crack contact occurs, the unknowns are the field singular behavior, the contact region and the distribution of contact tractions on the closed region of the crack. The latter class of unknowns does not exist in the case without crack closure. This type of complicated deformation of crack surfaces constitutes a non-linear problem that is too difficult to be treated with classical analytical procedures. Thus, a suitable numerical implementation is required when partial crack closure occurs.

In reference [16] constructed a lumped cracked beam model from the three-dimensional formulation of the general problem of elasticity with unilateral contact conditions on the crack lips. The problem of a beam with an edge crack subjected to a harmonic load was considered in [17]. The breathing crack behavior was simulated as a frictionless contact problem between the crack surfaces. Displacement constraints were applied to prevent penetration of the nodes of one crack surface into the other crack surface. In reference [18] studied the problem of a cantilever beam with an edge crack subjected to a harmonic load. The breathing crack behavior was represented via a frictionless contact model of the interacting surfaces. In [19] studied the effect of a helicoidal crack on the dynamic behavior of a rotating shaft. This study used a very accurate and simplified model that assumes linear stress and strain distributions to calculate the breathing mechanism. The determination of open and closed parts of the crack was performed through a non-linear iterative procedure.

This chapter presents the vibrational behavior of a beam with a non-propagating edge crack. To treat this problem, a two-dimensional beam finite element model is employed. The breathing crack is simulated as a full frictional contact problem between the crack surfaces, while the region around the crack is discretized into a number of conventional finite elements. This non-linear dynamic problem is solved using an incremental iterative procedure. This study is applied for the case of an impulsive loaded cantilever beam. Based on the derived time response, conclusions are extracted for the crack state (i.e. open or closed) over the time. Furthermore, the time response is analyzed by Fourier and continuous wavelet transforms to show the sensitivity of the vibrational behavior for both a transverse and slant crack of various depths and positions. Comparisons are performed with the corresponding vibrational behavior of the beam when the crack is considered as always open. To assess further the validity of this technique, the quasi-static problem of a three-dimensional rotating beam with a breathing crack is also presented. The formulation of this latter problem is similar to the former one. The main differences are: the inertia and damping terms are ignored, any possible sliding occurs in two dimensions and the iterative procedure is applied to load instead of time increment. The flexibility of the rotating beam and the crack state over time are presented for both a transverse and slant crack of various depths. The validation of the present study is demonstrated through comparisons with results available from the literature.

2. Finite element formulation

In the following, both a two and three-dimensional beam models with a non-propagating surface crack are presented. For both models the crack surfaces are assumed to be planar and smooth and the crack thickness negligible. The beam material properties are considered linear elastic and the displacements and strains are assumed to be small. The region around the crack is discretized into conventional finite elements. The breathing crack behavior is simulated as a full frictional contact problem between the crack surfaces, which is an inherently non-linear problem. Any possible sliding is assumed to obey Coulomb's law of friction, and penetration between contacting areas is not allowed. The non-linear dynamic problem is discussed for the two-dimensional model and the corresponding quasi-static for the three-dimensional model. Both problems are solved utilizing incremental iterative procedures. For completeness reasons, the contact analysis in three dimensions and the formulation and solution of the non-linear dynamic problem are presented below.

a. *Two-dimensional model*

Figure 1 illustrates a two-dimensional straight cantilever beam with a rectangular cross-section $b \times h$ and length L . A breathing crack of depth a exists at position L_c. The crack is located at the upper edge of the beam and forms an angle θ with respect to the $x-$ axis of the global coordinate system $x,\ y$. An impulsive load is applied transversally at point A (Figure 1).

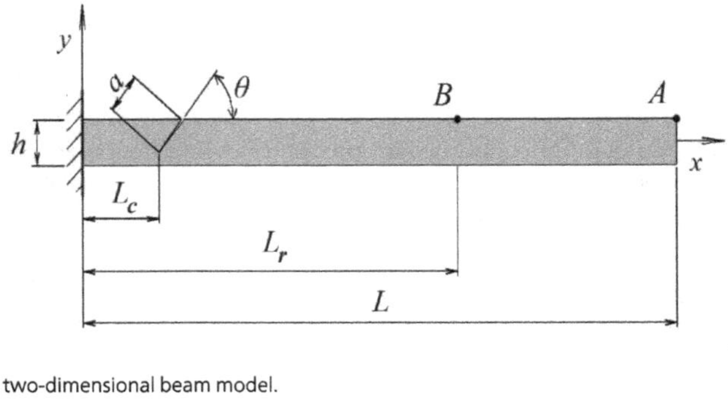

Figure 1. Cracked two-dimensional beam model.

In the finite element method (FEM) framework, the equilibrium equation governing the dynamic behavior of the model is:

$$M\ddot{U} + C\dot{U} + KU = R \tag{1}$$

where M , C , and K are the mass, damping, and stiffness matrices, respectively. The time-dependent vectors $\ddot{U}$, $\dot{U}$, U , and R denote the nodal accelerations, velocities, displacements, and external forces, respectively, in terms of a global Cartesian coordinate system x, y .

b. *Three-dimensional model*

Figure 2 depicts a three-dimensional cantilever beam with length $2L$ and circular cross-section of radius R . A breathing crack of depth a exists at the middle of the beam. The crack has either straight or curved front (Figures 2b and 2c). The slant crack forms angles θ_y , θ_z with respect to (x, y) and (x, z) planes, respectively. Two different load cases are separately applied at the tip of the cantilever beam, i.e., twisting moment T and bending moment M , respectively. The bending moment is applied in several aperture angles $M = M(\varphi)$, in order to simulate a rotating load on a fixed beam. The components $M_y = M_y(\varphi)$ and $M_z = M_z(\varphi)$ of the bending moment $M(\varphi)$, along the directions of axes y and z , respectively, are functions of the aperture angle.

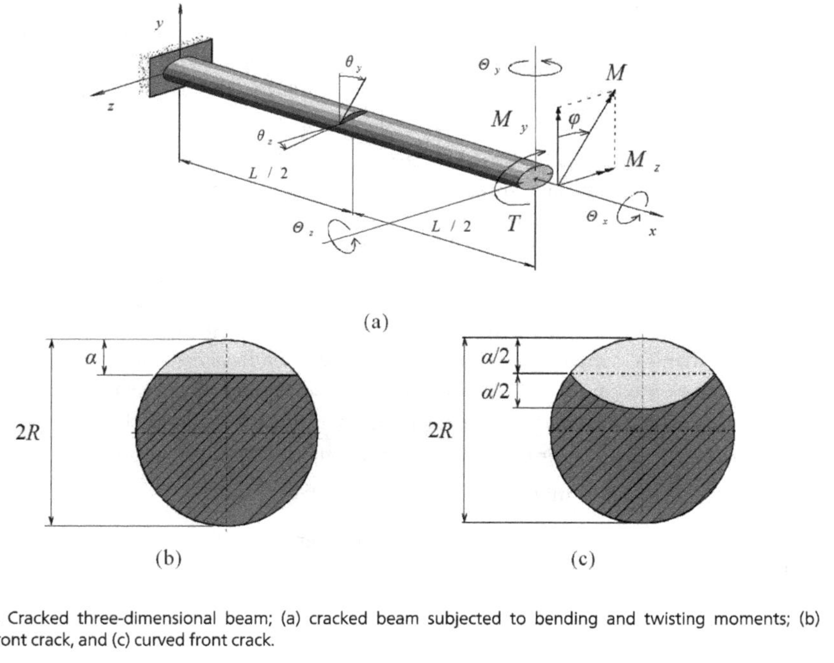

Figure 2. Cracked three-dimensional beam; (a) cracked beam subjected to bending and twisting moments; (b) straight-front crack, and (c) curved front crack.

The equilibrium equation governing the quasi-static behavior of the model is:

$$KU = R \tag{2a}$$

where $\boldsymbol{K}$ is the stiffness matrix, $\boldsymbol{U}$ denotes the nodal displacements and $\boldsymbol{R}$ represent the external forces, respectively, in terms of a global Cartesian coordinate system $x,\ y,\ z$.

The equilibrium equation governing the quasi-static behavior of the model is:

$$KU = R \tag{2b}$$

where $\boldsymbol{K}$ is the stiffness matrix, $\boldsymbol{U}$ denotes the nodal displacements and $\boldsymbol{R}$ represent the external forces, respectively, in terms of a global Cartesian coordinate system $x,\ y,\ z$.

c. *Crack modelling*

Considering the three-dimensional model, the crack is composed of two surfaces, which intersect on the crack front. Parts of these two surfaces may come into contact on an interface. The size of the interface can vary during the interaction between the load and the structure, but the interface is usually comprised of two parts, i.e., an adhesive part and a slipping part, depending on the friction conditions maintained between the contacting surfaces. In the open crack state, the corresponding part of the crack surface is subjected to traction-free conditions. The so-called slave–master concept that is widely used for the implementation of contact analysis is adopted in this work for prediction of the crack-surface interference. One of the two crack surfaces is considered as the master surface, with the other as the slave. Both master and slave crack surfaces are defined by the local coordinate systems $({}^{J}x_1,\ {}^{J}x_2,\ {}^{J}x_3)$, with J=I for master surface and J=II for slave surface. The axes ${}^{J}x_3$ define the direction of the unit outward normal vector of the corresponding surfaces. The nodes that belong to the master and slave surfaces are called the master and slave nodes, respectively. Contact segments that span master nodes cover the contact surface of the structure. Therefore, the above problem can be regarded as contact between a slave node and a point on a master segment. This point may be located at a node, an edge, or a point of a master segment. A slave node makes contact with only one point on the master segment, but one master segment can make contact with one or more slave nodes at each time. For each contact pair, the mechanical contact conditions are expressed in a local coordinate system in the direction of the average normal to the boundaries of the bodies. Symbols u_i and R_i , i=1, 2, 3 denote nodal displacement and force components, respectively, defined on the local coordinate systems $({}^{J}x_1,\ {}^{J}x_2,\ {}^{J}x_3)$, j=I, II . The subscripts that indicate nodal numbers are dropped for simplicity from this point forward.

Recalling the equilibrium condition, the force between the components is always expressed by the following equations:

$${}^{I}R_i + {}^{II}R_i = 0,\ i = 1,\ 2,\ 3 \tag{3}$$

In the open crack state, the following traction-free conditions are held between the components:

$$ {}^{I}R_{i} = {}^{II}R_{i} = 0, \; i = 1, 2, 3 \tag{4} $$

From the definition of adhesion, the displacement components on the corresponding crack surfaces are interconnected by the equations:

$$ {}^{I}u_{i} + {}^{II}u_{i} = 0, \; i = 1, 2 \tag{5} $$

When an initial gap g^{0} exists in the normal direction between the master and slave nodes of the corresponding node pair, the displacement component along the normal direction is:

$$ {}^{I}u_{3} + {}^{II}u_{3} = g^{0} \tag{6} $$

The slip state does not prohibit the existence of a gap between the crack surfaces, so equation (5) is still valid in this case. However, the tangential force component is defined in terms of friction as:

$$ {}^{I}R_{i} \pm \mu\, {}^{I}R_{3} = 0, \; i = 1, 2 \tag{7} $$

where μ is the coefficient of Coulomb friction. Concerning the corresponding two-dimensional contact analysis, the aforementioned approach is straightforwardly used neglecting one of the three directions.

d. *Incremental iterative procedure*

The simulation of the breathing crack behavior as a full frictional contact problem constitutes the present study as a non-linear dynamic problem. In the FEM framework, a non-linear dynamic problem described by an equation such equation (1) is solved using an implicit direct integration scheme [20]. According to this method, the solution time interval of interest $[0, T]$ is subdivided into N equal time increments Δt, where $\Delta t = T / N$. The variation of accelerations, velocities, and displacements within the time increment has a certain form and depends on the type of time integration scheme. Approximate solutions of equation (1) are sought at times $0, \Delta t, 2\Delta t, \ldots, t, t + \Delta t, \ldots, T$. The calculations performed to obtain the solution at time $t + \Delta t$ require that the solutions at previous times $0, \Delta t, 2\Delta t, \ldots, t$ are known. The initial conditions of accelerations, velocities, and displacements at time zero are also required. Thus, equation (1) is evaluated at time $t + \Delta t$ as:

$$ \boldsymbol{M}\,{}^{t+\Delta t}\ddot{\boldsymbol{U}} + \boldsymbol{C}\,{}^{t+\Delta t}\dot{\boldsymbol{U}} + \boldsymbol{K}\,{}^{t+\Delta t}\boldsymbol{U} = {}^{t+\Delta t}\boldsymbol{R} \tag{8} $$

where the left-hand subscripts denote the time.

The solution to this non-linear problem requires an iterative procedure. Employing the modified Newton-Raphson iteration method [20], the displacement vector ${}^{t+\Delta t}\boldsymbol{U}_{(k)}$ at time $t+\Delta t$ and iteration k is given by:

$$ {}^{t+\Delta t}\boldsymbol{U}_{(k)} = {}^{t+\Delta t}\boldsymbol{U}_{(k-1)} + \Delta\boldsymbol{U}^{(k)} \tag{9}$$

while equations (8) are written as:

$$ \boldsymbol{M}\,{}^{t+\Delta t}\ddot{\boldsymbol{U}}_{(k)} + \boldsymbol{C}\,{}^{t+\Delta t}\dot{\boldsymbol{U}}_{(k)} + {}^{t}\boldsymbol{K}_{T}\Delta\boldsymbol{U}^{(k)} = {}^{t+\Delta t}\boldsymbol{R} - {}^{t+\Delta t}\boldsymbol{F}_{(k-1)} \tag{10}$$

where the right-hand subscripts in brackets represent the iteration number, with $k=1, 2, 3, \ldots$. The symbols ${}^{t}\boldsymbol{K}_{T}$, ${}^{t+\Delta t}\boldsymbol{F}_{(k-1)}$ and $\Delta\boldsymbol{U}^{(k)}$ denote the tangent stiffness matrix, the nodal force vector, which is equivalent to the element stresses, and the incremental nodal displacement vector, respectively. The iterative method is called the modified Newton–Raphson method, since the tangent stiffness matrix is not calculated in every iteration, which is the case for the full method [20].

Employing for example an implicit time integration scheme, formulas are implemented that relate the nodal accelerations, velocities, and displacement vectors at time $t+\Delta t$ to those at previous times. Considering these formulas, equation (10) can be written in the following form [20]:

$$ {}^{t}\hat{\boldsymbol{K}}_{T}\Delta\boldsymbol{U}^{(k)} = \Delta\hat{\boldsymbol{R}}^{(k-1)} \tag{11}$$

where the matrix ${}^{t}\hat{\boldsymbol{K}}_{T}$ is a function of the tangent stiffness matrix, mass matrix, and damping matrix, while the vector $\Delta\hat{\boldsymbol{R}}^{(k-1)}$ contains the nodal force vector and contributions from the inertia and damping of the system. In the first iteration, the vectors $\Delta\hat{\boldsymbol{R}}^{(0)}$ and ${}^{t+\Delta t}\boldsymbol{U}_{(0)}$ are equal to the corresponding vectors of the last iteration at the previous time. In each iteration, the latest estimates of displacements are used to evaluate the vector $\Delta\hat{\boldsymbol{R}}^{(k-1)}$. Then, the incremental displacements $\Delta\boldsymbol{U}^{(k)}$ are obtained by solving equations (11), while the nodal displacements ${}^{t+\Delta t}\boldsymbol{U}_{(k)}$ are derived from equations (9). The iteration proceeds until the nodal displacements vector of the last iteration ${}^{t+\Delta t}\boldsymbol{U}_{(k)}$ are approximately equal to the corresponding vector of the previous iteration ${}^{t+\Delta t}\boldsymbol{U}_{(k-1)}$. This convergence criterion is expressed as

$$ \left| \frac{{}^{t+\Delta t}\boldsymbol{U}_{(k)} - {}^{t+\Delta t}\boldsymbol{U}_{(k-1)}}{{}^{t+\Delta t}\boldsymbol{U}_{(k)}} \right| \le \varepsilon \tag{12}$$

where ε is a small numerical quantity.

Considering that the problem has been solved for time t, and consequently, vectors ${}^{t}\boldsymbol{U}$ and ${}^{t}\boldsymbol{R}$ are known for the entire structure. To determine the corresponding displacement and forces vectors at time $t+\Delta t$, the equations (3)-(7) are written in incremental form as following:

$$^{t+\Delta t}({}^{\mathrm{I}}\Delta R_i) + {}^{t+\Delta t}({}^{\mathrm{II}}\Delta R_i) = 0,\ i=1,\ 2 \tag{13}$$

$$^{t+\Delta t}({}^{\mathrm{I}}\Delta R_i) = -{}^{t}({}^{\mathrm{II}}R_i),\ i=1,\ 2 \tag{14}$$

$$^{t}({}^{\mathrm{I}}u_1) + {}^{t+\Delta t}({}^{\mathrm{I}}\Delta u_1) = {}^{t}({}^{\mathrm{II}}u_1) + {}^{t+\Delta t}({}^{\mathrm{II}}\Delta u_1) \tag{15}$$

$$^{t}({}^{\mathrm{I}}u_2) + {}^{t+\Delta t}({}^{\mathrm{I}}\Delta u_2) = {}^{t}({}^{\mathrm{II}}u_2) + {}^{t+\Delta t}({}^{\mathrm{II}}\Delta u_2) - g^0 \tag{16}$$

$$^{t}({}^{\mathrm{I}}R_1) + {}^{t+\Delta t}({}^{\mathrm{I}}\Delta R_1) = \pm\mu({}^{t}({}^{\mathrm{I}}R_2) + {}^{t+\Delta t}({}^{\mathrm{I}}\Delta R_2)) \tag{17}$$

Assumption	Decision	
	Open	**Contact**
Open	${}^{\mathrm{II}}\Delta u_3m - {}^{\mathrm{I}}\Delta u_3m\,{}^{\mathrm{I}}u_{3m-1} - {}^{\mathrm{II}}u_{3m-1} + g^0$	${}^{\mathrm{II}}\Delta u_3m - {}^{\mathrm{I}}\Delta u_3m \le {}^{\mathrm{I}}u_{3m-1} - {}^{\mathrm{II}}u_{3m-1} + g^0$
Contact	${}^{\mathrm{I}}f_{3m-1} + {}^{\mathrm{I}}\Delta f_3m \ge 0$	${}^{\mathrm{I}}f_{3m-1} + {}^{\mathrm{I}}\Delta f_3m0$
	Adhesion	Slip
Adhesion	$\left\|{}^{\mathrm{I}}f_{im-1} + {}^{\mathrm{I}}\Delta f_im\right\| \left\|\mu({}^{\mathrm{I}}f_{3m-1} + {}^{\mathrm{I}}\Delta f_3m)\right\|, i=1,\ 2$	$\left\|{}^{\mathrm{I}}f_{im-1} + {}^{\mathrm{I}}\Delta f_im\right\| \ge \left\|\mu({}^{\mathrm{I}}f_{3m-1} + {}^{\mathrm{I}}\Delta f_3m)\right\|, i=1,\ 2$
Slip	$({}^{\mathrm{I}}f_{im-1} + {}^{\mathrm{I}}\Delta f_im)({}^{\mathrm{I}}\Delta f_im - {}^{\mathrm{II}}\Delta f_im)0, i=1,\ 2$	$({}^{\mathrm{I}}f_{im-1} + {}^{\mathrm{I}}\Delta f_im)({}^{\mathrm{I}}\Delta f_im - {}^{\mathrm{II}}\Delta f_im) \le 0, i=1,\ 2$

Table 1. Definition of contact status.

For reasons of simplicity, the iteration number has been omitted from equations (13)-(17). However, the formulation given below is repeated for all iterations. These equations are transformed to the global Cartesian coordinate system x, y, z and are then embedded and rearranged into equation (11). To determine the corresponding nodal displacements at time $t+\Delta t$, the contact conditions must first be satisfied. Therefore, the iterative procedure employed must be applied by initially using the convergent contact status (union of the adhesive, slipping and open parts of the crack surface) of the previous time t. The procedure initially assumes that the coplanar and normal incremental force components for a master surface at time $t+\Delta t$ are zero. Accurate values of the incremental forces can be estimated via the iterative procedure. The contact state for every node pair is examined according to Table 1. This table describes criteria to check whether violations involving geometrical compatibility and force continuity have occurred. Where necessary, appropriate changes from open to contact or from adhesion to slip states and vice versa are made to identify the equilibrium state of the contact conditions. The new contact condition is applied to the node pair closest to the change. If the change is from the open to the contact state, then the adhesion condition is adjusted. When the iterative procedure converges, the incremental nodal values ${}^{t+\Delta t}\Delta \boldsymbol{U}$ and ${}^{t+\Delta t}\Delta \boldsymbol{R}$ are known for the entire structure. After calculating the total nodal values, the procedure goes to the next step of the time increment and continues until the final time increment is reached. The problem solution is then attained.

3. Local flexibilities in cracked beams

A crack introduces local flexibilities in the stiffness of the structure due to strain energy concentration. Although local flexibilities representing the fracture in stationary structures are constant for open cracks, the breathing mechanism causes their time dependence. In a fixed direction, local flexibilities of rotating beams change also with time due to the breathing mechanism. Evidently, the vibrational response of a rotating cracked beam depends on the crack opening and closing pattern in one cycle.

Since the torsional and bending vibrations are dominant in rotating beams, in this chapter it is assumed that the corresponding local flexibilities are also dominant in the local flexibility matrix, neglecting the cross-coupling terms. Numerical results showed that the off-diagonal coefficients of this matrix are at least two orders of magnitude lower than the diagonal ones, and thus are considered negligible. Therefore, the presence of the crack can equivalently represented by a diagonal local flexibility matrix, independent of the crack contact conditions.

The exact relationship between the fracture characteristics and the induced local flexibilities is difficult to be determined by the strain energy approach, because, stress intensity factor expressions for this complex geometry are not available. For the computation of the local crack compliance, a finite element method was used. According to the point load displacement method, if at some node preferable lying on the tip of the beam is applied the external load vector $\{Q\}=\{T \;\; M_y \;\; M_z\}^T$, then at the same node the resulting rotations $\{\Theta\}=\{\Theta_x \;\; \Theta_y \;\; \Theta_z\}^T$ are:

$$\begin{bmatrix} c_x & 0 & 0 \\ 0 & c_y & 0 \\ 0 & 0 & c_z \end{bmatrix} \begin{Bmatrix} T \\ M_y \\ M_z \end{Bmatrix} = \begin{Bmatrix} \Theta_x \\ \Theta_y \\ \Theta_z \end{Bmatrix} \tag{18}$$

In equation (18), c_r , $r=x,\ y,\ z$ are the diagonal coefficients of the local flexibility matrix. Under the application of a particular load component Q_r at the r - direction, the above equation is then simplified as follows:

$$\Theta_r = c_r Q_r \tag{19}$$

where Θ_r , $r = x,\ y,\ z$ is the induced rotation, and c_r is the local flexibility component that corresponds to the particular loading mode Q_r . Equation (19) can be used for the computation of the local flexibility coefficients, as described in the following. When the original non-cracked beam is uploaded until the load Q_{ro} , a rotation is imposed in r - direction, such that equation (19) gives:

$$q_{ro} = c_{ro} Q_{ro} \tag{20}$$

where c_{ro} is the flexibility of the original structure. The deformation of the cracked beam when loaded at the same node gives:

$$\Theta_{rc} = c_{rc} Q_{rc} \tag{21}$$

where c_{rc} is the flexibility of the cracked beam, and Q_{rc} , Θ_{rc} the applied load and the resulting rotation, respectively. Between the flexibilities of the original and the fractured structure holds the condition

$$c_{rc} = c_r + c_{ro} \tag{22}$$

where c_r is the local flexibility due to the crack itself. Assuming that the applied load levels are of the same magnitude, i.e. $Q_{rc} = Q_{ro} = Q_r$, after some manipulation, equations (19)-(23) yield the local flexibility coefficient in the r - direction

$$c_r = \frac{\Theta_{rc} - \Theta_{ro}}{Q_r} \tag{23}$$

Equation (23) is used to compute the coefficients of the local flexibility matrix $[c]$ utilizing the FEM results. Tip loads Q_r , $r = x,\ y,\ z$ are applied independently, and the resulting rotations Θ_r are evaluated. The rotations of the original structure Q_{ro} , are evaluated for FEM models that do not present crack but have similar meshing with the cracked models. When fractured models are examined, FEM results are computed for several values of crack depth and different loading conditions.

4. Response analysis

The response derived from equations (9) and (11) cannot be examined directly to distinguish the breathing crack effects. For this reason, fast Fourier and continuous wavelet transforms are employed. These two popular transforms in signal analysis are briefly discussed below for reasons of completeness, and more information can be found in references [21, 22].

The fast Fourier transform (FFT) is a perfect tool for finding the frequency components in a signal of stationary nature. Unfortunately, FFT cannot show the time point at which a particular frequency component occurs. Therefore, FFT is not a suitable tool for a non-stationary signal, such as the impulsive response of the cracked cantilever beam considered in this study, which requires time-frequency representation. To overcome this FFT deficiency, the short time Fourier transform (STFT) could be adopted, which maps a signal into a two-dimensional function of time and frequency. This windowing technique analyzes only a small section of the signal at a time. However, the information about time and frequency that is obtained has a limited precision that is determined by the size of the window, which is the same for all frequencies.

Wavelet transforms are a novel and precise way to analyze signals and can overcome the problems that other signal transforms exhibit. The most important advantage of wavelet transformations is that they have changeable window dimensions. For low frequencies, the window is wide, while for high frequencies, it is narrow. Thus, maximum time frequency resolution is provided for all frequency intervals.

The continuous wavelet transform (CWT), as employed in this study, is defined mathematically as:

$$W f_{s,u} = \frac{1}{\sqrt{s}} \int_{-\infty}^{\infty} f(t) \psi^* \left(\frac{t-u}{s}\right) dt \tag{24}$$

where $f(t)$ is the signal for analysis, $\psi^*(t)$ is the complex conjugate of the mother wavelet $\psi(t)$ and s and u are real-valued parameters used to characterize the dilation and translation features of the wavelet.

The CWT has an inverse that permits recovery of the signal from its coefficient $W f_{s,u}$ and is defined as:

$$f(t) = \frac{1}{C_\psi} \int_{-\infty}^{\infty} \int_{-\infty}^{\infty} W f_{s,u} \psi\left(\frac{t-u}{s}\right) \frac{1}{s^2} ds du \tag{25}$$

5. Results and discussions

a. *Accuracy study*

i. *Two-dimensional model*

To demonstrate the accuracy of the presented study, a two-dimensional beam is considered with length $L = 1.5\text{m}$, cross-section $a/R = 1.0$, modulus of elasticity $E = 2.06 \times 10^{11}\text{Pa}$, mass density ρ=7650kg/m^3 and Poisson's ratio ν=0.29 . It is assumed that the beam contains a breathing crack of θ=90 , L_c/L =0.5 and a/h =0.5 . A transverse impulse loading is applied at point A of the beam (Figure 1). Based on the FFT of the transverse acceleration at point L_r/L =1 (Figure 1), the three lower dimensionless natural frequencies f_{ic}/f_{in} of the cantilever beam are evaluated and quoted at Table 2. Subscript i=1, 2, 3 denotes the order of natural frequency, while the subscript j represents the crack state of the beam. In particular, for $j=c$ the beam is cracked, while for $j=n$ the beam is non-cracked. It is derived from Table 2, that the results of the present study are generally close to the results of references [23] and [24]. It is noteworthy that the study of reference [23] has applied for a Timoshenko beam. Furthermore, the results from the work in [24] correspond to an internal crack of the same severity with the studied crack case. The internal crack constitutes a good approximation of a fully closed crack.

	f_{1c}/f_{1n}	f_{2c}/f_{2n}	f_{3c}/f_{3n}
Nandwana and Maiti [23]	1.000	0.992	1.014
Present study	0.986	0.992	0.998
% difference of natural frequency	1.41	-0.04	1.60
Kisa and Brandon [24]	0.980	0.925	0.999
Present study	0.986	0.992	0.998
% difference of natural frequency	-0.58	-6.71	0.08

Table 2. Three lower natural frequencies of the cantilever beam with a crack of θ=90°, L_c/L =0.5 and a/h =0.5 .

ii. *Three-dimensional model*

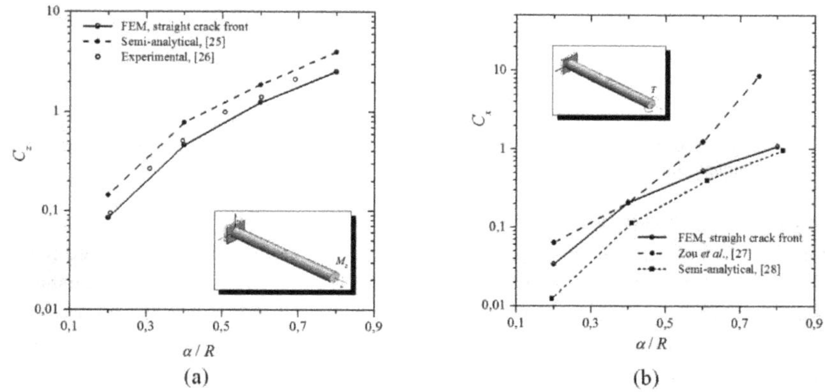

Figure 3. Flexibility of a transverse straight front crack under subjected to traction-free conditions; (a) bending flexibility coefficient C_z , and (b) torsional flexibility coefficient C_x .

To demonstrate further the accuracy of the presented study, a three-dimensional beam with length L =1.0m , radius R=0.05m , modulus of elasticity $E=2.06\times10^{11}$Pa , mass density $\rho=7650\text{kg/m}^3$ and Poisson's ratio $\nu=0.29$ is considered. The beam has a gaping straight front crack of various depths (Figure 3). The beam undergoes either pure bending M_z, i.e. $\phi=90^o$ or twisting moment T . For both loading cases, the dimensionless local flexibility coefficients, $C_{r=}(ER^3/(1-\nu^2))c_r$ with $r=x,\ y,\ z$, are calculated for various values of dimensionless crack depth a/h . The application of bending moment for this transversely fractured structure imposes the bending mode local flexibility C_z (Figure 3a), while the twisting moment imposes only the twisting coefficient is C_x (Figure 3b). As shown from Figures 3, the presented results are in good agreement with semi-analytical and experimental ones [25-27].

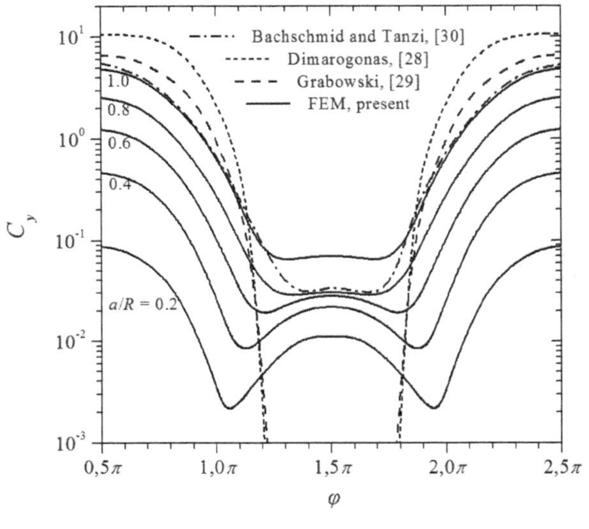

Figure 4. Dimensionless bending compliance variation of a transversely cracked beam as function of the aperture angle; (a) straight front crack, and (b) curved front crack.

Figure 4 presents the variation of the flexibility coefficient C_y when the transversely cracked beam is subjected to bending moment $M=M(\phi)$. The numerical results are plotted for a period that corresponds to one revolution of the beam. For better reasons of understanding, the local flexibility is plotted with phase lag $\pi/2$ that corresponds to the application only of the M_z component of bending moment or $\varphi=90^o$. That is, the evolution of the local flexibility is progressing from the full open condition of the crack. The same figure illustrates theoretical [28], experimental [29] and numerical [30] results from the literature for straight-fronted edge crack and crack depth $a/R=1.0$. As it is shown, the values of local flexibility for crack depth $a/R=1.0$, from FEM analysis are smaller than the other methods, when the crack is open. The experimental results are obtained for notched beam which is more flexible than the cracked beam. As the crack closes the FEM results are greater than the others. This is explained by the crack contact area (Figure 5), which differs and is generally smaller than the area assumed by the theoretical approaches for the same edge orientation. The impossibility of these approximations to predict the crack closure correctly is the main reason that these yield comparable results with the present ones only on the regions of partially opening portion of the crack, i.e., between $L_c/L = 0.5$ or $a/h = 0.75$. The variation of the coefficient C_y depends significantly on the crack depth. For small crack depths, the crack does not open regularly once per revolution, but contact is observed twice per revolution. The second contact state yields smaller compliance than the regular contact.

b. *Crack contact state*

Figure 5 illustrates predictions of the crack closure portion for a transverse crack with depth $a/R = 0.8$, when the three-dimensional beam is loaded in bending moment only, versus the aperture angle. Figure 5a shows the crack closure evolution for a straight front crack, and Figure 5b for a curved front crack, respectively. As the aperture angle increases, the crack closure portion increases in both cases examined. On full load reversal, a very small portion of the crack surface along the crack front remains always open. The shape of the contact surface and its portion clearly depends on the shape of the crack. This fact is expected to impose differences between the local flexibilities of different crack shapes. When the crack is slant, smaller portions of the crack surfaces are in contact.

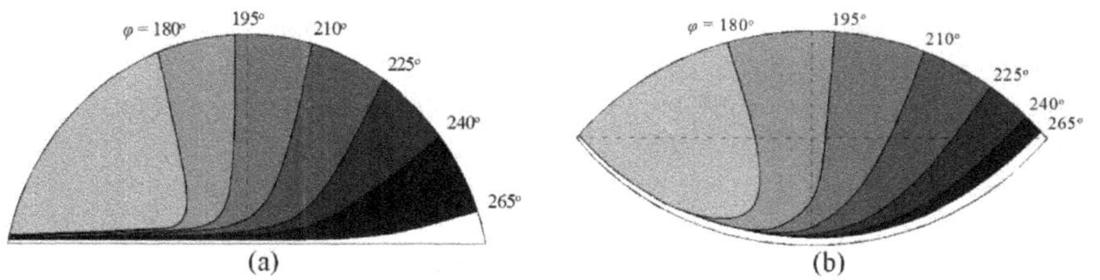

Figure 5. Evolution of contact area between the crack surfaces of a transverse crack under bending loading; (a) straight front crack, and (b) curved front crack.

Figure 6 depicts this situation for cracks having slope θ_z with respect to the z-axis ($\theta_y = 0^o$), loaded in bending M_z (or $\varphi = 90^o$). The crack slope θ_z seems to affect slightly the size and shape of the contact area between the crack surfaces for both of the crack fronts under investigation. The cases of crack orientation subjected to bending load that are not illustrated here present similar behavior with respect to the crack surface contact. For cracks subjected to twisting moment, the contact area does not generally change during the shaft rotation. Figure 7 depicts this fact when the crack slope is defined by the angles $\theta_y = 45^o$ and $\theta_z = 0^o$.

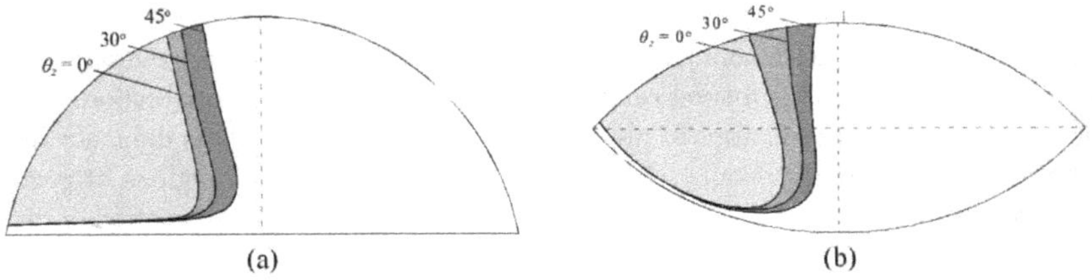

Figure 6. Contact area between crack surfaces for different crack slopes when the shaft is loaded in bending ($\theta_y=0$, $\varphi=90^o$); (a) straight front crack, and (b) curved front crack.

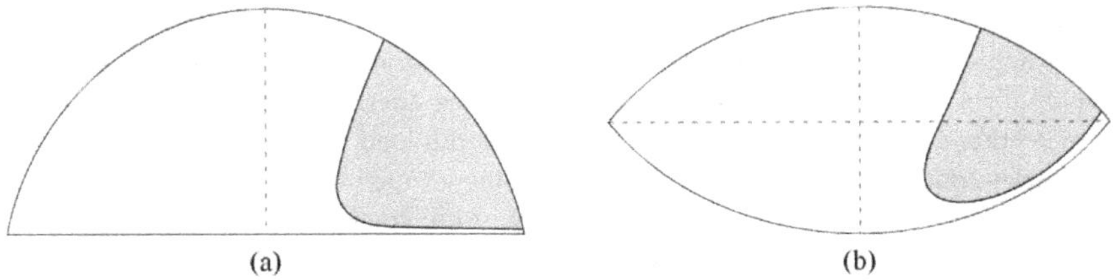

Figure 7. Contact area between crack surfaces when the shaft is loaded in torsion; (a) straight-fronted crack, and (b) penny shaped crack.

Further knowledge regarding the crack contact state can be also derived studying a strip under reversing loadings. Figure 8a illustrates the distinction between the open and partially closed crack configurations as defined by the crack opening displacements (CODs). In this figure the normalized displacements of the crack surface over the strip height *h*, close to the crack tip is shown. It is apparent that direct bending opens the crack, although load reversing causes partial crack closure. Solid lines represent crack opening mode and dashed lines partial crack closure, respectively. Negative values in this figure represent sign convention between the axes and not penetrating crack surfaces. Numerical results show that open crack displacements are two orders of magnitude higher than those corresponding to partially closed cracks. Figure 8b shows the transverse to crack line CODs, for two cracks with θ=135° and depths a/h =0.8 and a/h =1 , respectively. Under load reversing conditions and in the region close to crack tip, opening displacements develop, as in previous case. Crack orientation affects the deformation mode, yielding smaller displacements than those developed in transverse cracks. In the case of slant cracks, tangential stresses are shown to play a significant role in the crack surface interference process, producing small opening ligaments and destroying symmetry of displacements. CODs observed in slant cracks under load reversing are two orders of magnitude smaller than the corresponding displacements observed on normal loading inducing stress intensification.

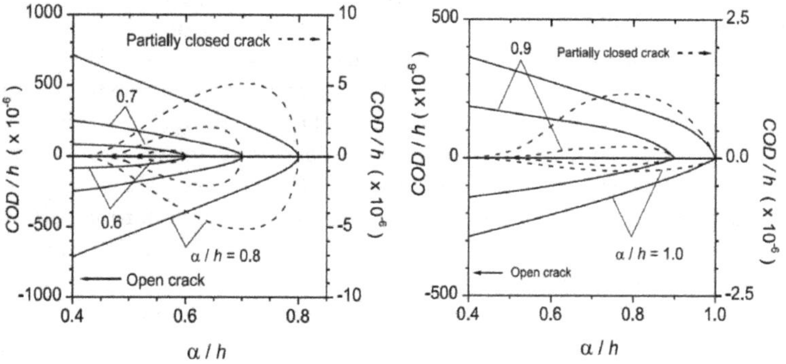

Figure 8. CODs for (a) transverse cracks, and (b) slant cracks.

c. *Time Response*

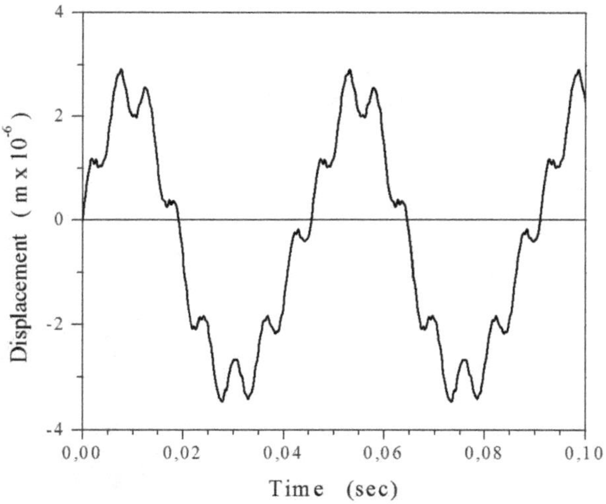

Figure 9. Time response when a crack is present at L_c / L =0.25 with depth a / h =0.75 and angle θ=90° .

Figure 9 depicts the impulsive displacement response in the transverse direction versus time t for the two-dimensional beam with a breathing crack of θ=90 , L_c / L =0.25 and a / h =0.75 . Consider a time period of response that spans from t=0sec to t=0.05sec . From the deformed mesh, it is assumed that during approximately the first half of the time period (from t=0sec to t=0.02sec) where the response sign is positive, the crack opens and closes continuously. On the contrary, during the second half of the period (from t=0.02sec to t=0.05sec) where the sign of the response is negative, the crack is open. For this period, the response is composed of a sinusoidal waveform that corresponds to that of the non-cracked beam, and several disturbances that correspond to lower frequencies than those of the non-cracked beam. For the first part of the indicated period (where the displacement is positive), these disturbances correspond to the existence of partial closure and contact along the crack surfaces, while the second part of the period (where the displacement is negative) corresponds to crack opening deformations. The same conclusions apply for the part of the response to the right of the considered time period. As it is deduced from Figure 9, the breathing of crack causes the beam to exhibit a highly nonlinear vibrational response. As a result, the portions of the beam on the left and right of the crack vibrate with different ways at the same time. In an attempt to clarify the intricacies involved in the response, Figure 10 illustrates the deformed shape and the detail of the corresponding crack opening deformation for a beam with either a small (a/h =0.25) or a deep (a/h =0.75) crack of angle θ=45 at position L_c/L =0.25 for time t=0.013sec at which the crack is instantaneously open. Apart from the instantaneous deformed shapes of the fractured beams, Figure 10a shows also the undeformed shape of the non-cracked beam (dotted lines) for comparison. The displacements in Figure 10a are magnified with a scale 10^4 , while in Figures 10b-10c are magnified with a scale 6×10^4 . It seems from Figure 10a that the instantaneous deformed shapes for the small and the deep crack approach mainly the fundamental and the second mode shapes of the non-cracked beam, respectively.

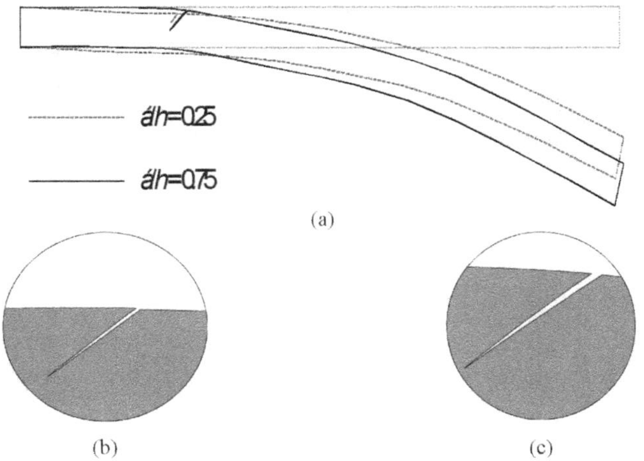

Figure 10. (a) Instantaneous deformed shape of a beam with an inclined crack (θ=45° , L_c/L =0.25) at time t=0.013sec and detail of the corresponding crack opening deformation for (b) a small (a/h =0.25) and (c) a deep (a/h =0.75) crack.

It is also visible from Figure 10a that for the deep crack the beam portion located on the left of the crack is nearly undeformed due to the dominance of the support conditions in the vibrational behavior of the beam at the considered time. Based on the Figures 10b-10c, it is observed that the distance between crack surfaces increases as the crack depth increase. This is reasonable since the small crack reduces slightly the local stiffness at crack position. This slight change of local stiffness matrix is time dependent and varies from zero to a maximum value depending on the breathing mechanism. On the contrary, for the deep crack the range of local stiffness change is wider than that of the small crack cases. For this reason, the effect of the crack is more important. The deformed shape of beam and the corresponding detail of the crack opening deformation for the two considered crack cases are also presented in Figure 11, which is magnified as Figure 10, for time t=0.023sec at which the crack is instantaneously closed. It seems from Figure11a that the instantaneous deformed shape for the small crack approaches mainly the fundamental mode of the non-cracked beam. In contrast, the instantaneous deformed shape for the deep crack approaches mainly the second mode shape of the non-cracked beam. It is noticeable from Figures 11b-11c that the crack surfaces are not completely closed due to the presence of the friction, which causes the development of shear stresses along crack surfaces. This stress component causes deformation of crack surfaces and develops a mechanism of energy absorption.

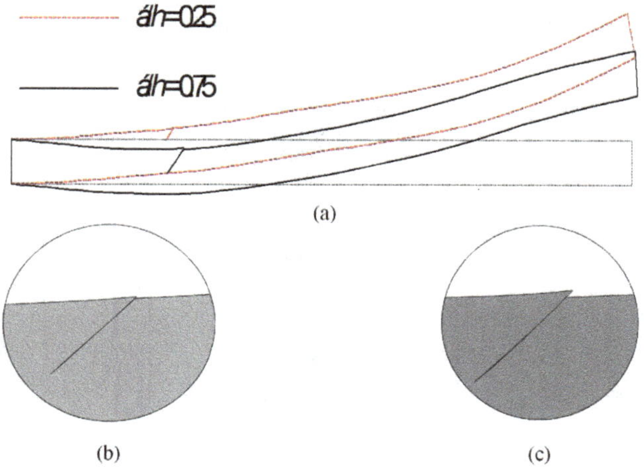

Figure 11. (a) Instantaneous deformed shape of a beam with an inclined crack (θ=45° , L_c/L =0.25) at time t=0.013sec and detail of the corresponding crack opening deformation for (b) a small (a/h =0.25) and (c) a deep (a/h =0.75) crack.

d. *Frequency Response*

Figure 12 depicts the FFT of the transverse acceleration response for the two-dimensional beam with a breathing crack of θ=30 , L_c/L =0.25 , and a/h =0.75 . The corresponding

FFT of transverse acceleration response with the crack always open is also plotted in this figure for comparison in frequency domain. For reasons of clarity, in this and the other figure of this subchapter, the corresponding response for the non-cracked beam is not plotted in this figure. The four vertical dash-dot lines represent the loci of the first four natural bending frequencies of the non-cracked beam. These natural frequencies are evaluated from the FFT of the transverse acceleration response and are 0.028, 0.174, 0.472 , and the 0.873kHz . The dash-dot-dot vertical lines in the figures represent the third natural bending frequency of the beam portion located on the right of the crack. For the present crack case, this portion of the beam has a length of $0.75L$, and its third natural bending frequency is 0.810kHz . This latter frequency is evaluated from the FFT of the transverse acceleration response. Figure 12 shows that the FFT of the open crack model exhibits five peaks at 0.026, 0.172, 0.448, 0.784 , and 0.850kHz . The fourth peak corresponds to the third natural bending frequency of the beam portion with length $0.75L$, while the remaining peaks correspond to the first four natural bending frequencies of the non-cracked beam. Corresponding peaks appear for the breathing crack model (Figure 12). The two FFTs illustrated in Figure 12 show that the first two frequencies of the open and breathing crack models are very close and are lower than those corresponding to the non-cracked beam. Shifts are observed for the remaining three frequencies. In particular, the third, fourth, and fifth frequencies of the breathing crack model are between the frequencies of the open cracked model and the non-cracked beam, as expected [8]. The absolute percentage differences for these three frequencies are 2.40% , 2.61% , and 0.93% , respectively.

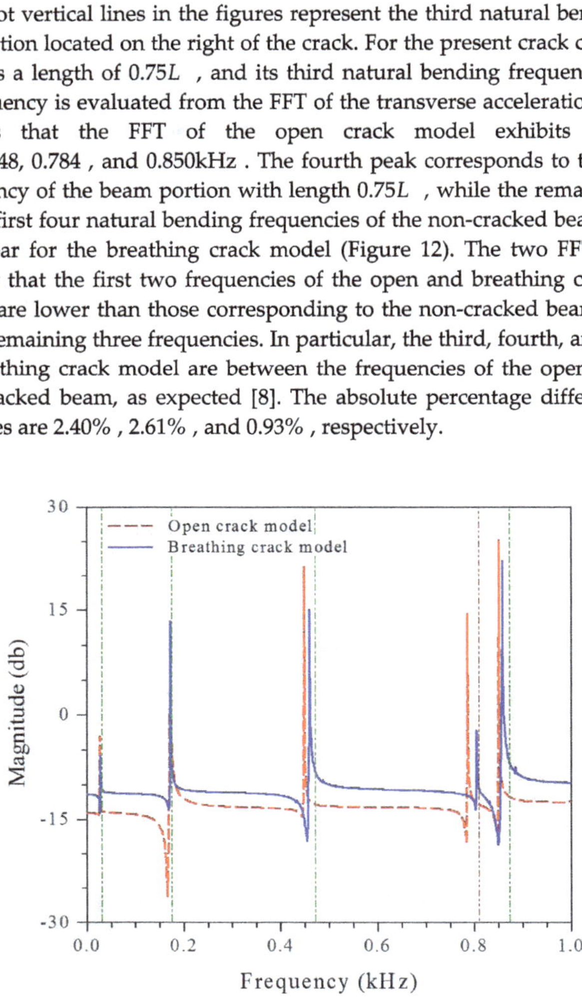

Figure 12. Frequency response of the transverse acceleration when a crack of depth $a/h = 0.25$ and angle $\theta = 90°$ is present at $L_c/L = 0.5$.

Figure 13 shows the sensitivity of beam vibrational behavior in terms of crack position. It seems that the crack position affects all but the first natural frequency. The absolute percentage differences are 12.87% , 8.53% , 7.29% , and 1.47% for the second through fifth natural frequencies, respectively.

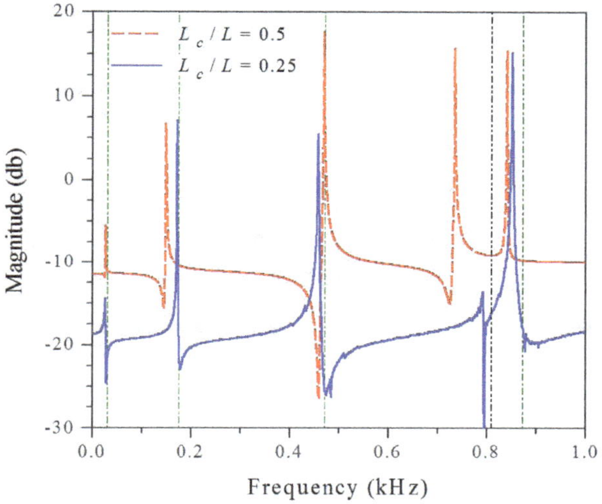

Figure 13. Frequency response of the transverse acceleration when a breathing crack of angle θ=45° and depth a/h =0.75 is present.

e. *Time-Frequency Response*

Figure 14a shows a contour map of the CWT for the transverse displacement response of the two-dimensional beam with a breathing crack of θ=90 , L_c/L =0.25 , and a/h =0.75 . In this and following contour map, the color at each point of the $s-u$ domain represents the magnitude of the wavelet coefficients $Wf_{s,u}$. A lighter color corresponds to larger coefficients, and a darker color to smaller ones. The symbol s denotes the scale space that is inversely proportional to the frequency domain, and u stands for the time space. The contour map of Figure 14a consists of three horizontal regions. The first region extends up to $s \approx 10$, the second one up to $s \approx 40$, and the third occupies the rest of the map. The third region is unchanged for $s128$. Each of these regions consists of a number of consecutive approximately equal vertical ridges. The number of ridges differs according to region.

Higher numbers of ridges appear in the first region while there are fewer in the third. A comparison of these three regions shows differences in the magnitude of the wavelet coefficients. Figure 14b shows the contour map of the CWT for the transverse displacement response for a fully open crack. This contour map consists of three visible horizontal regions. The first two regions are similar to those of the breathing crack model (Figure 14a) in respect to their extent and the number and extent of ridges. On the contrary, the magnitude of the wavelet coefficients appears differences. The third region consists of fewer and wider ridges than that of the breathing model (Figure 14a).

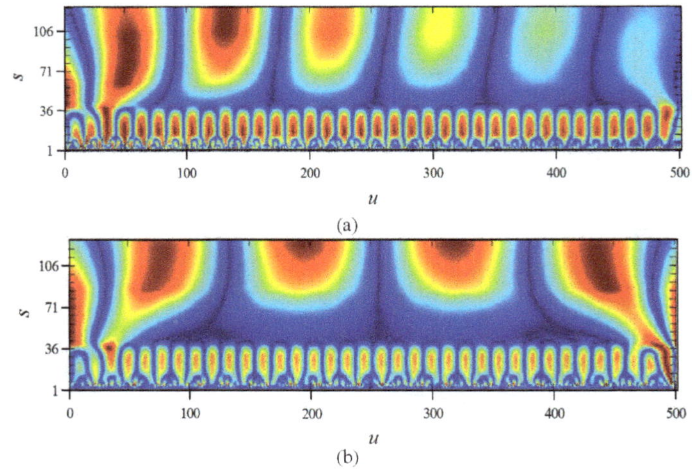

Figure 14. CWT of the transverse displacement history when (a) a breathing crack, (b) an open crack is present ($a/h = 0.75$, $\theta = 90°$, and $L_c/L = 0.25$).

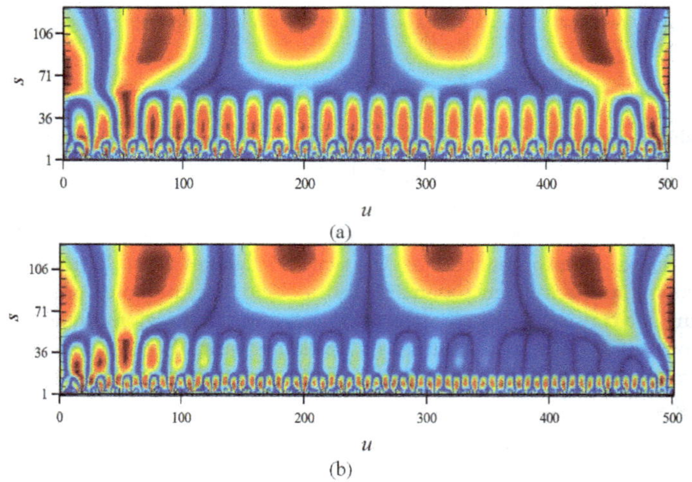

Figure 15. CWT of the transverse displacement history when a breathing crack of angle $\theta = 45°$ is present at $L_c/L = 0.5$; (a) $a/h = 0.25$, and (b) $a/h = 0.75$.

The higher number of ridges in the breathing model most likely represents changes to stiffness matrix due to the breathing of the crack. Similar conclusions regarding the magnitude of the wavelet coefficients, the extent of the regions, and the number and shape of ridges per region can be made from Figure 15, which shows the sensitivity analysis with respect to crack depth. Although, the effect of the crack depth in the morphology of the Figure 15 is visible, a qualitative analysis is required to correlate the natural frequencies obtained from the FFT to the extent of the horizontal regions and the number and the extent of ridges per region. Furthermore, the magnitude of the wavelet coefficients in the domain $s-u$ can be extracted. This analysis correlates the natural beam frequencies obtained by the FFT to the scales of contour maps. The magnitude of the wavelet coefficients at these scales can then be found from the contour maps for the entire time solution of interest, and conclusions for crack detection techniques can be extracted.

6. Conclusions

In this chapter, finite element procedures able to approach the vibrations of a beam with a breathing crack were presented. Although the developed models were discretized into a number of conventional finite elements, the breathing was treated as a full frictional contact problem between the crack surfaces. Quasi-static and non-linear dynamic analyses were performed aiming the prediction of vibration characteristics of cracked beams. The solutions were obtained using incremental iterative procedures. The results show good agreement with experimental or theoretical ones found in the open literature. The assesment of crack contact state in the different phases of the breathing mechanism gives comprehesive answers for the local flexibility variations and concequently for the vibrational response of a cracked beam. The derived time response of the two-dimensional beam was analyzed by various integral transforms including FFT and CWT. This model was assessed for the case of a cantilever beam subjected to an impulse loading. The results show the sensitivity of vibrational behavior with respect to crack characteristics. Although a qualitative analysis was required to interpret the results exactly, this study proved that FFT and CWT can be used as supplementary tools for crack detection techniques.

Author details

A. S. Bouboulas, S. K. Georgantzinos and N. K. Anifantis*

*Address all correspondence to: nanif@mech.upatras.gr

Machine Design Laboratory, Mechanical and Aeronautics Engineering Department, University of Patras, Greece

References

[1] Silva, J. M. M., & Gomes, A. J. M. A. (1990). Experimental dynamic analysis of cracked free-free beams. *Experimental Mechanics*, 30, 20-25.

[2] Doebling, S. W., Farrar, C. R., Prime, M. B., & Shevitz, D. W. (1998). A summary review of vibration based damage identification methods. *The Shock and Vibration Digest*, 30, 91-105.

[3] Christides, S., & Barr, A. D. S. (1984). One-dimensional theory of cracked Bernoulli-Euler beams. *International Journal of Mechanical Sciences*, 26, 639-648.

[4] Dimarogonas, A. D. (1976). *Vibration Engineering*, West Publishers, St Paul, Minesota.

[5] Chondros, T. G., & Dimarogonas, A. D. (1980). Identification of cracks in welded joints of complex structures. *Journal of Sound and Vibration*, 69, 531-538.

[6] Krawczuk, M., Żak, A., & Ostachowicz, W. (2000). Elastic beam finite element with a transverse elasto-plastic crack. *Finite Elements in Analysis and Design*, 34, 61-73.

[7] Bouboulas, A. S., & Anifantis, N. K. (2008). Formulation of cracked beam element for analysis of fractured skeletal structures. *Engineering Structures*, 30, 894-901.

[8] Gudmundson, P. (1983). The dynamic behavior of slender structures with cross-sectional cracks. *Journal of the Mechanics and Physics of Solids*, 31, 329-345.

[9] Cacciola, P., & Muscolino, G. (2002). Dynamic response of a rectangular beam with a known non-propagating crack of certain or uncertain depth. *Computers and Structures*, 80, 2387-2396.

[10] Benfratello, S., Cacciola, P., Impollonia, N., Masnata, A., & Muscolino, G. (2007). Numerical and experimental verification of a technique for locating a fatigue crack on beams vibrating under Gaussian excitation. *Engineering Fracture Mechanics*, 74, 2992-3001.

[11] Sholeh, K., Vafai, A., & Kaveh, A. (2007). Online detection of the breathing crack using an adaptive tracking technique. *Acta Mechanica*, 188, 139-154.

[12] Clark, R., Dover, W. D., & Bond, L. J. (1987). The effect of crack closure on the reliability of NDT predictions of crack size. *NDT International*, 20, 269-275.

[13] Abraham, O. N. L., & Brandon, J. A. (1995). The modelling of the opening and closure of a crack. *Journal of Vibration and Acoustics*, 117, 370-377.

[14] Douka, E., & Hadjileontiadis, L. J. (2005). Time-frequency analysis of the free vibration response of a beam with a breathing crack. *NDT&E International*, 38, 3-10.

[15] Loutridis, S., Douka, E., & Hadjileontiadis, L. J. (2005). Forced vibration behaviour and crack detection of cracked beams using instantaneous frequency. *NDT&E International*, 38, 411-419.

[16] Andrieux, S., & Varé, C. (2002). A 3D cracked beam model with unilateral contact. Application to rotors. *European Journal of Mechanics A/Solids*, 21, 793-810.

[17] Nandi, A., & Neogy, S. (2002). Modelling of a beam with a breathing edge crack and some observations for crack detection. *Journal of Vibration and Control*, 8, 673-693.

[18] Andreaus, U., Casini, P., & Vestroni, F. (2007). Nonlinear dynamics of a cracked cantilever beam under harmonic excitation. *International Journal of Nonlinear Mechanics*, 42, 566-575.

[19] Bachschmid, N., Tanzi, E., & Audebert, S. (2008). The effect of helicoidal cracks on the behavior of rotating shafts. *Engineering Fracture Mechanics*, 75, 475-488.

[20] Bathe, K. J. (1996). *Finite Element Procedures*, Prentice-Hall, Upper Saddle River, NJ.

[21] Brigham, E. O. (1973). *The Fast Fourier Transform: An Introduction to Its Theory and Application*, Englewood Cliffs, NJ, Prentice Hall.

[22] Rao, R. M., & Bopardikar, A. S. (1998). *Wavelet transforms- introduction to theory and applications*, Reading, MA, Addison Wesley Longman.

[23] Nandwana, B. P., & Maiti, S. K. (1997). Modelling of vibration of beam in presence of inclined edge or internal crack for its possible detection based on frequency measurements. *Engineering Fracture Mechanics*, 5, 193-205.

[24] Kisa, M., & Brandon, J. (2000). The effects of closure of cracks on the dynamics of a cracked cantilever beam. *Journal of Sound and Vibration*, 238, 1-18.

[25] Chasalevris, A. C., & Papadopoulos, C. A. (2006). Identification of multiple cracks in beams under bending. *Mechanical Systems and Signal Processing*, 20, 1631-1673.

[26] Bush, A. J. (1976). Experimentally determined stress-intensity factors for single-edge-crack round bars loaded in bending. *Experimental Mechanics*, 16-249.

[27] Zou, J., Chen, J., & Pu, Y. P. (2004). Wavelet time-frequency analysis of torsional vibrations in rotor system with a transverse crack. *Computers and Structures*, 82, 1181-1187.

[28] Dimarogonas, A. D., & Papadopoulos, C. A. (1983). Vibration of cracked shaft in bending. *Journal of Sound and Vibration*, 91, 583-593.

[29] Grabowski, B. (1979). The vibrational behavior of a turbine rotor containing a transverse crack. *ASME Design Engineering Technology Conference* [79-DET], Paper.

[30] Bachschmid, N., & Tanzi, E. (2004). Deflections and strains in cracked shafts due to rotating loads: a numerical and experimental analysis. *International Journal of Rotating Machinery*, 10, 283-291.

Chapter 10

A Simplified Analytical Method for High-Rise Buildings

Hideo Takabatake

Additional information is available at the end of the chapter

1. Introduction

High-rise buildings are constructed everywhere in the world. The height and size of high-rise buildings get larger and larger. The structural design of high-rise buildings depends on dynamic analysis for winds and earthquakes. Since today performance of computer progresses remarkably, almost structural designers use the software of computer for the structural design of high-rise buildings. Hence, after that the structural plane and outline of high-rise buildings are determined, the structural design of high-rise buildings which checks structural safety for the individual structural members is not necessary outstanding structural ability by the use of structural software on the market. However, it is not exaggeration to say that the performance of high-rise buildings is almost determined in the preliminary design stages which work on multifaceted examinations of the structural form and outline. The structural designer is necessary to gap exactly the whole picture in this stage. The static and dynamic structural behaviors of high-rise buildings are governed by the distributions of transverse shear stiffness and bending stiffness per each storey. Therefore, in the preliminary design stages of high-rise buildings a simple but accurate analytical method which reflects easily the structural stiffness on the whole situation is more suitable than an analytical method which each structural member is indispensable to calculate such as FEM.

There are many simplified analytical methods which are applicable for high-rise buildings. Since high-rise buildings are composed of many structural members, the main treatment for the simplification is to be replaced with a continuous simple structural member equivalent to the original structures. This equivalently replaced continuous member is the most suitable to use the one-dimensional rod theory.

Since the dynamic behavior of high-rise buildings is already stated to govern by the shear stiffness and bending stiffness determined from the structural property. The deformations of high-rise buildings are composed of the axial deformation, bending deformation, transverse shear deformation, shear-lag deformation, and torsional deformation. The problem is to be how to take account of these deformations under keeping the simplification.

There are many rod theories. The most simple rod theory is Bernonlli-Euler beam theory which may treat the bending deformation excluding the transverse shear deformation. The Bernonlli-Euler beam theory is unsuitable for the modeling of high-rise buildings.

The transverse resistance of the frame depends on the bending of each structural member consisted of the frame. Therefore, the transverse deformation always occurs corresponding to the transverse stiffness κGA. Since the transverse shear deformation is independent of the bending deformation of the one-dimensional rod, this shear deformation cannot neglect as for equivalent rod theory. This deformation behavior can be expressed by Timoshenko beam theory. Timoshenko beam theory may consider both the bending and the transverse shear deformation of high-rise buildings. The transverse deformation in Timoshenko beam theory is assumed to be linear distributed in the transverse cross section.

Usual high-rise buildings have the form of the three-dimensional structural frame. Therefore the structures produce the three dimensional behaviors. The representative dissimilarity which is differ from behavior of plane frames is to cause the shear-lag deformation. The shear-lag deformation is noticed in bending problem of box form composed of thin-walled closed section.

Reissner [1] presented a simplified beam theory including the effect of the shear-lag in the Bernonlli-Euler beam for bending problem of box form composed of thin-walled member. In this theory the shear-lag is considered only the flange of box form. This phenomenon appears in high-rise buildings the same as wing of aircrafts. Especially the shear-lag is remarkable in tube structures of high-rise buildings and occurs on the flange sides and web ones of the tube structures. The shear-lag occurs on all three-dimensional frame structures to a greater or lesser degree. Thus the one-dimensional rod theory which is applicable to analyze simply high-rise buildings is necessary to consider the longitudinal deformation, bending deformation, transverse shear deformation, shear-lag deformation, and torsional deformation. In generally, high-rise buildings have doubly symmetric structural forms from viewpoint the balance of facade and structural simplicity. Therefore the torsional deformation is considered to separate from the other deformations. Takabatake [2-6] presented a one-dimensional rod theory which can consider simply the above deformations. This theory is called the one-dimensional extended rod theory.

The previous works for continuous method are surveyed as follows: Beck [7] analyzed coupled shear walls by means of beam model. Heidenbrech et al. [8] indicated an approximate analysis of wall-frame structures and the equivalent stiffness for the equivalent beam. Dynamic analysis of coupled shear walls was studied by Tso et al. [9], Rutenberg [10, 11], Danay et al. [12], and Bause [13]. Cheung and Swaddiwudhipong [14] presented free vibration of frame shear wall structures. Coull et al. [15, 16] indicated simplified analyses of tube structures subjected to torsion and bending. Smith et al. [17, 18] proposed an approximate method for deflections and natural frequencies of tall buildings. However, the aforementioned continuous approaches have not been presented as a closed-form solution for tube structures with variable stiffness due to the variation of frame members and bracings. In this chapter high-rise buildings are expressed as tube structures in which three dimensional frame structures are included naturally.

2. Formulation of the one-dimensional extended rod theory for high-rise buildings

Frame tubes with braces and/or shear walls are replaced with an equivalent beam. Assuming that in-plane floor's stiffness is rigid, the individual deformations of outer and inner tubes in tube-in-tube are restricted. Hence, the difference between double tube and single tube depends on only the values of bending stiffness, transverse shear stiffness, and torsional stiffness. Therefore, for the sake of simplicity, consider a doubly symmetric single tube structure, as shown in Figure 1. Cartesian coordinate system, x, y, z is employed, in which the axis x takes the centroidal axis, and the transverse axes y and z take the principal axes of the tube structures. Since the lateral deformation and torsional deformation for a doubly symmetric tube structure are uncouple, the governing equations for these deformations can be formulated separately for simplicity.

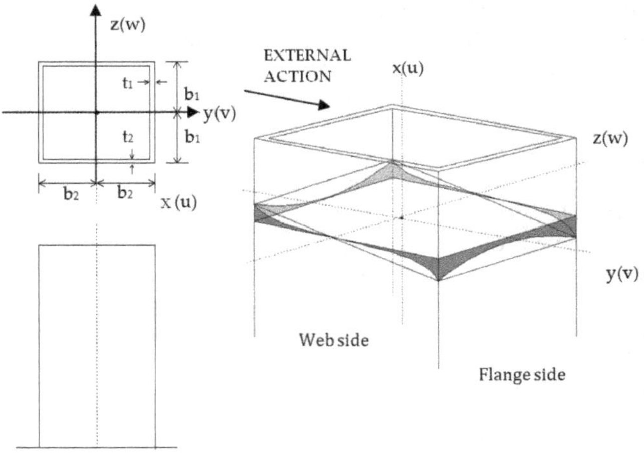

Figure 1. Doubly symmetric tube structure

2.1. Governing equations for lateral forces

Consider a motion of the tube structure subjected to lateral external forces such as winds and earthquakes acting in the y-direction, as shown in Figure 1. The deformation of the tube structures is composed of axial deformation, bending, transverse shear deformation, and shear-lag, in which the in-plane distortion of the cross section is neglected due to the in-plane stiffness of the slabs. The displacement composes $\overline{U}(x, y, z, t)$, $\overline{V}(x, y, z, t)$, and $\overline{W}(x, y, z, t)$ in the x-, y-, and z-directions on the middle surface of the tube structures as

$$\overline{U}(x, y, z, t)=u(x, t)+y\phi(x, t)+\varphi^*(y, z)u^*(x, t) \tag{1}$$

$$\overline{V}(x, y, z, t)=v(x, t) \tag{2}$$

$$\overline{W}(x, y, z, t)=0 \tag{3}$$

in which u and v= longitudinal and transverse displacement components in the x-and y-directions on the axial point, respectively; ϕ= rotational angle on the axial point along the z-axis; u^*= shear-lag coefficient in the flanges; $\varphi^*(x, y)$= shear-lag function indicating the distribution of shear-lag. These displacements and shear-lag coefficient are defined positive as the positive direction of the coordinate axes. However, the rotation is defined positive as counterclockwise along the z axis, as shown in Figure 2. The shear-lag function for the flange sections is used following function given by Reissner [1] and for the web sections sine distribution [5, 6] is assumed:

$$\varphi^*(y, z)=\pm[1-(\frac{z}{b_1})^2] \text{ for flange} \tag{4}$$

$$\varphi^*(y, z)=\sin\left(\frac{\pi y}{b_2}\right) \text{ for web} \tag{5}$$

in which the positive of ± takes for the flange being the positive value of the y-axis and vice versa b_1 and b_2 are hafe width of equivalent flange and web sections, as shown in Figure 1.

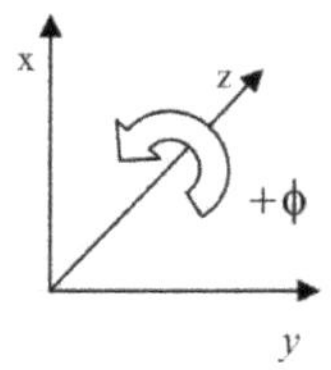

Figure 2. Positive direction of rotation ϕ

The governing equation of tube structures is proposed by means of the following Hamilton's principle.

$$\delta I = \delta \int_{t_1}^{t_2} (T - U - V) dt = 0 \tag{6}$$

in which T = the kinetic energy; U = the strain energy; V = the potential energy produced by the external loads; and δ = the variational operator taken during the indicated time interval.

Using linear relationship between strain and displacement, the following expressions are obtained.

$$\varepsilon_x = \frac{\partial \overline{U}}{\partial x} = u' + y\phi' + \varphi^* u^{*\prime} \tag{7}$$

$$\gamma_{xy} = \frac{\partial \overline{U}}{\partial y} + \frac{\partial \overline{V}}{\partial x} = \phi + \varphi^*_{,y} u^* + v' \tag{8}$$

$$\gamma_{xz} = \frac{\partial \overline{U}}{\partial z} + \frac{\partial \overline{W}}{\partial x} = \varphi^*_{,z} u^* \tag{9}$$

in which dashes indicate the differentiation with respect x and the differentiations with respect y and z are expressed as

$$\varphi^*_{,y} = \frac{\partial \varphi^*}{\partial y} \tag{10}$$

$$\varphi^*_{,z} = \frac{\partial \varphi^*}{\partial z} \tag{11}$$

The relationships between stress and strain are used well-known engineering expression for one-dimentional structural member of the frame structure.

$$\sigma_x = E \varepsilon_x \tag{12}$$

$$\tau_{xy} = G \gamma_{xy} \tag{13}$$

$$\tau_{xz} = G \tau_{xz} \tag{14}$$

in which E is Yound modulus and G shear modulus.

Assuming the above linear stress-strain relation, the strain energy U is given by

$$U = \frac{1}{2} \int_0^L \left[EA(u')^2 + EI(\phi')^2 + EI^*(u^{*\prime})^2 + \kappa G F^*(u^*)^2 + 2ES^*\phi' u^{*\prime} + \kappa GA(v' + \phi)^2 \right] dx \tag{15}$$

in which L = the total height of the tube structure; k = the shear coefficient; and A, I, I^*, S^*, and F^* = the sectional stiffnesses. These sectional stiffnesses vary discontinuously with respect to x for a variable tube structure and are defined as

$$A=\iint dydz=\sum A_c \tag{16}$$

$$I=\iint y^2 dydz \tag{17}$$

$$A^* = \iint \varphi^* dydz = \iint [\varphi_f^* + \varphi_w^*]dydz \tag{18}$$

$$I^* = \iint (\varphi^*)^2 dydz = 2t_2\int_{-b_1}^{b_1}(\varphi_f^*)^2 dz + 2t_1\int_{-b_2}^{b_2}(\varphi_w^*)^2 dy \tag{19}$$

$$S^* = \iint y\varphi^* dydz = 2t_2\int_{-b_1}^{b_1} b_2\varphi_f^* dz + 2t_2\int_{-b_2}^{b_2} y\varphi_w^* dy \tag{20}$$

$$F^* = \iint (\varphi_{,z}^*)^2 dydz + \iint (\varphi_{,y}^*)^2 dydz = \iint (\varphi_{f,z}^* + \varphi_{w,z}^*)^2 dydz + \iint (\varphi_{f,y}^* + \varphi_{w,y}^*)^2 dydz \tag{21}$$

in which $\sum A_c$ = the total cross-sectional area of columns per story.

The kinetic energy, T, for the time interval from t_1 to t_2 is

$$T = \int_{t_1}^{t_2}\left\{\frac{1}{2}\int_0^L [\rho A(\dot{u})^2 + \rho I(\dot{\phi})^2 + \rho I^*(\dot{u}^*)^2 + 2\rho S^*\dot{\phi}\dot{u}^* + \rho A(\dot{v})^2]dx\right\}dt \tag{22}$$

in which the dot indicates differentiation with respect to time and ρ = mass density of the tube structure. Now assuming that the variation of the displacements and rotation at $t=t_1$ and $t=t_2$ is negligible, the variation δT may be written as

$$\delta T = -\int_{t_1}^{t_2}\left\{\int_0^L [\rho A\ddot{u}\delta u + \rho I\ddot{\phi}\delta\phi + \rho I^*\ddot{u}^*\delta u^* + \rho S^*(\ddot{\phi}\delta u^* + \ddot{u}^*\delta\phi) + \rho A\ddot{v}\delta v]dx\right\}dt \tag{23}$$

When the external force at the boundary point (top for current problem) prescribed by the mechanical boundary condition is absent, the variation of the potential energy of the tube structures becomes

$$\delta V = -\int_0^L \left[\iint (p_x\delta\overline{U} + p_y\delta\overline{V})dydz\right]dx + \int_0^L (c_u\dot{u}\delta u + c_v\dot{v}\delta v)dx \tag{24}$$

in which p_x and p_y= components of external loads in the x- and y-directions per unit are, respectively; c_u , and c_v= damping coefficients for longtitudinal and transverse motions, respectively. The substitution of Eqs. (1) and (2) into Eq. (24) yields

$$\delta V = -\int_0^L \left(P_x \delta u + m\delta\phi + m^* \delta u^* + P_y \delta v - c_u \dot{u} \delta u - c_v \dot{v} \delta v\right) dx \tag{25}$$

in which P_x , P_y , m , and m^* are defined as

$$P_x = \iint p_x dydz \tag{26}$$

$$P_y = \iint p_y dydz \tag{27}$$

$$m = \iint p_x ydydz \tag{28}$$

$$m^* = \iint p_x \varphi^* dydz = 0 \tag{29}$$

Since for a doubly symmetric tube structure the distribution of the shear-lag function on the flange and web surfaces confronting each other with respect to z axis is asymmetric, m^* vanishes. Hence, Eq. (25) reduces to

$$\delta V = -\int_0^L (P_x \delta u + m\delta\phi + P_y \delta v - c_u \dot{u} \delta u - c_v \dot{v} \delta v) dx \tag{30}$$

Substituting Eqs. (15), (28), and (30) into Eq. (6), the differential equations of motion can be obtained

$$\delta u : \rho A\ddot{u} + c_u \dot{u} - (EAu')' - P_x = 0 \tag{31}$$

$$\delta v : \rho A\ddot{v} + c_v \dot{v} - [\kappa GA(v' + \phi)]' - P_y = 0 \tag{32}$$

$$\delta\phi : \rho I\ddot{\phi} + \rho S^* \ddot{u}^* - (EI\phi' + ES^* u^{*\prime})' + \kappa GA(v' + \phi) - m = 0 \tag{33}$$

$$\delta u^* : \rho I^* \ddot{u}^* + \rho S^* \ddot{\phi} - (EI^* u^{*\prime} + ES^* \phi')' + \kappa GF^* u^* = 0 \tag{34}$$

together with the associated boundary conditions at $x=0$ and $x=L$.

$$u \quad = 0 \text{ or } EAu' = 0 \tag{35}$$

$$v \;=0 \text{ or } \kappa\, GA(v' + \phi) = 0 \tag{36}$$

$$\phi \;=0 \; or \; EI\phi' + ES^{*}u^{*\prime} = 0 \tag{37}$$

$$u^{*} = 0 \text{ or } EI^{*}u^{*\prime} + ES^{*}\phi' = 0 \tag{38}$$

2.2. Governing equations for torsional moment

The displacement components for current tube structures subjected to torsional moments, m_x , around the x-axis are expressed by

$$\overline{U} = \overline{w}(y, z)\theta'(x, t) \tag{39}$$

$$\overline{V} = -z\theta \tag{40}$$

$$\overline{W} = y\theta \tag{41}$$

in which $\overline{W}$ = the displacement component in the z-direction on the tube structures; θ = torsional angle; and $\overline{w}(y, z)$ = warping function. Using the same manner as the aforementioned development, the differential equation of motion for current tube structures can be obtained

$$\delta\theta : \rho I_p \ddot{\theta} - (GJ\theta')' - m_x = 0 \tag{42}$$

together with the association boundary conditions

$$\theta = 0 \text{ or } GJ\theta' = m_{xL} \tag{43}$$

at $x = 0$ and L , in which GJ = the torsional stiffness.

2.3. Sectional constants

The sectional constants are defined by Eqs. (16) to (21). For doubly symmetric single-tube structures as shown in Figure 1, these sectional constants are simplified as follow.

$$A^{*} = 0 \tag{44}$$

$$I^{*} = \frac{8}{15}A_f + \frac{1}{2}A_w \tag{45}$$

$$S^* = \frac{2}{3} b_2 A_f + \frac{b_2}{\pi} A_w \tag{46}$$

$$F^* = \frac{4}{3(b_1)^2} A_f + \frac{\pi^2}{2b_2^2} A_w \tag{47}$$

in which A_f = the total cross-sectional area of columns in the flanges and webs, respectively, per story. For the cross-section of tube structures, as shown in Figure 1, A_f and A_w are given as $A_f = 4t_1 b_1$ and $A_w = 4t_2 b_2$.

2.4. Equivalent transverse shear stiffness κGA

When the tube structure are composed of frame and bracing, the equivalent transverse shear stiffness κGA for each story is given by

$$\kappa GA = \sum (\kappa GA)_{frame} + \sum (\kappa GA)_{brace} \tag{48}$$

in which $\sum$ is taken the summation of equivalent transverse shear stiffnesses of web frame and of braces per story. The shear stiffnesses of web frame and web double-brace $\sum (\kappa GA)_{frame}$ and $\sum (\kappa GA)_{brace}$ for each side of the web surfaces, respectively, are given by

$$\frac{1}{(\kappa GA)_{frame}} = \frac{h\left(\frac{1}{\sum K_c} + \frac{1}{\sum K_b}\right)}{12E} + \frac{1}{\sum_c (\kappa G A_{cw})} + \frac{h}{\sum_b (\ell \kappa G A_{bw})} \tag{49}$$

$$(\kappa GA)_{brace} = \frac{h}{\ell} k_{brace} A_B E_B \cos^2\theta_B \tag{50}$$

in which the first term on the right side of Eq. (49) indicates the deformation of the frame with the stiffnesses of columns and beams K_c and K_b , respectively; the second and third terms indicate the shear deformation of only the columns and beams in the current web-frame, respectively. A_{cw} and A_{bw} = the web's cross-sectional area of a column and of a beam, respectively. $\sum_c$ and $\sum_b$ = the sums of columns and beams, respectively, in a web-frame at the current story of the frame tube. If the shear deformations of columns and beams are neglected, these terms must vanish. Furthermore, ℓ = the span length; A_B = the cross-sectional area of a brace; E_B = the Young's modulus; and θ_B = the incline of the brace. The coefficient k_{brace} indicates the effective number of brace and takes $k_{brace} = 1$ for a brace resisting only tension and $k_{brace} = 2$ for two brace resisting tension and compression, as shown in Figure 3.

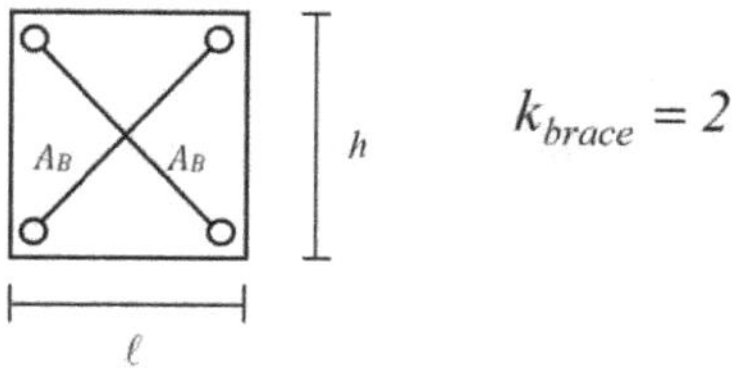

Figure 3. Brace resisting tension and compression

2.5. Equivalent bending stiffness *EI*

The equivalent bending stiffness EI for each storey is determined from the total sum of the moment of inertia about the z-axis of each column located on the storey.

$$EI = \sum_{i=1}^{n} E_i \left[I_{0i} + A_i e_i^2 \right] \tag{51}$$

in which E_i , I_{0_i} and A_i = Young modulus, moment of inertia and the cross section of the ith column; and e_i = the distance measured from the z-axis.

3. Static analysis by the finite defference method

3.1. Expression of static analysis

The governing equations for the one-dimensional extended rod theory are differential equations with variable coefficients due to the variation of structural members and forms in the longitudinal direction. Furthermore, although the equations of motion and boundary conditions for the vertical displacement u are uncoupled from the other displacement components, the governing equations take coupled from concerning variables v , ϕ , and u^* .

Takabatake [2, 3] presented the uncoupled equations as shown in section 7 by introducing positively appropriate approximations into the coupled equations and proposed a closed-form solution. For usual tube structures this method produces reasonable results. However the analytical approach deteriorates on the accuracy of numerical results for high-rise buildings with the rapid local variations of transverse shear stiffness and/or braces. Especially the difference appears on the distributions of not dynamic deflection but story acceleration and storey shear force. It is limit to express these rapid variations by a functional expression. So, the above governing equations are solved by means of the finite difference method.

The equations of motion and boundary conditions for the longitudinal displacement u are uncoupled from the other displacement components. So we consider only the lateral motion given by the governing equation coupled about the lateral displacements v , rotational angle ϕ , and shear-lag displacement u^* .

Using ordinary central finite differences, the finite difference expressions of the current equilibrium equations, obtained from the equations of motion Eqs. (32)-(34), may be written, respectively, as follows:

$$\left[-\frac{\kappa GA}{\Delta^2}+\frac{(\kappa GA)'}{2\Delta}\right]v_{i-1}+\frac{\kappa GA}{2\Delta}\phi_{i-1}+\frac{2\kappa GA}{\Delta^2}v_i -(\kappa GA)'\phi_i-\left[\frac{\kappa GA}{\Delta^2}+\frac{(\kappa GA)'}{2\Delta}\right]v_{i+1}-\frac{\kappa GA}{2\Delta}\phi_{i+1}=P_{yi} \tag{52}$$

$$-\frac{\kappa GA}{2\Delta}v_{i-1}+\left[-\frac{EI}{\Delta^2}+\frac{(EI)'}{2\Delta}\right]\phi_{i-1}+\left[-\frac{ES^*}{\Delta^2}+\frac{(ES^*)'}{2\Delta}\right]u_{i-1}^*+\left(\frac{2EI}{\Delta^2}+\kappa GA\right)\phi_i +\frac{2ES^*}{\Delta^2}u_i^*+\frac{\kappa GA}{2\Delta}v_{i+1}+\left[-\frac{EI}{\Delta^2}+\frac{(EI)'}{2\Delta}\right]\phi_{i+1}+\left[-\frac{ES^*}{\Delta^2}-\frac{(ES^*)'}{2\Delta}\right]u_{i+1}^*=m_{zi} \tag{53}$$

$$\left[-\frac{ES^*}{\Delta^2}+\frac{(ES^*)'}{2\Delta}\right]\phi_{i-1}+\left[-\frac{EI^*}{\Delta^2}+\frac{(EI^*)'}{2\Delta}\right]u_{i-1}^*+\frac{2ES^*}{\Delta^2}\phi_i+\left(\frac{2EI^*}{\Delta^2}+\kappa GF^*\right)u_i^* +\left[-\frac{ES^*}{\Delta^2}-\frac{(ES^*)'}{2\Delta}\right]\phi_{i+1}+\left[-\frac{EI^*}{\Delta^2}-\frac{(EI^*)'}{2\Delta}\right]u_{i+1}^*=0 \tag{54}$$

in which Δ = the finite difference mesh; v_{i-1} , v_i , v_{i+1} ,... represent displacements at the (*i*-1)th, *i*th, and (*i*+1)th mesh points, respectively, as shown in Figure 4; and P_{yi} and m_{zi}= the lateral load and moment, respectively, at the *i*th mesh point. In the above equations, the rigidities κGA , EI ,... at the pivotal mesh point i are taken as the mean value of the rigidities of current prototype tube structures located in the mesh region, in which the mesh region is defined as each half height between the mesh point i and the adjoin mesh points, i-1 and i+1, namely from $(x_i+x_{i-1})/2$ to $(x_i+x_{i+1})/2$, as shown in Figure 5. Hence, the stiffness $k_{(i)}$ at a mesh point i is evaluated

$$k_{(i)}=\frac{a_{i1}k_{i1}+a_{i2}k_{i2}+\cdots+a_{in}k_{in}}{h_{(i)}} \tag{55}$$

in which a_{i1} , a_{i2} ,.., a_{in} and k_{i1} , k_{i2} ,..., k_{in}= the effective story heights and story rigidities, located in the mesh region, respectively; and $h_{(i)}$= the current mesh region for the pivotal mesh point i. The first mesh region in the vicinity of the base is defined as region from the base to the mid-height between the mesh points 1 and 2.

Now, the boundary conditions for a doubly-symmetric tube structure are assumed to be fixed at the base and free, except for the shear-lag, at the top. The shear-lag at the top is considered for two cases: free and constrained. Hence, from Eqs. (36) to (38)

$$v = 0 \text{ at } x = 0 \tag{56}$$

$$\phi = 0 \text{ at } x = 0 \tag{57}$$

$$u^* = 0 \text{ at } x = 0 \tag{58}$$

$$v' + \phi = 0 \text{ at } x = L \tag{59}$$

$$ES^* u^{*'} + EI\phi' = 0 \text{ at } x = L \tag{60}$$

$$EI^* u^{*'} + ES\phi' = 0 \text{ at } x = L \text{ (Shear−lag is free.)} \tag{61a}$$

$$u^* = 0 \text{ at } x = L \text{(Shear − lag is constraint.)} \tag{61b}$$

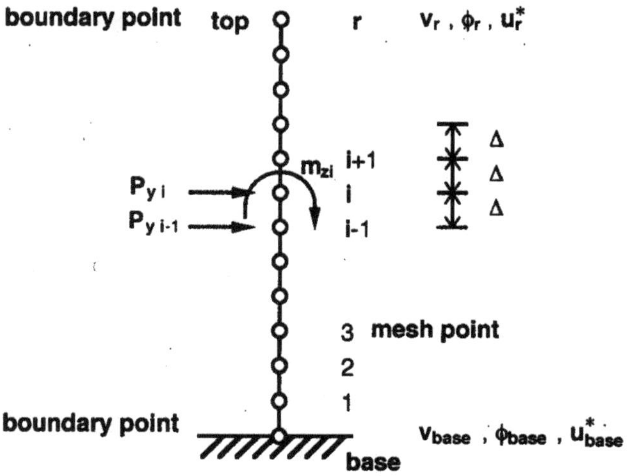

Figure 4. Mesh point in finite difference method [6]

Let us consider the finite difference expression for the boundary conditions (56) - (61). Since tube structures are replaced with an equivalent cantilever in the one-dimensional extended rod theory, the inner points for finite difference method take total numbers m as shown in Figure 6, in which the mesh point m locates on the boundary point at $x = L$.

Since the number of each boundary condition of the base and top for v , ϕ and u^* is one, respectively, the imaginary number of the boundary mesh in finite differences can be taken one for each displacement component at each boundary.

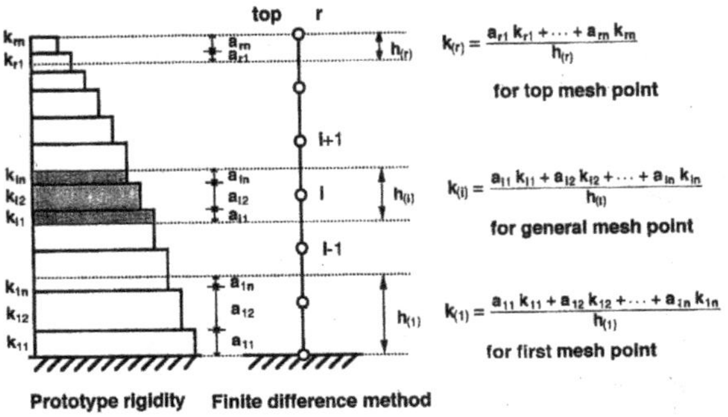

Figure 5. Equivalent rigidity in finite difference method [6]

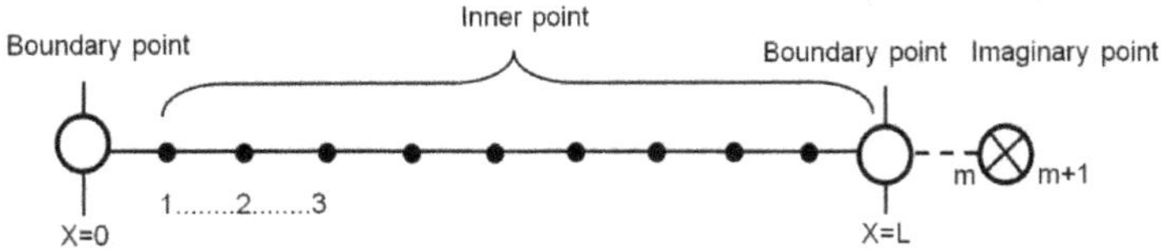

Figure 6. Inner points and imaginary point

The finite differences expressions for the boundary conditions (56)-(58) at the base ($x=0$) are

$$v_{base} = 0 \tag{62}$$

$$\phi_{base} = 0 \tag{63}$$

$$u^*_{base} = 0 \tag{64}$$

in which v_{base} , ϕ_{base} , and u^*_{base} represent quantities at the base.

On the other hand, using central difference method, the finite difference expressions for the boundary conditions (59), (60), and (61a), in case where the shear-lag is free at the top(x=L), are expressed as

$$[-v_{m-1}+v_{m+1}]\frac{1}{2\Delta}+\phi_m=0 \tag{65}$$

$$EI[-\phi_{m-1}+\phi_{m+1}]\frac{1}{2\Delta}+ES^*[-u^*_{m-1}+u^*_{m+1}]\frac{1}{2\Delta}=0 \tag{66}$$

$$ES^*[-\phi_{m-1}+\phi_{m+1}]\frac{1}{2\Delta}+EI^*[-u^*_{m-1}+u^*_{m+1}]\frac{1}{2\Delta}=0 \tag{67}$$

in which the mesh point m locates on the boundary point at the free end of $x=L$; the mesh point $m+1$ is imaginary point adjoining the mesh point m ; and the mesh point $m-1$ is inner point adjoining the mesh point m . Solving the above eqations for the variables v_{m+1} , ϕ_{m+1} , u^*_{m+1} at the imaginary point $m+1$, we have

$$v_{m+1}=v_{m-1}-2\Delta\cdot\phi_m \tag{68}$$

$$\phi_{m+1}=\phi_{m-1} \tag{69}$$

$$u^*_{m+1}=u^*_{m-1} \tag{70}$$

On the other hand, the finite difference expressions for boundary conditions (59), (60), and (61b), in case where the shear-lag is constraint at the top, use the central diference for v and ϕ but backward difference for u^*, because u^*_{m+1} is unsolvable in the use of the central difference.

$$[-v_{m-1}+v_{m+1}]\frac{1}{2\Delta}+\phi_m=0 \tag{71}$$

$$EI[-\phi_{m-1}+\phi_{m+1}]\frac{1}{2\Delta}+ES^*[-u^*_{m-1}+u^*_{m}]\frac{1}{\Delta}=0 \tag{72}$$

$$u^*_m=0 \tag{73}$$

Solving the above eqations for the variables v_{m+1} , ϕ_{m+1} , u^*_m , we have

$$v_{m+1}=v_{m-1}-2\Delta\cdot\phi_m \tag{74}$$

$$\phi_{m+1} = \phi_{m-1} + \frac{2ES^*}{EI} u^*_{m-1} \tag{75}$$

$$u^*_m = 0 \tag{76}$$

Static solutions are obtained by solving a system of linear, homogeneous, simultaneous algebraic equation (77) with respect to unknown displacement components at the internal mesh points. In finite difference method the equilibrium equations are formulated on each inner point from *1* to *m*.

$$\boldsymbol{A}\ \boldsymbol{v} = \boldsymbol{P} \tag{77}$$

in which the matrix **A** is the total stiffness matrix summed the individual stiffness matrix at each mesh point. **v** and **P** are the total displacement vector and total external load vector, respectively.

Figure 7 shows stencil of equilibrium equations at a general inner point *i*. Figure 8 shows stencil of equilibrium equations at inner point *1* adjoining the base. Figure 9 shows stencil of equilibrium equations at inner point *i=m* for the case that the shear-lag is free at the top. Figure 10 shows stencil of equilibrium equations at inner point *i=m* for the case that the shear-lag is constrained at the top.

3.2. Axial forces of columns

Let us consider the axial forces of columns. The axial stress σ_x of the tube structure is given from Eq. (12) by

$$\sigma_x = E\left[y\phi'(x, t) + \varphi^*(z)u^{*'}(x, t)\right] \tag{78}$$

Hence, the axial force N_i in the *i*th column with the column's sectional area A_i is

$$N_i = E\left\{y\phi' A_i \pm \left[z - \frac{z^3}{3b_z^2}\right]_{z1}^{z2} t_i u^{*'}\right\} \tag{79}$$

for columns in flange surfaces,

$$N_i = E\left[\frac{1}{2}(y_2 + y_1)A_i\phi'\right] \tag{80}$$

for columns in web surfaces, and

$$N_i = E\left\{ y\phi' A_i \pm \left[z - \frac{z^3}{3b_z^2} \right]_{z1}^{z2} t_i u^{*'} \right\} + E\left[\frac{1}{2}\left(y_2^2 - y_1^2\right)\phi' \right] t_i \qquad (81)$$

for corner columns, in which y_1 , y_1 and y_1 , z_2= lower and upper coordinate values of the half between the ith column and both adjacent columns, respectively, and t_i= the cross-sectional area A_i of the ith column divided by the sum of half spans between the ith column and the both adjacent columns.

	i-1			i			i+1				P
	v_{i-1}	ϕ_{i-1}	u^*_{i-1}	v_i	ϕ_i	u^*_i	v_{i+1}	ϕ_{i+1}	u^*_{i+1}		
δv	$-\frac{\kappa GA}{\Delta^2} + \frac{(\kappa GA)'}{2\Delta}$	$+\frac{\kappa GA}{2\Delta}$	0	$+\frac{2\kappa GA}{\Delta^2}$	$-(\kappa GA)'$	0	$-\frac{\kappa GA}{\Delta^2} - \frac{(\kappa GA)'}{2\Delta}$	$-\frac{\kappa GA}{2\Delta}$	0	=	P_{yi}
$\delta\phi$	$-\frac{\kappa GA}{2\Delta}$	$-\frac{EI}{\Delta^2} + \frac{(EI)'}{2\Delta}$	$-\frac{ES^*}{\Delta^2} + \frac{(ES^*)'}{2\Delta}$	0	$\frac{2EI}{\Delta^2} + \kappa GA$	$\frac{2ES^*}{\Delta^2}$	$\frac{\kappa GA}{2\Delta}$	$-\frac{EI}{\Delta^2} - \frac{(EI)'}{2\Delta}$	$-\frac{ES^*}{\Delta^2} - \frac{(ES^*)'}{2\Delta}$	=	m_i
δu^*	0	$-\frac{ES^*}{\Delta^2} + \frac{(ES^*)'}{2\Delta}$	$-\frac{EI^*}{\Delta^2} + \frac{(EI^*)'}{2\Delta}$	0	$\frac{2ES^*}{\Delta^2}$	$\frac{2EI^*}{\Delta^2} + \kappa GF^*$	0	$-\frac{ES^*}{\Delta^2} - \frac{(ES^*)'}{2\Delta}$	$-\frac{EI^*}{\Delta^2} - \frac{(EI^*)'}{2\Delta}$	=	0

Figure 7. Stencil of equilibrium equations at inner point *i*

	i-1 (0)			i (1)			i+1 (2)				P
	v_{i-1}	ϕ_{i-1}	u^*_{i-1}	v_i	ϕ_i	u^*_i	v_{i+1}	ϕ_{i+1}	u^*_{i+1}		
δv	$\frac{\kappa GA}{\Delta^2} + \frac{(\kappa GA)'}{2\Delta}$	$-\frac{\kappa GA}{2\Delta}$	0	$-\frac{2\kappa GA}{\Delta^2}$	$(\kappa GA)'$	0	$\frac{\kappa GA}{\Delta^2} - \frac{(\kappa GA)'}{2\Delta}$	$\frac{\kappa GA}{2\Delta}$	0	=	P_{yi}
$\delta\phi$	$-\frac{\kappa GA}{2\Delta}$	$-\frac{EI}{\Delta^2} + \frac{(EI)'}{2\Delta}$	$-\frac{ES^*}{\Delta^2} + \frac{(ES^*)'}{2\Delta}$	0	$\frac{2EI}{\Delta^2} + \kappa GA$	$\frac{2ES^*}{\Delta^2}$	$\frac{\kappa GA}{2\Delta}$	$-\frac{EI}{\Delta^2} - \frac{(EI^*)'}{2\Delta}$	$-\frac{ES^*}{\Delta^2} - \frac{(ES^*)'}{2\Delta}$	=	m_i
δu	0	$-\frac{ES^*}{\Delta^2} + \frac{(ES^*)'}{2\Delta}$	$-\frac{EI^*}{\Delta^2} + \frac{(EI^*)'}{2\Delta}$	0	$\frac{2ES^*}{\Delta^2}$	$\frac{2EI^*}{\Delta^2} + \kappa GF^*$	0	$-\frac{ES^*}{\Delta^2} - \frac{(ES^*)'}{2\Delta}$	$-\frac{EI^*}{\Delta^2} - \frac{(EI^*)'}{2\Delta}$	=	0

Figure 8. Stencil of equilibrium equations at inner point *1*

	$i-1\ (=m-1)$			$i\ (=m)$			$i+1\ (m+1)$				P
	v_{i-1}	ϕ_{i-1}	u^*_{i-1}	v_i	ϕ_i	u^*_i	v_{i+1}	ϕ_{i+1}	u^*_{i+1}		
δv	$-\frac{\kappa GA}{\Delta^2}+\frac{(\kappa GA)'}{2\Delta}+c17$	$\frac{\kappa GA}{2\Delta}+c18$	$0+c19$	$\frac{2\kappa GA}{\Delta^2}$	$-(\kappa GA)'-c17(2\Delta)$	0	$-\frac{\kappa GA}{\Delta^2}-\frac{(\kappa GA)'}{2\Delta}$ c17	$-\frac{\kappa GA}{2\Delta}$ c18	0 c19	=	P_{yi}
$\delta\phi$	$-\frac{\kappa GA}{2\Delta}+c27$	$-\frac{EI}{\Delta^2}+\frac{(EI)'}{2\Delta}+c28$	$-\frac{ES^*}{\Delta^2}+\frac{(ES^*)'}{2\Delta}+c29$	0	$\frac{2EI}{\Delta^2}+\kappa GA-c27(2\Delta)$	$\frac{2ES^*}{\Delta^2}$	$\frac{\kappa GA}{2\Delta}$ c27	$-\frac{EI}{\Delta^2}-\frac{(EI^*)'}{2\Delta}$ c28	$-\frac{ES^*}{\Delta^2}-\frac{(ES^*)'}{2\Delta}$ c29	=	m_i
δu^*	$0+c37$	$-\frac{ES^*}{\Delta^2}+\frac{(ES^*)'}{2\Delta}+c38$	$-\frac{EI^*}{\Delta^2}+\frac{(EI^*)'}{2\Delta}+c39$	0	$\frac{2ES^*}{\Delta^2}-c37(2\Delta)$	$\frac{2EI^*}{\Delta^2}+\kappa GF^*$	0 c37	$-\frac{ES^*}{\Delta^2}-\frac{(ES^*)'}{2\Delta}$ c38	$-\frac{EI^*}{\Delta^2}-\frac{(EI^*)'}{2\Delta}$ c39	=	0

Figure 9. Stencil of equilibrium equations at inner point *i=m* for the case that the shear-lag is free at the top

	$i-1$			i			$i+1$ (F)				P
	v_{i-1}	ϕ_{i-1}	u^*_{i-1}	v_i	ϕ_i	u^*_i	v_{i+1}	ϕ_{i+1}	u^*_{i+1}		
δv	$-\frac{\kappa GA}{\Delta^2}+\frac{(\kappa GA)'}{2\Delta}+c17$	$\frac{\kappa GA}{2\Delta}+c18$	$0+c18(\frac{2ES^*}{EI})$	$\frac{2\kappa GA}{\Delta^2}$	$-(\kappa GA)'-c17(2\Delta)$	0	$-\frac{\kappa GA}{\Delta^2}-\frac{(\kappa GA)'}{2\Delta}$ c17	$-\frac{\kappa GA}{2\Delta}$ c18	0 c18	=	P_{yi}
$\delta\phi$	$-\frac{\kappa GA}{2\Delta}+c27$	$-\frac{EI}{\Delta^2}+\frac{(EI)'}{2\Delta}+c28$	$-\frac{ES^*}{\Delta^2}+\frac{(ES^*)'}{2\Delta}+c28(\frac{2ES^*}{EI})$	0	$\frac{2EI}{\Delta^2}+\kappa GA-c27(2\Delta)$	$\frac{2ES^*}{\Delta^2}$	$\frac{\kappa GA}{2\Delta}$ c27	$-\frac{EI}{\Delta^2}-\frac{(EI^*)'}{2\Delta}$ c28	$-\frac{ES^*}{\Delta^2}-\frac{(ES^*)'}{2\Delta}$ c29	=	m_i
δu^*	$0+c37$	$-\frac{ES^*}{\Delta^2}+\frac{(ES^*)'}{2\Delta}+c38$	$-\frac{EI^*}{\Delta^2}+\frac{(EI^*)'}{2\Delta}+c38(\frac{2ES^*}{EI})$	0	$\frac{2ES^*}{\Delta^2}-c37(2\Delta)$	$\frac{2EI^*}{\Delta^2}+\kappa GF^*$	0 c37	$-\frac{ES^*}{\Delta^2}-\frac{(ES^*)'}{2\Delta}$ c38	$-\frac{EI^*}{\Delta^2}-\frac{(EI^*)'}{2\Delta}$ c39	=	0

Figure 10. Stencil of equilibrium equations at inner point *i=m* for the case that the shear-lag is constrained at the top

4. Free transverse vibrations by finite difference method

Consider free transverse vibrations of the current doubly-symmetric tube structures by means of the finite difference method. Now, $v(x,\ t)$, $\phi(x,\ t)$, and $u^*(x,\ t)$ are expressed as

$$v(x,\ t)=\bar{v}(x,\ t)\exp\{i\omega t\} \tag{82}$$

$$\phi(x,\ t)=\bar{\phi}(x,\ t)\exp\{i\omega t\} \tag{83}$$

$$u^{*}(x,\ t)=\bar{u}^{*}(x,\ t)\exp\{i\omega t\} \tag{84}$$

Substituting the above equations into the equations of motion for the free transverse vibration obtained from Eqs. (32)-(39), the equations for free vibrations become

$$\delta v:\omega^{2}m\bar{v}+[\kappa GA(\bar{v}'+\bar{\phi})]'=0 \tag{85}$$

$$\delta\phi:\omega^{2}\rho I\bar{\phi}+\omega^{2}\rho S^{*}\bar{u}^{*}+(EI\bar{\phi})'-\kappa GA(\bar{v}'+\bar{\phi})+(ES^{*}\bar{u}^{*\prime})'=0 \tag{86}$$

$$\delta u^{*}:\omega^{2}\rho S^{*}\bar{\phi}+\omega^{2}\rho I^{*}\bar{u}^{*}+(EI^{*}\bar{u}^{*\prime})'+(ES^{*}\bar{\phi}')-\kappa GF^{*}\bar{u}^{*}=0 \tag{87}$$

The finite difference expressions of the above equations reduce to eigenvalue problem for $\bar{v}$, $\bar{\phi}$, and $\bar{u}^{*}$.

$$[\mathbf{A}-\omega^{2}\mathbf{B}]\ \boldsymbol{v}=0 \tag{88}$$

Here the matrix **A** is the total stiffness matrix as given in Eq. (78). On the other hand, the matrix **B** is total mass matrix which is the sum of individual mass matrix. The individual mass matrix at the *i*th mesh point is given in Figure 11. The *i*th natural frequencies ω_i can be obtained from the *i*th eigenvalue.

i		
v_i	ϕ_i	u^{*}_i
$-\rho A$	0	0
0	$-\rho I$	$-\rho S^{*}$
0	$-\rho S^{*}$	$-\rho I^{*}$

Figure 11. Individual mass matrix at mesh point *i*

5. Forced transverse vibrations by finite difference method

Forced lateral vibration of current tube structures may be obtained easily by means of modal analysis for elastic behavior subject to earthquake motion. Applying the finite difference method into Eq. (32), the equation of motion of current tube structures with distributed properties may be changed to discreet structure with degrees of freedom three times the total number of mesh points because each mesh point has three freedoms for the displacement components. Hence, Eq. (32) for current tube structure, subjected to earthquake acceleration $\ddot{v}_0$ at the base may be written in the matrix form as

$$[M]\{\ddot{v}\}+[c_v]\{\dot{v}\}-[(\kappa GA(v'+\phi))']=-[M]\{1\}\{\ddot{v}_0\} \tag{89}$$

in which $[M]$= mass matrix; $[c_v]$= the damping coefficient matrix; and $\{\ddot{v}\}$, $\{\dot{v}\}$, and $\{v\}$ are the relative acceleration vector, the relative velocity vector and relative displacement vector, respectively, measured from the base. $\{1\}$= unit vector. It is assumed that the dynamic deflection vector $\{v\}$ and the rotational angle vector $\{\phi\}$ may be written as

$$\{v\}=\sum_{j=1}^{n}\beta_j\{v\}_j q_j(t) \tag{90}$$

$$\{\phi\}=\sum_{j=1}^{n}\beta_j\{\phi\}_j q_j(t) \tag{91}$$

in which β_j= the j-th participation coefficient; $\{v\}_j$ and $\{\phi\}_j$= the j-th eigenfunctions for v and ϕ, respectively; $q_j(t)$= the j-th dynamic response depending on time t; and n= the total number of degrees of freedom taken into consideration here. Substituting Eqs. (90) and (91) into Eq. (89) and multiplying the reduced equation by $\{v\}_i^T$, we have

$$\begin{aligned}&\{v\}_i^T[M]\{v\}_i\beta_i\ddot{q}_i(t)+\{v\}_i^T[c_v]\{v\}_i\beta_i\dot{q}_i(t)-\\&[\{v\}_i^T(\kappa GA)\{v\}_i''q_i+\{v\}_i^T(\kappa GA)'\{v\}_i'q_i]\beta_i=-\{v\}_i^T[M]\{1\}\ddot{v}_0\end{aligned} \tag{92}$$

Now, Eq. (86) may be rewritten as

$$\omega^2[M]\{v\}_i-[\kappa GA(\{v\}_i'+\{\phi\}_i)]'=0 \tag{93}$$

Multiplying the above equation by $\{v\}_i^T$ and substituting the reduced equation into Eq. (93), we have

$$\{v\}_i^T[M]\{v\}_i\beta_i\ddot{q}_i(t)+\{v\}_i^T[c]\{v\}_i\beta_i\dot{q}_i(t)+\beta_i\omega^2\{v\}_i^T[M]\{v\}_i q_i=-\{v\}_i^T[M]\{1\}\ddot{v}_0 \tag{94}$$

Here, the damping coefficient matrix $[c_v]$ and the participation coefficients β_i are assumed to satisfy the following expressions:

$$\frac{\{v\}_i^T[c_v]\{v_i\}}{\{v\}_i^T[M]\{v_i\}}=2h_i\omega_i \tag{95}$$

$$\beta_i=\frac{\{v\}_i^T[M]\{1\}}{\{v\}_i^T[M]\{v_i\}} \tag{96}$$

in which h_i is the ith damping constant. Thus, Eq. (94) may be reduced to

$$\ddot{q}_i(t)+2h_i\omega_i\dot{q}_i(t)+\omega_i^2q_i(t)=-\ddot{v}_0 \tag{97}$$

The general solution of Eq. (98) is

$$q_i(t)=\exp(-h_i\omega_it)(C_1\sin\omega_{Di}t+C_2\cos\omega_{Di}t)-\frac{1}{\omega_{Di}}\int_0^t\exp[-h_i\omega_i(t-\tau)]\sin\omega_{Di}(t-\tau)\ddot{v}_0d\tau \tag{98}$$

in which $\omega_{Di}=\omega_i\sqrt{1-h_i^2}$ and C_1 and C_2 are constants determined from the initial conditions. The Duhamel integral in Eq. (98) may be calculated approximately by means of Paz [19] or Takabatake [2].

6. Numerical results by finite difference method

6.1. Numerical models

Numerical models for examining the simplified analysis proposed here have are shown in Figure 12. These numerical models are determined to find out the following effects: (1) the effect of the aspect ratio of the outer and inner tubes; (2) the effect of omitting the corners; and (3) the effect of bracing. **Model T1** is a doubly symmetric single frame-tube prepared for comparison with the numerical results of the doubly symmetric frame-double-tube. **T7** and **T8** are made up steel reinforced concrete frame-tubes, and the other models are steel frame-tubes. The total number of stories is 30. The difference between models **T2** to **T5** concerns the number of story and span attached bracing. The members of the single and double tubes are shown in Figures 13 and 14.

In the numerical computation, the following assumptions are made:

1. the static lateral force is a triangularly distributed load, as shown in Figures 13 and 14;

2. the dynamic loads are taken from El Centro 1940 NS, Taft 1952 EW, and Hachinohe 1968 NS, in which each maximum acceleration is 200 m/s^2;

3. the damping ratio for the first mode of the frame-tubes is h_1= 0.02, and the higher damping ratio for the n-th mode is $h_1 = h_1\omega_n / \omega_1$;

4. the weight of each floor is 9.807 kN/m^{-2} and the mass of the frame-tube is considered to be only floor's weight;

5. in the modal analysis, the number of modes for the participation coefficients is taken five into consideration as five.

MODEL (1)	PLAN (2)	BRACING (3)	CONSTRACTION (4)
T 1			STEEL
T 2	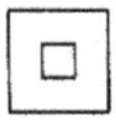		STEEL
T 3	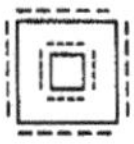	15, 16 STORY	STEEL
T 4		9, 10 STORY	STEEL
T 5		15, 16 STORY	STEEL
T 6		15, 16, 29, 30 STORY	STEEL
T 7			STEEL REINFORCED CONCRETE
T 8			STEEL REINFORCED CONCRETE

Figure 12. Numerical models [6]

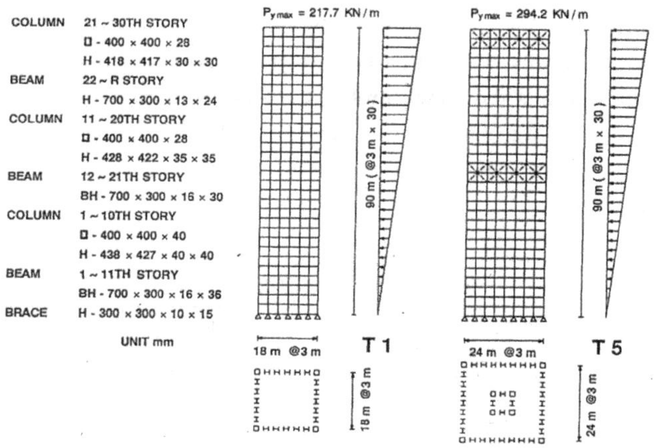

Figure 13. Member of numerical models T1 and T5 [6]

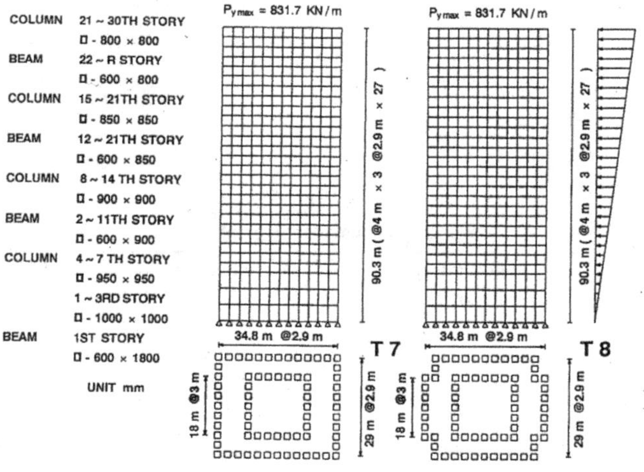

Figure 14. Member of numerical models T7 and T8 [6]

6.2. Static numerical results

First, the static numerical results are stated. Tables 1 and 2 show the maximum values of the static lateral displacements and shear-lags, calculated from the present theory, NASTRAN and DEMOS, in which a discrepancy between results obtained from NASTRAN and DE-

MOS is negligible, in practice. The ratios are those of the values obtained from the present theory to the corresponding values from the three-dimensional frame analysis using NASTRAN and DEMOS. The distributions of the static lateral displacements are shown in Figure 15. These numerical results show the simplified analysis is in good agreement, in practice, with the results of three-dimensional frame analysis using NASTRAN and DEMOS. Since the shear-lag is far smaller than the transverse deflections, as shown in Table 2 the discrepancy in shear-lag is negligible in practice.

Maximum static lateral deflection (m)			
Model (1)	Present theory (2)	Frame analysis (3)	Ratio(2)/(3) (4)
T1	0.441	0.430	1.026
T2	0.327	0.343	0.953
T3	0.307	0.318	0.965
T4	0.299	0.319	0.937
T5	0.312	0.330	0.945
T6	0.329	0.311	1.058
T7	0.151	0.158	0.956
T8	0.157	0.166	0.946

Table 1. Maximum values of static lateral deflections [6]

Maximum shear-lag (m)			
Model (1)	Present theory (2)	Frame analysis (3)	Ratio(2)/(3) (4)
T1	0.0145	0.0079	1.835
T2	0.0149	0.0085	1.753
T7	0.0090	0.0052	1.731
T8	0.0103	0.0028	3.679

Table 2. Maximum values of static shear-lags [6]

Figure 16 shows the distribution of axial forces of model **T2**. A discrepancy between the results obtained from the proposed theory and those from three-dimensional frame analysis is found. However, this discrepancy is within 10 % and is also allowable for practical use because the axial forces in tube structures are designed from the axial forces on the flange surfaces, being always larger than those on the web surfaces.

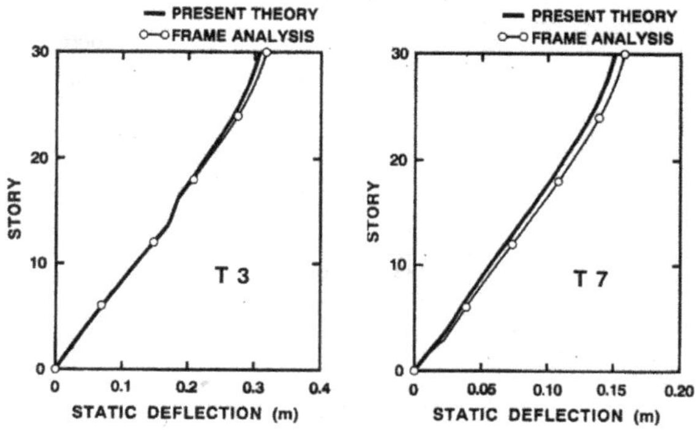

Figure 15. Distribution of static lateral deflection [6]

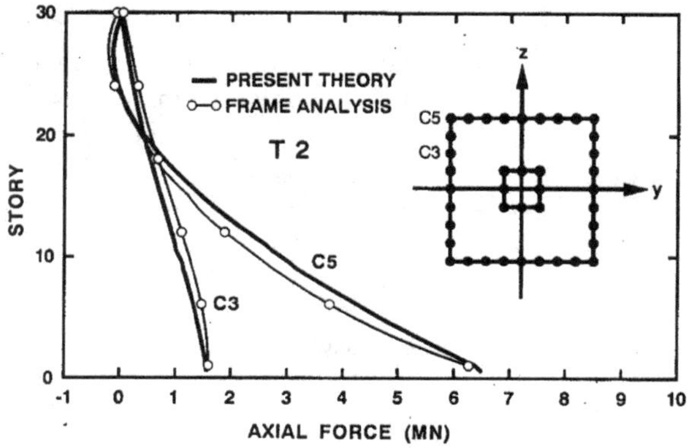

Figure 16. Axial force [6]

		Natural frequency (rad/s)				
Model (1)	Analytical methods (2)	First (3)	Second (4)	Third (5)	Fourth (6)	Fifth (7)
T1	Present theory	1.998	6.077	10.942	15.759	20.397
	Frame analysis	2.062	6.211	11.048	15.907	20.648
	Ratio	0.969	0.978	0.990	0.991	0.988
T2	Present theory	2.080	6.255	11.150	15.977	20.614

Model (1)	Analytical methods (2)	Natural frequency (rad/s) First (3)	Second (4)	Third (5)	Fourth (6)	Fifth (7)
	Frame analysis	2.058	6.223	11.076	16.020	20.826
	Ratio	1.011	1.005	1.007	0.997	0.990
T3	Present theory	2.137	6.197	11.694	15.924	21.740
	Frame analysis	2.138	6.290	11.687	16.198	22.062
	Ratio	1.000	0.985	1.001	0.983	0.985
T4	Present theory	2.192	6.186	11.112	16.805	21.334
	Frame analysis	2.152	6.296	11.224	16.813	21.756
	Ratio	1.019	0.983	0.990	1.000	0.981
T5	Present theory	2.126	6.266	11.559	16.073	21.390
	Frame analysis	2.100	6.260	11.464	16.136	21.636
	Ratio	1.012	1.001	1.008	0.996	0.989
T6	Present theory	2.055	6.045	11.631	16.053	22.019
	Frame analysis	2.147	6.369	11.848	16.662	22.610
	Ratio	0.957	0.949	0.982	0.963	0.974
T7	Present theory	3.458	9.920	17.924	25.863	32.626
	Frame analysis	3.462	10.037	17.983	26.246	34.675
	Ratio	0.999	0.988	0.997	0.985	0.941
T8	Present theory	3.401	9.811	17.825	25.786	32.696
	Frame analysis	3.382	9.856	17.729	26.028	34.561
	Ratio	1.006	0.995	1.005	0.991	0.946

Table 3. Natural frequencies [6]. Note. Ratio = present theory/frame analysis

6.3. Free vibration results

Secondly, consider the natural frequencies. Table 3 shows the natural frequencies of the above-mentioned numerical models. It follows that, in practical use, the simplified analysis gives in excellent agreement with the results obtained from the three-dimensional frame analysis using NASTRAN and DEMOS. Since the transverse stiffness of the bracing is far larger than for frames, the transverse stiffness of current frame-tube with braces varies discontinuously, particularly at the part attached to the bracing. However, such discontinuous and local variation due to bracing can be expressed by the present theory.

6.4. Dynamic results

Thirdly, let us present dynamic results. The maximum values of dynamic deflections, story shears, and overturning moments are shown in Tables 4-6, respectively. Figure 17 shows the distribution of the maximum dynamic deflections and of the maximum story shear forces of model **T7** for El Centro 1949 NS. Figure 18 indicates the distribution of the absolute accelerations and of the maximum overturning moments for model **T7**. Thus, the proposed approximate theory is in good agreement with the results of the three-dimensional frame analysis using NASTRAN and DEMOS in practice. These excellent agreements may be estimated from participation functions as shown in Figure 19. The present one-dimensional extended rod theory used the finite difference method can always express discontinuous and local behavior caused by the part attached to the bracing.

Maximum dynamic lateral deflection (m)					
Model (1)	Earthquake type (2)		Present theory (3)	Frame analysis (4)	Ratio(3)/(4) (5)
T1	El Centro	NS	0.263	0.293	0.898
	Hachinohe	NS	0.453	0.411	1.102
	Taft	EW	0.213	0.208	1.024
T2	El Centro	NS	0.311	0.291	1.069
	Hachinohe	NS	0.420	0.419	1.002
	Taft	EW	0.212	0.213	0.995
T3	El Centro	NS	0.327	0.329	0.994
	Hachinohe	NS	0.460	0.465	0.989
	Taft	EW	0.205	0.206	0.995
T4	El Centro	NS	0.318	0.324	0.981
	Hachinohe	NS	0.515	0.469	1.098
	Taft	EW	0.196	0.201	0.975
T5	El Centro	NS	0.327	0.315	1.038
	Hachinohe	NS	0.454	0.429	1.058
	Taft	EW	0.207	0.212	0.976
T6	El Centro	NS	0.297	0.321	0.925
	Hachinohe	NS	0.424	0.437	0.970
	Taft	EW	0.211	0.207	1.019
T7	El Centro	NS	0.155	0.150	1.033
	Hachinohe	NS	0.202	0.195	1.036
	Taft	EW	0.221	0.215	1.028
T8	El Centro	NS	0.158	0.154	1.026
	Hachinohe	NS	0.232	0.235	0.987
	Taft	EW	0.219	0.210	1.043

Table 4. Maximum dynamic lateral deflections [6]

The above-mentioned numerical computations are obtained from that the total number of mesh points, including the top, is 60. Figure 20 shows the convergence characteristics of the static and dynamic responses for model **T7**, due to the number of mesh points. The convergence is obtained by the number of mesh points, being equal to the number of stories of the tube structures.

Maximum story shear (kN)					
Model (1)	Earthquake type (2)		Present theory (3)	Frame analysis (4)	Ratio(3)/(4) (5)
T1	El Centro	NS	5482	6659	0.823
	Hachinohe	NS	12494	10003	1.249
	Taft	EW	4972	4835	1.028
T2	El Centro	NS	11464	11082	1.035
	Hachinohe	NS	17260	17309	0.997
	Taft	EW	8414	8071	1.043
T3	El Centro	NS	12239	11768	1.040
	Hachinohe	NS	18937	19378	0.977
	Taft	EW	9248	9012	1.026
T4	El Centro	NS	14749	12258	1.203
	Hachinohe	NS	27498	21084	1.304
	Taft	EW	9316	9307	1.001
T5	El Centro	NS	11484	11180	1.027
	Hachinohe	NS	18172	17515	1.038
	Taft	EW	8865	8659	1.024
T6	El Centro	NS	11562	12160	0.951
	Hachinohe	NS	17632	20270	0.870
	Taft	EW	9807	9150	1.072
T7	El Centro	NS	38746	37167	1.042
	Hachinohe	NS	56153	53642	1.047
	Taft	EW	56731	54819	1.035
T8	El Centro	NS	41306	40109	1.030
	Hachinohe	NS	59595	57957	1.028
	Taft	EW	49004	44718	1.096

Table 5. Maximum story shears [6]

Maximum overturning moment (MN m)					
Model (1)	Earthquake type (2)		Present theory (3)	Frame analysis (4)	Ratio(3)/(4) (5)
T1	El Centro	NS	3.233	3.991	0.810
	Hachinohe	NS	6.099	5.599	1.089
	Taft	EW	2.731	2.863	0.954

Maximum overturning moment (MN m)					
Model (1)	Earthquake type (2)		Present theory (3)	Frame analysis (4)	Ratio(3)/(4) (5)
T2	El Centro	NS	7.118	6.531	1.090
	Hachinohe	NS	9.829	9.413	1.044
	Taft	EW	4.928	4.849	1.016
T3	El Centro	NS	8.090	7.972	1.015
	Hachinohe	NS	11.277	11.122	1.014
	Taft	EW	5.129	5.070	1.012
T4	El Centro	NS	8.140	7.727	1.053
	Hachinohe	NS	13.827	10.983	1.259
	Taft	EW	4.984	5.021	0.993
T5	El Centro	NS	7.879	7.315	1.077
	Hachinohe	NS	10.778	10.250	1.052
	Taft	EW	5.053	5.007	1.009
T6	El Centro	NS	6.566	8.070	0.814
	Hachinohe	NS	9.593	11.431	0.839
	Taft	EW	4.968	5.091	0.976
T7	El Centro	NS	22.148	21.574	1.027
	Hachinohe	NS	29.914	28.929	1.034
	Taft	EW	32.186	31.675	1.016
T8	El Centro	NS	21.662	21.659	1.000
	Hachinohe	NS	32.693	33.462	0.977
	Taft	EW	28.779	28.246	1.019

Table 6. Maximum overturning moments [6]

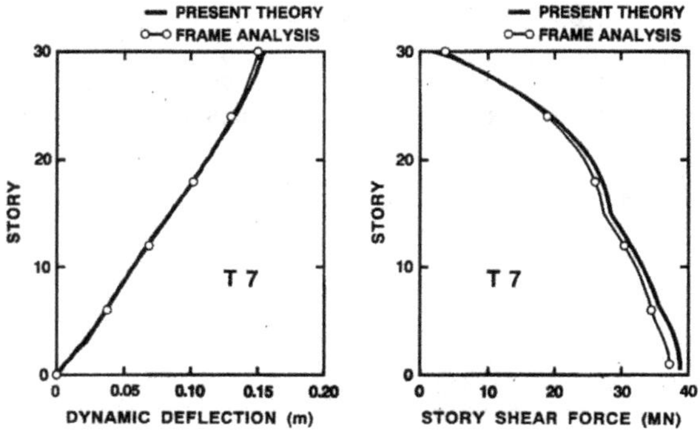

Figure 17. Distribution of dynamic lateral deflection and story shear force [6]

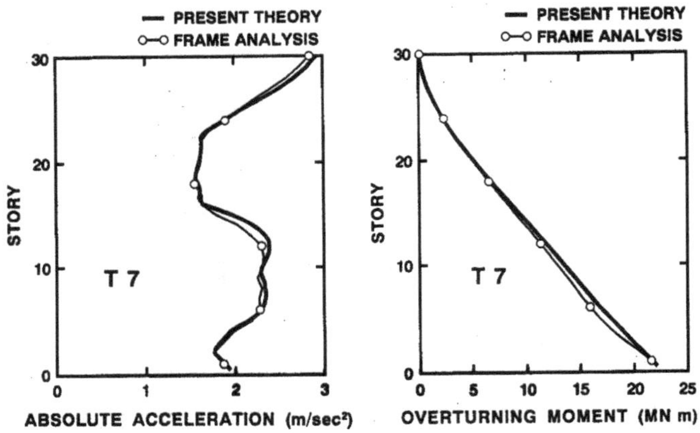

Figure 18. Distribution of absolute acceleration and overturning moment [6]

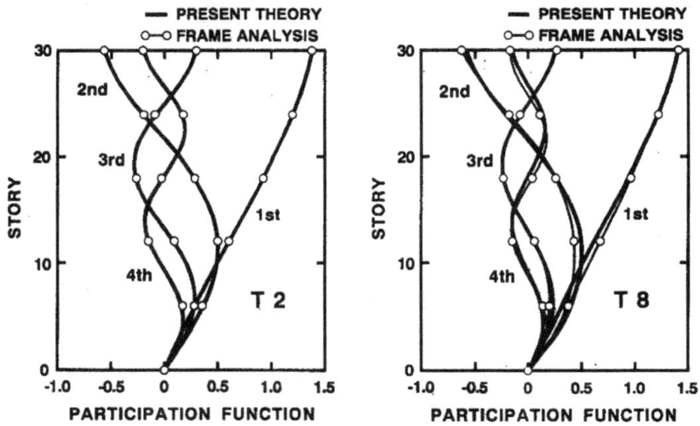

Figure 19. Participation functions [6]

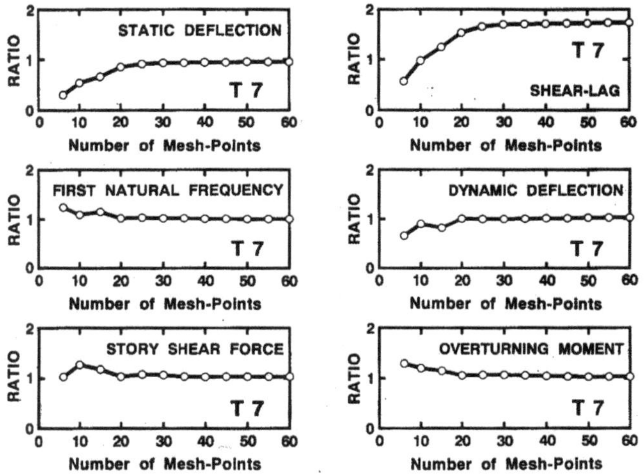

Figure 20. Convergence characteristics [6]

7. Natural frequencies by approximate method

7.1. Simplification of governing equation

In the structural design of high-rise buildings, structure designers want to grasp simply the natural frequencies in the preliminary design stages. Takabatake [3] presented a general and simple analytical method for natural frequencies to meet the above demands. This section explains about this simple but accurate analytical method.

The one-dimensional extended rod theory for the transverse motion takes the coupled equations concerning v , ϕ , and u^* , as given in Eqs. (32) to (34). Now consider the equation of motion expressed in terms with the lateral deflection. Neglecting the differential term of the transverse shear stiffness, κGA , in Eq. (32), the differential of rotational angle with respect to x may be written as

$$\phi = -v'' + \frac{1}{\kappa GA}(-P_y + \rho A\ddot{v} + c_v\dot{v}) \tag{99}$$

From (33) and (34), u^* becomes

$$u^* = \frac{1}{\kappa GF^*}\frac{I^*}{S^*}[\rho\hat{I}\ddot{\phi} - E\hat{I}\phi'' + \kappa GA(v' + \phi)] \tag{100}$$

in which $\hat{I}$ is defined as

$$\hat{I} = I\left[1 - \frac{(S^*)^2}{II^*}\right] \tag{101}$$

Differentiating Eq. (33) with respect to x and substituting Eqs. (32), (99), and (100) into the result, the equation of motion expressed in terms with the transverse deflection may be written as

$$\begin{aligned}
&EIv'''' + \rho A\ddot{v} + c_v\dot{v} - P_y - \rho I\ddot{v}'' - \frac{EI}{\kappa GA}\left(-P''_y + \rho A\ddot{v}'' + c_v\dot{v}''\right) - \frac{EI^*}{\kappa GF^*}\left(-P''_y + \rho A\ddot{v}'' + c_v\dot{v}''\right)\\
&+\left(\frac{\rho I}{\kappa GA} + \frac{\rho I^*}{\kappa GF^*}\right)\left(-P_y + \rho A\ddot{v} + c_v\dot{v}\right)^{\cdot\cdot} + \frac{\rho I^*}{\kappa GF^*}E\hat{I}\ddot{v}'''' - \frac{\rho I^*}{\kappa GF^*}\frac{E\hat{I}}{\kappa GA}\left(-P''_y + \rho A\ddot{v}'' + c_v\dot{v}''\right)^{\cdot\cdot''}\\
&+\frac{\rho I^*}{\kappa GF^*}\frac{\rho\hat{I}}{\kappa GA}\left(-P_y + \rho A\ddot{v} + c_v\dot{v}\right)^{\cdot\cdot\cdot\cdot} - \frac{\rho I^*}{\kappa GF^*}\rho\hat{I}\overset{\cdot\cdot\cdot\cdot}{v}'' + \frac{EI^*}{\kappa GF^*}\rho\hat{I}\ddot{v}''''\\
&-\frac{EI^*}{\kappa GF^*}\frac{\rho\hat{I}}{\kappa GA}\left(-P_y + \rho A\ddot{v} + c_v\dot{v}\right)\\
&-\frac{EI^*}{\kappa GF^*}E\hat{I}v'''''' + \frac{EI^*}{\kappa GF^*}\frac{E\hat{I}}{\kappa GA}\left(-P_y + \rho A\ddot{v} + c_v\dot{v}\right)'''' = 0
\end{aligned} \tag{102}$$

Eq. (102) is a sixth-order partial differential equation with variable coefficients with respect to x . In order to simplify the future development, considering only bending, transverse shear deformation, shear lag, inertia, and rotatory inertia terms in Eq. (102), a simplified governing equation is given

$$EIv'''' + \rho A\ddot{v} + c_v\dot{v} - P_y - \rho I\ddot{v}'' - \frac{EI}{\kappa GA}\left(1 + \underline{\frac{\kappa GAI^*}{I\kappa GF^*}}\right)\left(-P''_y + \rho A\ddot{v}'' + c_v\dot{v}''\right) = 0 \tag{103}$$

The equation neglecting the underlined term in Eq. (103) reduces to the equation of motion of Timoshenko beam theory, for example, Eq. (9.49) Craig [20]. Since Eq. (103) is very simple equation, the free transverse vibration analysis is developed by means of Eq. (103). To simplify the future expression, the following notation is introduced

$$\left(\underline{\underline{\kappa GA}}\right) = \kappa GA\frac{1}{1 + \dfrac{\kappa GA\,I^*}{\kappa GF^*I}} \tag{104}$$

Hence, Eq. (104) may be rewritten

$$EI\,v'''' + \rho\,A\ddot{v} + c_v\dot{v} - P_y - \rho I\,\ddot{v}'' - \frac{EI}{\underline{\underline{(\kappa GA)}}}\left(-P_y'' + \rho A\ddot{v}'' + c_v\dot{v}''\right) = 0 \tag{105}$$

The aforementioned equation suggests that in the simplified equation the transverse shear stiffness κGA must be replaced with the modified transverse shear stiffness $\underline{\underline{(\kappa GA)}}$.

7.2. Undamped free transverse vibrations

Let us consider undamped free transverse vibration of high-rise buildings. The equation for undamped free transverse vibrations is written from Eq. (105) as

$$v'''' - \frac{\rho A}{EI}\left[\frac{\rho I}{\rho A} + \frac{EI}{\underline{\underline{(\kappa GA)}}}\right]\ddot{v}'' + \frac{\rho A}{EI}\ddot{v} = 0 \tag{106}$$

Using the separation method of variables, $v(x,\,t)$ is expressed as

$$v(x,t) = \Phi(x)e^{iwt} \tag{107}$$

Substituting the above equation into Eq. (106), the equation for free vibrations becomes

$$\Phi + b\Phi'''' + c\,\Phi = 0 \tag{108}$$

in which the coefficients, b and c, are defined as

$$b = \frac{(k\,L)^4}{L^2}\left(\frac{1}{\hat{\lambda}_0}\right)^2 \tag{109}$$

$$c = -k^4 \tag{110}$$

in which k^4 and $\hat{\lambda}_0$ are defined as

$$k^4 = \frac{\rho A}{EI}\omega^2 \tag{111}$$

$$\left(\frac{1}{\hat{\lambda}_0}\right)^2 = \left(\frac{1}{\lambda_0^*}\right)^2 + \left(\frac{1}{\lambda_0}\right)^2 \tag{112}$$

in which λ_0^* and λ_0 are defined as

$$\lambda_0^* = L\sqrt{\frac{\rho A}{\rho I}} \tag{113}$$

$$\lambda_0 = L\sqrt{\frac{\left(\underline{\underline{\kappa G A}}\right)}{EI}} \tag{114}$$

λ_0^* and λ_0 are pseudo slenderness ratios of the tube structures, depending on the bending stiffness and the transverse shear stiffness, respectively. Since for a variable tube structure the coefficients, b and c, are variable with respect to x, it is difficult to solve analytically Eq. (108). So, first we consider a uniform tube structure where these coefficients become constant. The solution for a variable tube structure will be presented by means of the Galerkin method.

Thus, since for a uniform tube structure Eq. (108) becomes a fourth-order differential equation with constant coefficients, the general solution is

$$\Phi = C_1 \cos\lambda_1 x + C_2 \sin\lambda_1 x + C_3 \sin h\lambda_2 x + C_4 \cosh\lambda_2 x \tag{115}$$

in which C_1 to C_4 are integral constants and λ_1 and λ_2 are defined as

$$\lambda_1 = \frac{(kL)^2}{L}\alpha_1^* \tag{116}$$

$$\lambda_2 = \frac{(kL)^2}{L}\alpha_2^* \tag{117}$$

in which α_1^* and α_2^* are

$$\alpha_1^* = \frac{1}{\hat{\lambda}_0}\sqrt{\frac{1+\alpha}{2}} \tag{118}$$

$$\alpha_2^* = \frac{1}{\hat{\lambda}_0}\sqrt{\frac{-1+\alpha}{2}} \tag{119}$$

in which

$$\alpha = \sqrt{1+\frac{4}{(kL)^4\left(\frac{1}{\hat{\lambda}_0}\right)^4}} \tag{120}$$

Meanwhile, ϕ' for the free transverse vibration is from Eq. (99)

$$\phi' = -v'' + \frac{\rho A}{\underline{\underline{(\kappa GA)}}}\ddot{v} \tag{121}$$

in which κGA is replaced with $\underline{\underline{(\kappa GA)}}$. The substitution of Eqs. (108) and (111) into the aforementioned equation yields

$$\phi'(x,t) = -\left[\Phi(x) + k^4\frac{EI}{\underline{\underline{(\kappa GA)}}}\Phi(x)\right]e^{iwt} \tag{122}$$

The integration of the aforementioned equation becomes

$$\phi(x,t) = -\left[\Phi'(x) + k^4\frac{EI}{\underline{\underline{(\kappa GA)}}}\int\Phi(x)\,dx\right]e^{i\omega t} \tag{123}$$

The boundary conditions for the current tube structures are assumed to be constrained for all deformations at the base and free for bending moment, transverse shear and shear-lag at the top. Hence the boundary conditions are rewritten from Eqs. (35) to (38) as

$$v = 0 \text{ at } x = 0 \tag{124}$$

$$\phi = 0 \text{ at } x = 0 \tag{125}$$

$$u^* = 0 \text{ at } x = 0 \tag{126}$$

$$v' + \phi = 0 \text{ at } x = L \tag{127}$$

$$E\,S^*u^{*'} + E\,I\,\phi' = 0 \text{ at } x = L \tag{128}$$

$$EI^*u^{*'} + ES^*\phi' = 0 \text{ at } x = L \tag{129}$$

Eqs. (128) and (129) reduce to

$$\phi' = 0 \quad \text{at} \quad x = L \tag{130a}$$

$$u^{*'} = 0 \text{ at } x = L \tag{130b}$$

Hence the boundary conditions for current problem become as Eqs. (124), (125), (127), and (130a). Using Eq. (107), these boundary conditions are rewritten as

$$\Phi = 0 \text{ at } x = 0 \tag{131}$$

$$\Phi'(x) + k^4 \frac{EI}{(\kappa GA)} \int \Phi(x)\,dx = 0 \text{ at } x = 0 \tag{132}$$

$$\int \Phi(x)\,dx = 0 \text{ at } x = L \tag{133}$$

$$\Phi''(x) + k^4 \frac{EI}{(\kappa GA)} \Phi(x) = 0 \text{ at } x = L \tag{134}$$

Substituting Eq. (115) into the aforementioned boundary conditions, the equation determining a nondimensional constant $(k_n L)^2$ corresponding to the nth natural frequency is obtained as

$$\left[(\alpha_1^*)^2 - \frac{1}{\lambda_0^2}\right](\bar{k}_2 \cos\lambda_1 L - \bar{k}_1 \sin\lambda_1 L) + \left[(\alpha_2^*)^2 + \frac{1}{\lambda_0^2}\right](\sinh\lambda_2 L + \bar{k}_2 \cosh\lambda_2 L) = 0 \tag{135}$$

in which $\bar{k}_1$ and $\bar{k}_2$ are defined as

$$\bar{k}_1 = -\frac{\alpha_2^* + \dfrac{1}{\alpha_2^* \lambda_0^2}}{\alpha_1^* + \dfrac{1}{\alpha_1^* \lambda_0^2}} \tag{136}$$

$$\bar{k}_2 = -\frac{-\dfrac{\bar{k}_1}{\alpha_1^*}\cos\lambda_1 L + \dfrac{1}{\alpha_2^*}\cosh\lambda_2 L}{-\dfrac{1}{\alpha_1^*}\sin\lambda_1 L + \dfrac{1}{\alpha_2^*}\sinh\lambda_2 L} \tag{137}$$

The value of $(k_n L)^2$ is determined from Eq. (135) as follows:

STEP 1. From Eqs. (113) and (114), determine λ_0^* and λ_0.

STEP 2. From Eq. (112), determine $\hat{\lambda}_0$.

STEP 3. Assume the value of $(k_n L)^2$.

STEP 4. From Eq. (120), determine α.

STEP 5. From Eqs. (116) to (119), determine λ_1, λ_2, α_1^* and α_2^*, respectively.

STEP 6. From Eqs. (136) and (137), calculate $\bar{k}_1$ and $\bar{k}_2$.

STEP 7. Substitute these vales into Eq. (135) and find out the value of $(k_n L)^2$ satisfying Eq. (135) with trial and error.

Hence, the value of $(k_n L)^2$ depends on the slenderness ratios, λ_0^* and λ_0, of the uniform tube structure. So, for practical uses, the value of $(k_n L)^2$ for the given values λ_0^* and λ_0 can be presented previously as shown in Figure 21. Numerical results show that the values of $(k_n L)^2$ depend mainly on λ_0 and are negligible for the variation of λ_0^*. When λ_0 increases, the value of $(k_n L)^2$ approaches the value of the well-known Bernoulli-Euler beam. The practical tube structures take a value in the region from $\lambda_0 = 0.1$ to $\lambda_0 = 5$.

Thus, substituting the value of $(k_n L)^2$ into Eq. (111), the nth natural frequency, ω_n, of the tube structure is

$$\omega_n = \frac{(k_n L)^2}{L^2}\sqrt{\frac{EI}{\rho A}} \tag{138}$$

Using Figure 21, the structural engineers may easily obtain from the first to tenth natural frequencies and also grasp the relationships among these natural frequencies.

The nth natural function, Φ_n, corresponding to the nth natural frequency is

$$\Phi_n(x) = -\bar{k}_2 \cos \lambda_1 x + \bar{k}_1 \sin \lambda_1 x + \sin h\lambda_2 x + \bar{k}_2 \cos\ h\lambda_2 \tag{139}$$

Now, neglecting the effect of the shear lag, the solutions proposed here agree with the results for a uniform Timoshenko beam presented by Herrmann [21] and Young [22].

7.3. Natural frequency of variable tube structures

The natural frequency of a uniform tube structure has been proposed in closed form. For a variable tube structure the proposed results give the approximate natural frequency by replacing the variable tube structure with a pseudo uniform tube structure having an appropriate reference stiffness.

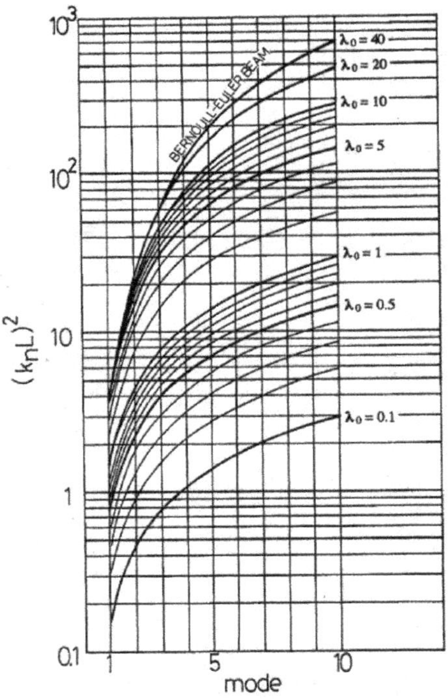

Figure 21. Values of $(k_nL)^2$ [3]

On the other hand, the natural frequency for a variable tube structure is presented by means of the Galerkin method. So, Eq. (108) may be rewritten as

$$EI\Phi'''' - \omega^2\left[\rho A\Phi - \rho A\left(\frac{L}{\hat{\lambda}_0}\right)^2\Phi''\right] = 0 \tag{140}$$

Φ is expressed by a power series expansion as follows

$$\Phi(x) = \sum_{n=1}^{\infty} c_n\ \Phi_n \tag{141}$$

in which c_n = unknown coefficients; and $\Phi_n(x)$ = functions satisfying the specified boundary conditions of the variable tube structure. Approximately, $\Phi_n(x)$ take the natural function of the pseudo uniform tube structure, as given in Eq. (139). Applying Eq. (141) into Eq. (140), the Galerkin equations of Eq. (140) become

$$\delta c_m : \sum_{n=1}^{\infty} c_n\left(A_{mn} - \omega^2 B_{mn}\right) = 0 \tag{142}$$

in which the coefficients, A_{mn} and B_{mn}, are defined as

$$A_{mn} = \int_0^L EI\Phi_n''''\ \Phi_m\, dx \tag{143}$$

$$B_{mn} = \int_0^L \rho A\Phi_n\Phi_m\, dx - \int_0^L \rho A\left(\frac{L}{\hat{\lambda}_0}\right)\Phi_n''''\Phi_m dx \tag{144}$$

Hence, the natural frequency of the variable tube structure is obtained from solving eigen-value problem of Eq. (142).

7.4. Numerical results for natural frequencies

The natural frequencies for doubly symmetric uniform and variable tube structures have been presented by means of the analytical and Galerkin methods, respectively. In order to examine the natural frequencies proposed here, numerical computations were carried out for a doubly symmetric steel frame tube, as shown in Figure 22. This frame tube equals to the tube structure used in the static numerical example in the Section 6, except for with or without bracing at 15 and 16 stories. The data used are as follows: the total story is 30; each story height is 3 m; the total height, L , is 90 m; the base is rigid; Young's modulus E of the material used is 2.05 x 10^{11} N/m^2. The weight per story is 9.8 kN/m^2 x 18 m x 18 m = 3214 kN.

The cross sections of columns and beams in the variable frame tube shown in Figure 22 vary in three steps along the height. On the other hand, the uniform frame tube is assumed to be the stiffness at the midheight ($L/2$) of the variable tube structure.

Table 7 shows the natural frequencies of the uniform and variable frame tubes, in which the approximate solution for the variable frame tube indicates the value obtained from replacing the variable frame tube with a pseudo uniform frame tube having the stiffnesses at the lowest story. The results obtained from the proposed method show excellent agreement with the three-dimensional frame analysis using FEM code NASTRAN. The approximate solution for the variable frame tube is also applicable to determine approximately the natural frequencies in the preliminary stages of the design.

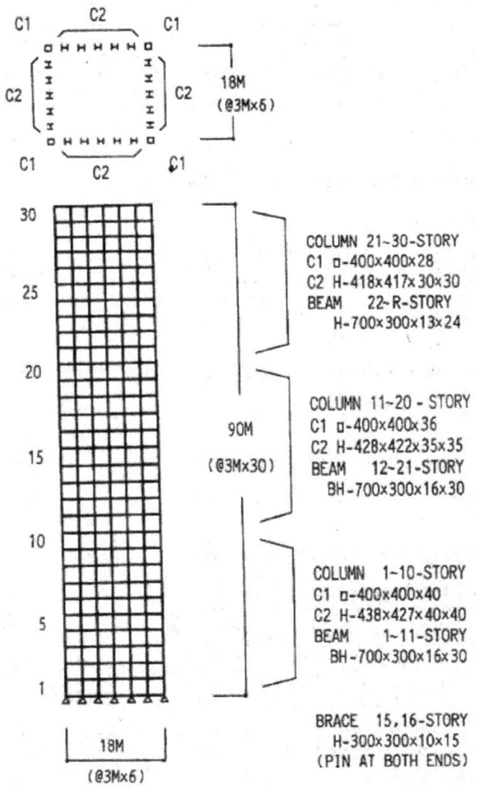

Figure 22. Numerical model of frame tube [3]

NATURAL FREQUENCIES (rad/s)					
	UNIFORM FRAME TUBE		VARIABLE FRAME TUBE		
Mode (1)	Analytical solution (2)	Frame theory (3)	Approximate solution (4)	Galerkin method (5)	Frame theory (6)
First	1.956	2.011	2.059	1.906	2.044
Second	5.907	6.196	6.218	6.203	6.141
Third	10.458	11.121	10.994	11.056	10.908
Fourth	14.708	15.822	15.460	15.341	15.683
Fifth	19.053	20.672	20.024	20.002	20.359

Table 7. Natural frequencies of uniform and variable frame tubes [3]

8. Expansion of one-dimensional extended rod theory

In order to carry out approximate analysis for a large scale complicated structure such as a high-rise building in the preliminary design stages, the use of equivalent rod theory is very effective. Rutenberg [10], Smith and Coull [23], Tarjan and Kollar [24] presented approximate calculations based on the continuum method, in which the building structure stiffened by an arbitrary combination of lateral load-resisting subsystems, such as shear walls, frames, coupled shear walls, and cores, are replaced by a continuum beam. Georgoussis [25] proposed to asses frequencies of common structural bents including the effect of axial deformation in the column members for symmetrical buildings by means of a simple shear-flexure model based on the continuum approach. Tarian and Kollar [24] presented the stiffnesses of the replacement sandwich beam of the stiffening system of building structures.

Takabatake et al. [2-6, 26-28] developed a simple but accurate one-dimensional extended rod theory which takes account of longitudinal, bending, and transverse shear deformation, as well as shear-lag. In the preceding sections the effectiveness of this theory has been demonstrated by comparison with the numerical results obtained from a frame analysis on the basis of FEM code NASTRAN for various high-rise buildings, tube structures and mega structures.

The equivalent one-dimensional extended rod theory replaces the original structure by a model of one-dimensional rod with an equivalent stiffness distribution, appropriate with regard to the global behavior. Difficulty arises in this modeling due to the restricted number of freedom of the equivalent rod; local properties of each structural member cannot always be properly represented, which leads to significant discrepancy in some cases. The one-dimensional idealization is able to deal only with the distribution of stiffness and mass in the longitudinal direction, possibly with an account of the averaged effects of transverse stiffness variation. In common practice, however, structures are composed of a variety of members or structural parts, often including distinct constituents such as a frame-wall or coupled wall with opening. Overall behavior of such a structure is significantly affected by the local distribution of stiffness. In addition, the individual behavior of each structural member plays an important role from the standpoint of structural design. So, Takabatake [29, 30] propose two-dimensional extended rod theory as an extension of the one-dimensional extended rod theory to take into account of the effect of transverse variations in individual member stiffness.

Figure 23 illustrates the difference between the one- and two-dimensional extended rod theories in evaluating the local stiffness distribution of structural components. In the two-dimensional approximation, structural components with different stiffness and mass distribution are continuously connected. On the basis of linear elasticity, governing equations are derived from Hamilton's principle. Use is made of a displacement function which satisfies continuity conditions across the boundary surfaces between the structural components.

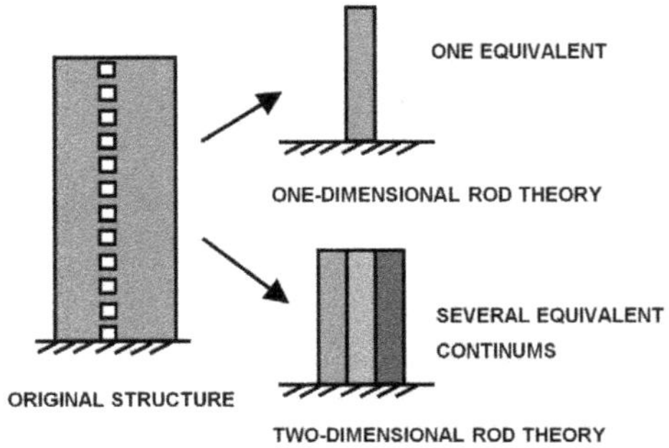

Figure 23. The difference between one- and two-dimensional rod theories [29]

Two-dimensional extended rod theory has been presented for simply analyzing a large or complicated structure such as a high-rise building or shear wall with opening. The principle of this theory is that the original structure comprising various different structural components is replaced by an assembly of continuous strata which has stiffness equivalent to the original structure in terms of overall behavior. The two-dimensional extended rod theory is an extended version of a previously proposed one-dimensional extended rod theory for better approximation of the structural behavior. The efficiency of this theory has been demonstrated from numerical results for exemplified building structures of distinct components. This theory may be applicable to soil-structure interaction problems involving the effect of multi-layered or non-uniform grounds.

On the other hand, the exterior of tall buildings has frequently the shape with many setback parts. On such a building the local variation of stress is considered to be very remarkable due the existence of setback. This nonlinear phenomenon of stress distribution may be explained by two-dimensional extended rod theory but not by one-dimensional extended rod theory. In order to treat exactly the local stress variation due to setback, the proper boundary condition in the two-dimensional extended rod theory must separate into two parts. One part is the mechanical boundary condition corresponding to the setback part and the other is the continuous condition corresponding to longitudinally adjoining constituents. Thus, Takabatake et al. [30] proved the efficiency of the two-dimensional extended rod theory to the general structures with setbacks.

Two-dimensional extended rod theory has been presented for simply analyzing a large or complicated structure with setback in which the stiffness and mass due to the existence of setback vary rapidly in the longitudinal and transverse directions. The effectiveness of this theory has been demonstrated from numerical results for exemplified numerical models. The transverse-wise distribution of longitudinal stress for structures with setbacks has been clarified to behave remarkable nonlinear behavior. Since the structural form of high-rise buildings with setbacks is frequently adapted in the world, the incensement of stress distribution occurred locally due to setback is very important for structural designers. The present theory may estimate such nonlinear stress behaviors in the preliminary design stages. The further development of the present theory will be necessary to extend to the three-dimensional extended rod theory which is applicable to a complicated building with three dimensional behaviors due to the eccentric station of many earthquake-resistant structural members, such as shear walls with opening.

9. Current problem of existing high-rise buildings

High-rise buildings have relative long natural period from the structural form. This characteristic is considered to be the most effective to avoid structural damages due to earthquake actions. However, when high-rise buildings subject to the action of the earthquake wave included the excellent long period components, a serious problem which the lateral deflection is remarkably large is produced in Japan. This phenomenon is based on resonance between the long period of high-rise buildings and the excellent long period of earthquake wave.

The 2011 Tohoku Earthquake (M 9.0) occurred many earthquake waves, which long period components are distinguished, on everywhere in Japan. These earthquake waves occur many physical and mental damages to structures and people living in high-rise buildings. The damage occurs high-rise buildings existing on all parts of Japan which appears long distance from the source. People entertain remarkable doubt about the ability to withstand earthquakes of high-rise buildings. This distrust is an urgent problem to people living and working in high-rise buildings. Existing high-rise buildings are necessary to improve urgently earthquake resistance. This section presents about an urgency problem which many existing high-rise buildings face a technical difficulty.

Let us consider dynamic behavior for one plane-frame of a high-rise building, as shown in Figure 24.

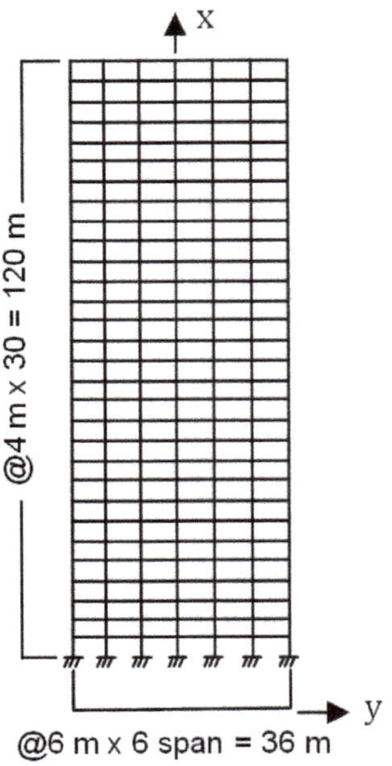

Figure 24. Numerical model

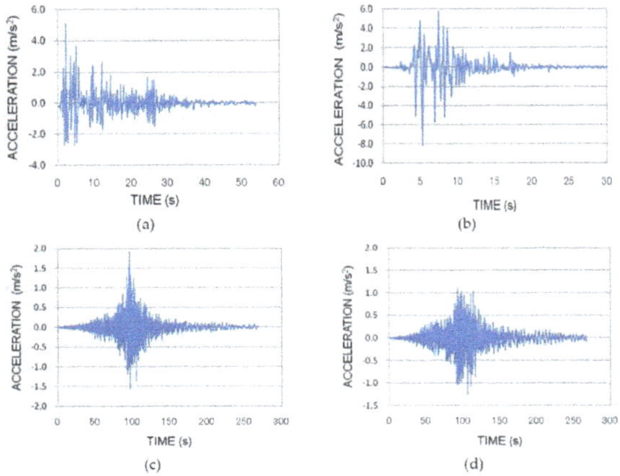

Figure 25. Time histories of acceleration (a) EL-Centro 1940 NS, (b) JMA Kobe 1995 NS, (c) Shinjuku 2011 NS, and (d) Urayasu 2011 NS

This plane frame is composed of uniform structural members. The sizes of columns and beams are □-800 x 800 x 25 (BCP) and H-400 x 300 x 11 x 18 (SN400B), respectively. The inertia moment of beams takes twice due to take into account of slab stiffness. This plane frame is a part of a three-dimensional frame structure with the span 6 m between adjacent plane-frames. The width and height are 36 m and 120 m, respectively. The main data used in numerical calculations are given in Table 8. Four kinds of earthquake waves are given in Table 9. El-Centro 1940 NS is converted the velocity to 0.5 m/s; JMA Kobe 1995 NS is the original wave with the maximum velocity 0.965 m/s; Shinjuku 2011 NS is the original wave with the maximum velocity 0.253 m/s; and Urayasu 2011 NS is the original wave with the maximum velocity 0.317 m/s. Figures 25(a) to 25(d) indicate time histories of accelerations for the four earthquake waves. In these earthquake waves, Shinjuku 2011 NS and Urayasu 2011 NS are obtained from K-net system measured at the 2011 Tohoku Earthquake. These earthquake waves are considered as earthquake waves included the excellent long periods. The excellent periods obtained from the Fourier spectrum of Shinjuku 2011 NS and Urayasu 2011 NS earthquake waves are 1.706 s and 1.342 s, respectively. The maximum acceleration and maximum velocity of these earthquake waves are shown in Table 9.

Structure shape	**Width:**	@6 m x 6 = 36 m
	Height:	@4 m x 30 floors = 120 m
Weight per floor (kN/m²)		12
Young modulus E (N/m²)		2.06×10^{11}
Shear modulus G (N/m²)		7.92×10^{10}
Mass density ρ (N/m³)		7850
Damping constant		0.02
Poisson ratio		0.3

Table 8. Main data for numerical model

Figure 26(a) shows the dynamic maximum lateral displacement subjected to the four kinds of earthquake waves. The maximum dynamic lateral displacement subject to Urayasu 2011 NS is remarkable larger than in the other earthquake waves. Figures 26(b) and (c) indicate the maximum shear force and overturning moment of the plane high-rise building subject to these earthquake actions, respectively. Earthquake wave Urayasu 2011 NS which includes long period components influence remarkable dynamic responses on the current high-rise building.

Earthquake Wave	Maximum Acceleration m/s²	Maximum Velocity m/s
EL-CENTRO 1940 NS	5.11	0.500
JMA KOBE 1995 NS	8.18	0.965
SHINJUKU 2011 NS	1.92	0.253
URAYASU 2011 NS	1.25	0.317

Table 9. Maximum acceleration and maximum velocity of each earthquake wave

It is very difficult to sort out this problem. If the existing structure stiffens the transverse shear rigidity of overall or selected stories, the dynamic responses produced by the earthquake wave included excellently long period decrease within initial design criteria for dynamic calculations. However, inversely the dynamic responses produced by both EL-Centro 1940 NS with the maximum velocity 0.5 m/s and JMA Kobe 1995 NS exceed largely over the initial design criteria. The original design is based on flexibility which is the most characteristic of high-rise buildings. This flexibility brings an effect which lowers dynamic responses produced by earthquake actions excluding long period components. Now, changing the structural stiffness from relatively soft to hard, this effect is lost and the safety of the high-rise building becomes dangerous for earthquake waves excluding the long period components.

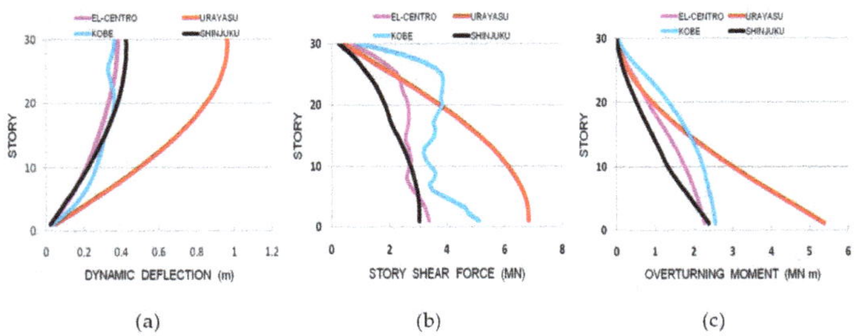

Figure 26. Distribution of dynamic responses (a) dynamic lateral deflection, (b) story shear force, and (c) overturning moment

Author has not in this stage a clear answer to this problem. This problem includes two situations. The first point is to find out the appropriate distribution of the transverse stiffness. The variation of the transverse stiffness is considered to stiffen or soften. In general, existing high-rise buildings are easily stiffening then softening. However, there is a strong probability that the stiffening of the transverse shear stiffness exceeds the allowable limit for the lateral deflection, story shear force, and overturning moment in the dynamic response subjected to earthquake waves used in original structural design. Therefore, the softening of the transverse shear stiffness used column isolation for all columns located on one or more selected story is considered to be effective. It is clarified from author's numerical computations that the isolated location is the most effective at the midheight. The second point is to find out an effective seismic retrofitting to existing high-rise buildings without the movement of people living and working in the high-rise building. These are necessary to propose urgently these measures for seismic retrofitting of existing high-rise buildings subject to earthquake waves included excellently long wave period. This subject will be progress to ensure comfortable life in high-rise buildings by many researchers.

10. Conclusions

A simple but accurate analytical theory for doubly symmetric frame-tube structures has been presented by applying ordinary finite difference method to the governing equations proposed by the one-dimensional extended rod theory. From the numerical results, the present theory has been clarified to be usable in the preliminary design stages of the static and dynamic analyses for a doubly symmetric single or double frame-tube with braces, in practical use. Furthermore, it will be applicable to hyper high-rise buildings, e.g. over 600m in the total height, because the calculation is very simple and very fast. Next the approximate method for natural frequencies of high-rise buildings is presented in the closed-form solutions. This method is very simple and effective in the preliminary design stages. Furthermore, the two-dimensional extended rod theory is introduced as for the expansion of the one-dimensional extended rod theory. Last it is stated to be urgently necessary seismic retrofitting for existing high-rise buildings subject to earthquake wave included relatively long period.

Author details

Hideo Takabatake*

Address all correspondence to: hideo@neptune.kanazawa-it.ac.jp

Department of Architecture, Kanazawa Institute of Technology, Institute of Disaster and Environmental Science, Japan

References

[1] Reissner, E. (1946). Analysis of shear lag in box beams by the principle of minimum potential energy. *Quarterly of Applied Mathematics*, 4(3), 268-278.

[2] Takabatake, H., Mukai, H., & Hirano, T. (1993). Doubly symmetric tube structures- I: Static analysis. *Journal of Structural Engineering ASCE*, 119(7), 1981-2001.

[3] Takabatake, H., Mukai, H., & Hirano, T. (1993). Doubly symmetric tube structures- II: Static analysis. *Journal of Structural Engineering ASCE*, 119(7), 2002-2016.

[4] Takabatake, H., Mukai, H., & Hirano, T. (1996). Erratum for "Doubly symmetric tube structures- I: Static analysis. *Journal of Structural Engineering ASCE*, 122(2), 225.

[5] Takabatake, H., Takesako, R., & Kobayashi, M. (1995). A simplified analysis of doubly symmetric tube structures. *The Structural Design of Tall Buildings*, 4(2), 137-153.

[6] Takabatake, H. (1996). A simplified analysis of doubly symmetric tube structures by the finite difference method. *The Structural Design of Tall Buildings*, 5(2), 111-128.

[7] Beck, H. (1962). Contribution to the analysis of coupled shear walls. *Journal of the American Concrete Institute*, 59(8), 1055-1069.

[8] Heidenbrech, A. C., & Smith, B. S. (1973). Approximate analyses of tall wall-frame structures. *Journal of the Structural Division ASCE*, 99(2), 199-221.

[9] Tso, W. K., & Chan, H. (1971). Dynamic analysis of plane coupled shear walls. *Journal of Engineering Mechanics Division ASCE*, 97(1), 33-48.

[10] Rutenberg, A. (1975). Approximate natural frequencies for coupled shear walls. *Earthquake Engineering and Structural Dynamics*, 4(1), 95-100.

[11] Rutenberg, A. (1977). Dynamic properties of asymmetric wall-frame structures. *Earthquake Engineering and Structural Dynamics*, 5(1), 41-51.

[12] Danay, A., Gluck, J., & Geller, M. (1975). A generalized continuum method for dynamic analysis of asymmetric tall buildings. *Earthquake Engineering and Structural Dynamics*, 4(2), 179-203.

[13] Bause, A. K., Nagpal, A. K., Bajaj, R. S., & Guiliani, A. K. (1979). Dynamic characteristics of coupled shear walls. *Journal of the Structural Division ASCE*, 105(8), 1637-1652.

[14] Cheung, Y. K., & Swaddiwudhipong, S. (1979). Free vibration of frame shear wall structures on flexible foundations. *Earthquake Engineering and Structural Dynamics*, 7(4), 355-367.

[15] Coull, A., & Smith, B. S. (1973). Torsional analyses of symmetric structures. *Journal of the Structural Division ASCE*, 99(1), 229-233.

[16] Coull, A., & Bose, B. (1975). Simplified analysis of frame-tube structures. *Journal of the Structural Division ASCE*, 101(11), 2223-2240.

[17] Smith, B. S., Kuster, M., & Hoenderkamp, J. C. D. (1984). Generalized method for estimating drift in high-rise structures. *Journal of Structural Engineering ASCE*, 110(7), 1549-1562.

[18] Smith, B. S., & Crowe, E. (1986). Estimating periods of vibration of tall buildings. *Journal of Structural Engineering ASCE*, 112(5), 1005-1019.

[19] Paz, M. (2006). Structural Dynamics. , 3rd Edn. Van Nostrand Reinhold New York., 74-75.

[20] Craig, R. R. (1981). Structural dynamics. John Wiley and Sons, New York.

[21] Herrmann, G. (1955). Forced motions of Timoshenko beams. *J. Appl. Mech. Trans. ASME*, 22(2), 53-56.

[22] Young, D. (1962). Continuos systems: Handbook of engineering mechanics. , W. Flügge, ed., McGraw-Hill. New York, N.Y. , 1-34.

[23] Smith, B. S., & Coull, A. (1991). Tall building structures: analysis and design:. John Willy & Sons, New York.

[24] Tarjian, G., & Kollar, L. P. (2004). Approximate analysis of building structures with identical stories subjected to earthquake. *International Journal of Solids and Structures*, 41(5), 1411-1433.

[25] Georgoussis, G. K. (2006). A simple model for assessing periods of vibration and modal response quantities in symmetrical buildings. *The Structural Design of Tall and Special Buildings*, 15(2), 139-151.

[26] Takabatake, H., & Nonaka, T. (2001). Numerical study of Ashiyahama residential building damage in the Kobe Earthquake. *Earthquake Engineering and Structural Dynamics*, 30(6), 879-897.

[27] Takabatake, H., Nonaka, T., & Tanaki, T. (2005). Numerical study of fracture propagating through column and brace of Ashiyahama residential building in Kobe Earthquake. *The Structural Design of Tall and Special Buildings*, 14(2), 91-105.

[28] Takabatake, H., & Satoh, T. (2006). A simplified analysis and vibration control to super-high-rise buildings. *The Structural Design of Tall and Special Buildings*, 15(4), 363-390.

[29] Takabatake, H. (2010). Two-dimensional rod theory for approximate analysis of building structures. *Earthquakes and Structures*, 1(1), 1-19.

[30] Takabatake, H., Ikarashi, F., & Matsuoka, M. (2011). A simplified analysis of super building structures with setback. *Earthquakes and Structures*, 2(1), 43-64.

Chapter 11

An Analysis of the Beam-to-Beam Connections Effect and Steel-Concrete Interaction Degree Over the Composite Floors Dynamic Response

José Guilherme Santos da Silva,
Sebastião Arthur Lopes de Andrade,
Pedro Colmar Gonçalves da Silva Vellasco,
Luciano Rodrigues Ornelas de Lima,
Elvis Dinati Chantre Lopes and
Sidclei Gomes Gonçalves

Additional information is available at the end of the chapter

1. Introduction

Nowadays steel and composite (steel-concrete) building structures are more and more becoming the modern landmarks of urban areas. Designers seem to continuously move the safety border, in order to increase slenderness and lightness of their structural systems. However, more and more steel and composite floors are carried out as light weight structures with low frequencies and low damping. These facts have generated very slender composite floors, sensitive to dynamic excitation, and consequently changed the serviceability and ultimate limit states associated to their design.

A direct consequence of this new design trend is a considerable increase in problems related to unwanted composite floor vibrations. For this reason, the structural floors systems become vulnerable to excessive vibrations produced by impacts such as human rhythmic activities. On the other hand, the increasing incidence of building vibration problems due to human activities led to a specific design criterion to be addressed in structural design [1-7]. This was the main motivation for the development of a design methodology centred on the steel-concrete composite floors non-linear dynamic response submitted to loads due to human rhythmic activities.

Considering all aspects mentioned before, the main objective of this paper is to investigate the beam-to-beam connections effect (rigid, semi-rigid and flexible) and the influence of steel-concrete interaction degree (from total to various levels of partial interaction) over the non-linear dynamic behaviour of composite floors when subjected to human rhythmic activities [1,2]. This way, the dynamic loads were obtained through experimental tests with individuals carrying out rhythmic and non-rhythmic activities such as stimulated and non-stimulated jumping and aerobic gymnastics [7]. Based on the experimental results, human load functions due to rhythmic and non-rhythmic activities are proposed [7].

The investigated structural model was based on a steel-concrete composite floor spanning 40m by 40m, with a total area of 1600m^2. The structural system consisted of a typical composite floor of a commercial building. The composite floor studied in this work is supported by steel columns and is currently submitted to human rhythmic loads. The structural system is constituted of composite girders and a 100mm thick concrete slab [1,2].

The proposed computational model adopted the usual mesh refinement techniques present in finite element method simulations, based on the ANSYS program [8]. This numerical model enabled a complete dynamic evaluation of the investigated steel-concrete composite floor especially in terms of human comfort and its associated vibration serviceability limit states.

Initially, all the composite floor natural frequencies and vibration modes were obtained. In sequence, based on an extensive parametric study, the floor dynamic response in terms of peak accelerations was obtained and compared to the limiting values proposed by several authors and design codes [6,9]. An extensive parametric analysis was developed focusing in the evaluation of the beam-to-beam connections effect and the influence of steel-concrete interaction degree over the investigated composite floor non-linear dynamic response, when submitted to human rhythmic activities.

The structural system peak accelerations were compared to the limiting values proposed by several authors and design standards [6,9]. The current investigation indicated that human rhythmic activities could induce the steel-concrete composite floors to reach unacceptable vibration levels and, in these situations, lead to a violation of the current human comfort criteria for these specific structures.

2. Dynamic Loading Induced by Human Rhythmic Activities

The description of the dynamic loads generated by human activities is not a simple task. The individual characteristics in which each individual perform the same activity and the existence of external excitation are key factors in defining the dynamic action characteristics. Numerous investigations were made aiming to establish parameters to describe such dynamic actions [1-6].

Several investigations have described the loading generated by human activities as a Fourier series, which consider a static part due to the individual weight and another part due to the dynamic load [1-6]. The dynamic analysis is performed equating one of the activity harmonics to the floor fundamental frequency, leading to resonance.

This study have considered the dynamic loads obtained by Faisca [7], based on the results achieved through a long series of experimental tests with individuals carrying out rhythmic and non-rhythmic activities. The dynamic loads generated by human rhythmic activities, such as jumps, aerobics and dancing were investigated by Faisca [7].

The loading modelling was able to simulate human activities like aerobics, dancing and free jumps. In this paper, the Hanning function was used to represent the human dynamic actions. The Hanning function was used since it was verified that this mathematical representation is very similar to the signal force obtained through experimental tests developed by Faisca [7].

The mathematical representation of the human dynamic loading using the Hanning function is given by Equation (1) and illustrated in Figure 1. The required parameters for the use of Equation (1) are related to the activity period, T, contact period with the structure, T_c, period without contact with the model, T_s, impact coefficient, K_p, and phase coefficient, CD. Figure 2 and the Table 1 illustrate the phase coefficient variation, CD, for human activities studied by Faisca [7], considering a certain number of individuals and later extrapolated for large number of peoples. Table 2 presents the experimental parameters used for human rhythmic activities representation and Figure 3 presents examples of dynamic action related to human rhythmic activities investigated in this work.

$$F(t) = CD\left\{K_p P\left[0.5 - 0.5\cos\left(\frac{2\pi}{T_c}t\right)\right]\right\} \qquad (1)$$

$$F(t) = 0$$

When $t \leq T_c$

When $T_c \leq t \leq T$

Where:

F(t): dynamic loading (N);

t: time (s);

T: activity period (s);

T_c: activity contact period (s);

P: person's weight (N);

K_p: impact coefficient;

CD: phase coefficient

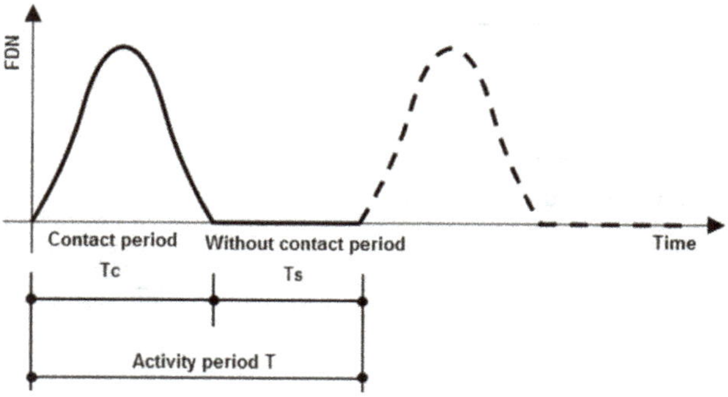

Figure 1. Representation of the dynamic loading induced by human rhythmic activities.

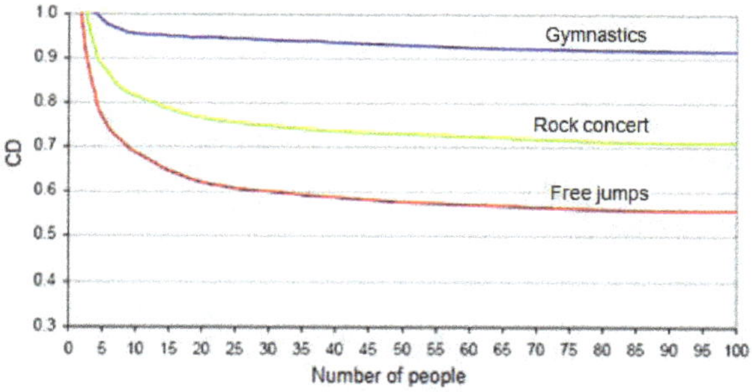

Figure 2. Variation of the phase coefficient CD for human rhythmic activities [7].

People number	Aerobics gymnastics	Free Jumps
1	1	1
3	1	0.88
6	0.97	0.74
9	0.96	0.70
12	0.95	0.67
16	0.94	0.64
24	0.93	0.62
32	0.92	0.60

Table 1. Numeric values adopted for the phase coefficient CD [7].

Activity	T (s)	T_c (s)	K_p
Free Jumps	0.44±0.15	0.32±0.09	3.17±0.58
Aerobics	0.44±0.09	0.34±0.09	2.78±0.60
Show	0.37±0.03	0.37±0.03	2.41±0.51

Table 2. Experimental parameters used for human rhythmic activities representation [7].

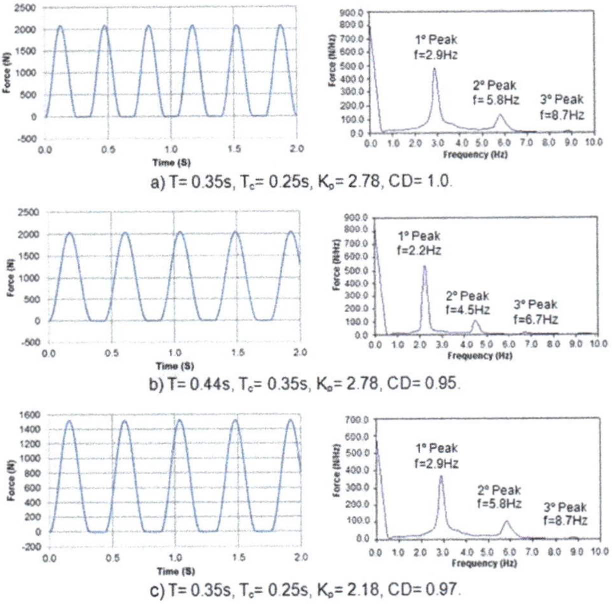

Figure 3. Dynamic loading induced by human rhythmic activities.

3. Investigated Structural Model

The investigated structural model was based on a steel-concrete composite floor spanning 40m by 40m, with a total area of 1600m^2. The structural system consisted of a typical composite floor of a commercial building. The floor studied in this work is supported by steel columns and is currently submitted to human rhythmic loads. The model is constituted of composite girders and a 100mm thick concrete slab [1,2], see Figures 4 and 5.

The steel sections used were welded wide flanges (WWF) made with a 345MPa yield stress steel grade. A 2.05x10^{5}MPa Young's modulus was adopted for the steel beams. The concrete slab has a 30MPa specified compression strength and a 2.6x10^4 MPa Young's Modulus. Table 3 depicted the geometric characteristics of the steel beams and columns.

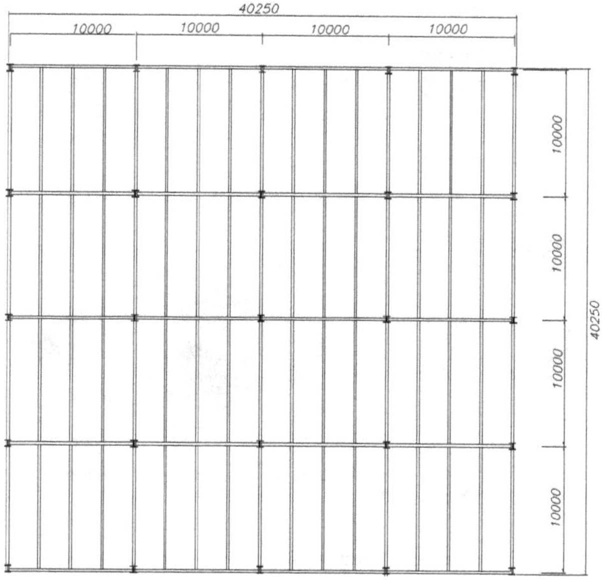

Figure 4. Structural model: composite floor (steel-concrete). Dimensions in (mm).

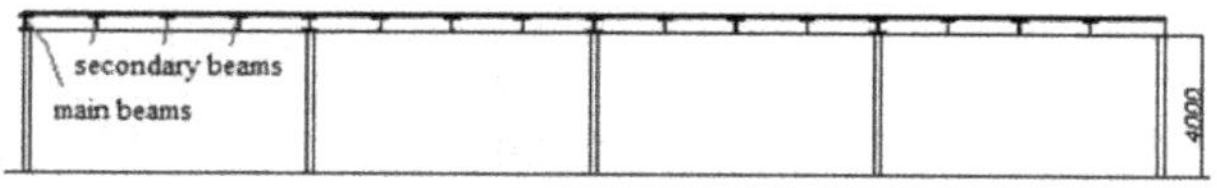

Figure 5. Cross section of the generic models. Dimensions in (mm).

Profile Type	Height (d)	Flange Width (b_f)	Top Flange Thickness (t_f)	Bottom Flange Thickness (t_f)	Web Thickness (t_w)
Main Beams (W610x140)	617	230	22.2	22.2	13.1
Secondary Beams (W460x60)	455	153	13.3	13.3	8.0
Columns (HP250x85)	254	260	14.4	14.4	14.4

Table 3. Geometric characteristics of the building composite floor (mm).

The human-induced dynamic action was applied on the aerobics area, see Figure 6. The composite floor dynamic response, in terms of peak accelerations values, were obtained on the nodes A to H, in order to verify the influence of the dynamic loading on the adjacent slab floors, as illustrated in Figure 8. In this investigation, the dynamic loadings were applied to the structural model corresponding to the effect of thirty two individuals practising aerobics.

The live load considered in this analysis corresponds to one person for each 4.0m^2 (0.25 person/m^2), according to reference [5]. The load distribution was considered symmetrically centred on the slab panels, as depicted in Figure 8. It is also assumed that an individual person weight is equal to 800N (0.8kN) [5]. In this study, the damping ratio, ξ=1% (ξ = 0.01) was considered for all cases [5].

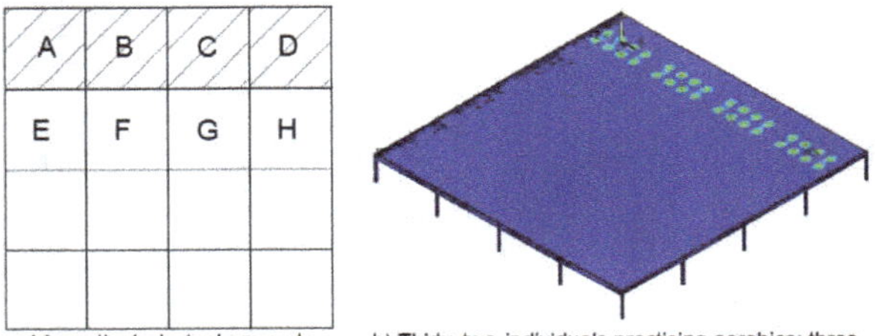

a) Investigated steel-concrete composite floor areas: top view

b) Thirty two individuals practising aerobics: three-dimensional view (3D view)

Figure 6. Dynamic loading: thirty two individuals practising aerobics on the investigated floor.

4. Finite Element Modelling

The proposed computational model, developed for the composite floor dynamic analysis, adopted the usual mesh refinement techniques present in finite element method simulations implemented in the ANSYS program [8]. The present investigation considered that both materials (steel and concrete) have an elastic behaviour. The finite element model is illustrated in Figure 7.

In this computational model, all "I" steel sections, related to beams and columns, were represented by three-dimensional beam elements (BEAM44 [8]) with tension, compression, torsion and bending capabilities. These elements have six degrees of freedom at each node: translations in the nodal x, y, and z directions and rotations about x, y, and z axes, see Figure 8.

On the other hand, the reinforced concrete slab was represented by shell finite elements (SHELL63 [8]). This finite element has both bending and membrane capabilities. Both in-plane and normal loads are permitted. The element has six degrees of freedom at each node: translations in the nodal x, y, and z directions and rotations about the nodal x, y, and z axes, see Figure 8.

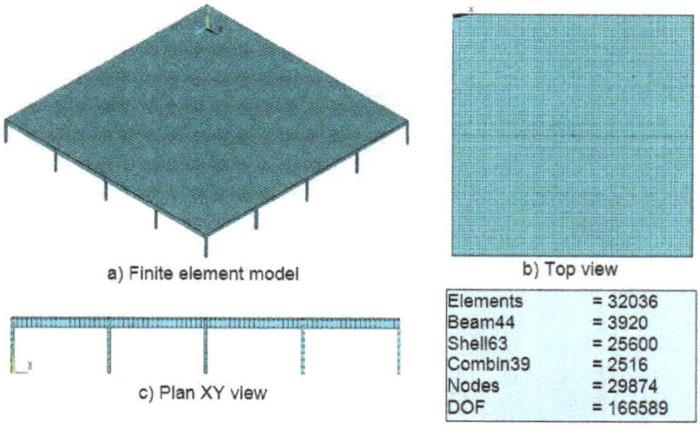

Figure 7. Steel-concrete composite floor finite element model mesh and layout.

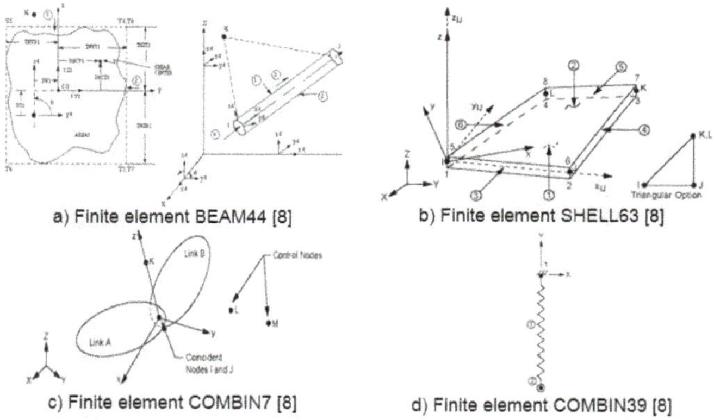

Figure 8. Finite elements used in the computational modelling.

The structural behaviour of the beam-to-beam connections (rigid, semi-rigid and flexible) present in the investigated composite floor was simulated by non-linear spring elements (COMBIN7 and COMBIN39 [8]), see Figure 8, which incorporates the geometric nonlinearity and the hysteretic behaviour effects. The moment versus rotation curve related to the adopted semi-rigid connections was based on experimental data [10], see Figure 9.

When the complete interaction between the concrete slab and steel beams was considered in the analysis, the numerical model coupled all the nodes between the beams and slab, to prevent the occurrence of any slip. On the other hand, to enable the slip between the concrete slab and the "I" steel profiles, to represent the partial interaction (steel-concrete) cases, the modelling strategy used non-linear spring elements (COMBIN39 [8]), see Figure 8, simulating the shear connector actions. The adopted shear connector force versus displacement curves were also based on experimental tests [11,12], see Figure 10.

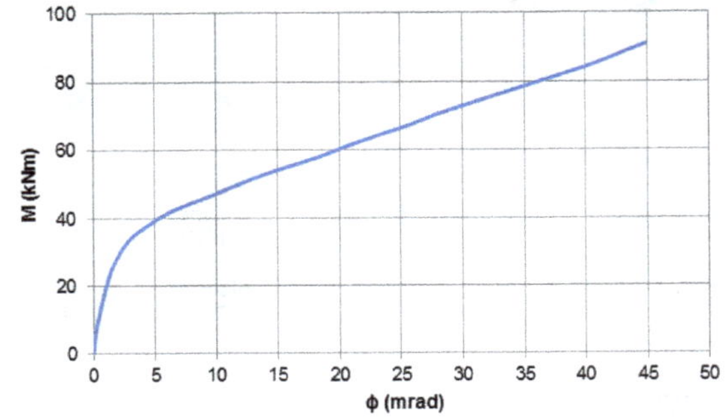

Figure 9. Moment versus rotation curve: beam-to-beam semi-rigid connections [10].

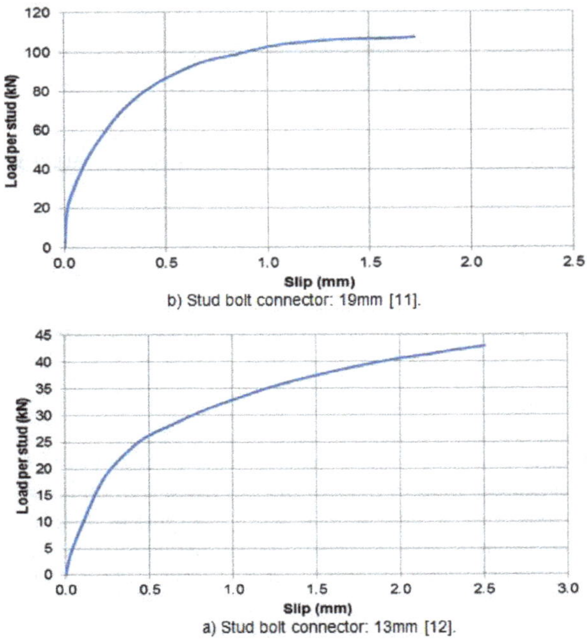

Figure 10. Force versus slip curve: shear connectors.

5. Dynamic Analysis

For practical purposes, a non-linear time-domain analysis was performed throughout this study. This section presents the evaluation of the composite floor vibration levels when submitted to human rhythmic activities. The composite floor dynamic response was determined through an analysis of its natural frequencies and peak accelerations. The results of the dynamic analysis were obtained from an extensive parametric analysis, based on the finite element method using the ANSYS program [8].

In order to evaluate quantitative and qualitatively the obtained results according to the proposed methodology, the composite floor peak accelerations were calculated and compared to design recommendations limiting values [6,9]. This comparison was made to access a possible occurrence of unwanted excessive vibration levels and human discomfort.

5.1. Natural Frequencies and Vibration Modes

The steel-concrete composite floor natural frequencies were determined with the aid of the numeric simulations, see Tables 4 and 5. The structural behaviour of the beam-to-beam connections (rigid, semi-rigid and flexible joints) and the stud connectors (from total to various levels of partial interaction cases) present in the investigated structural model were simulated objectifying to verify the influence of these connections and the steel-concrete interaction degree on the composite floor dynamic response.

Frequencies	Total Interaction			Partial Interaction (50%)		
(Hz)	Rigid	Semi-rigid	Flexible	Rigid	Semi-rigid	Flexible
f_{01}	6.57	6.14	6.00	6.32	5.91	5.76
f_{02}	6.69	6.41	6.30	6.45	6.19	6.05
f_{03}	7.03	6.52	6.37	6.76	6.27	6.31
f_{04}	7.04	6.71	6.58	6.77	6.46	6.31
f_{05}	7.11	6.97	6.85	6.87	6.72	6.58
f_{06}	7.28	7.10	6.98	7.01	6.83	6.68

Table 4. Composite floor natural frequencies (Beam-to-beam semi-rigid connections: S_j = 12kNmm/rad. Stud 13mm: S_j = 65kN/mm).

Considering the investigated composite floor natural frequencies, a small difference between the numeric results obtained with the use of total interaction or partial interaction (50%) can be observed. The largest difference between the natural frequencies was approximately equal to 5% to 7%, as presented in Tables 4 and 5 and illustrated in Figure 11.

Another interesting fact concerned that when the joints flexibility (rigid to flexible) and steel-concrete interaction degree (from total to partial) decreases the composite floor natural frequencies become smaller, see Tables 4 and 5. This conclusion is very important due to the fact that the structural system becomes more susceptible to excessive vibrations induced by human rhythmic activities.

Frequencies (Hz)	Total Interaction			Partial Interaction (50%)		
	Rigid	Semi-rigid	Flexible	Rigid	Semi-rigid	Flexible
f_{01}	6.63	6.18	6.06	6.39	5.98	5.84
f_{02}	6.75	6.46	6.36	6.52	6.26	6.13
f_{03}	7.10	6.58	6.43	6.84	6.35	6.19
f_{04}	7.11	6.77	6.65	6.85	6.54	6.40
f_{05}	7.17	7.02	6.91	6.94	6.79	6.67
f_{06}	7.35	7.16	7.05	7.08	6.91	6.78

Table 5. Composite floor natural frequencies (Beam-to-beam semi-rigid connections: S_j = 12kNmm/rad. Stud 19mm: S_j = 200kN/mm).

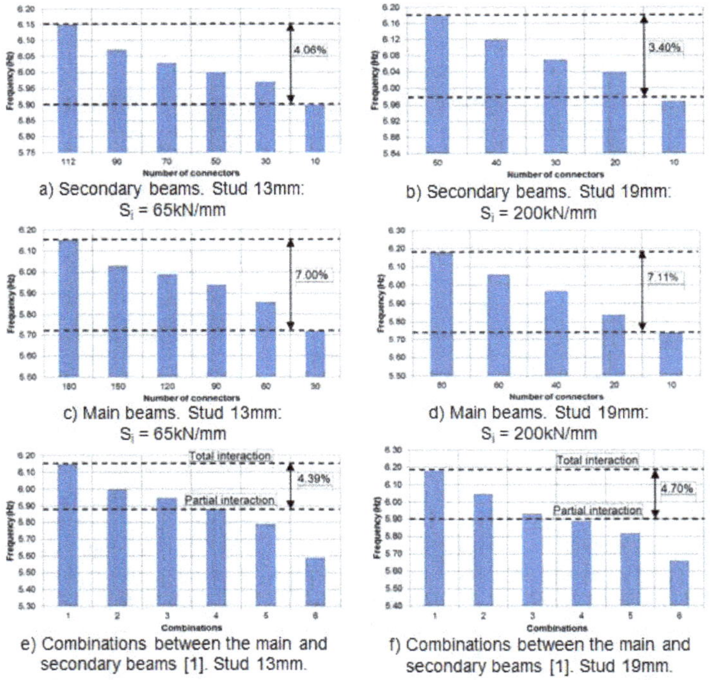

Figure 11. Steel-concrete composite floor fundamental frequency (f_{01}) variation.

In sequence, Figure 12 presents the composite floor vibration modes when total and partial interaction situations were considered in the numerical analysis. It must be emphasized that the composite floor vibration modes didn't present significant modifications when the connections flexibility and steel-concrete interaction was changed. It must be emphasized that the structural model presented vibration modes with predominance of flexural effects, as illustrated in Figure 12.

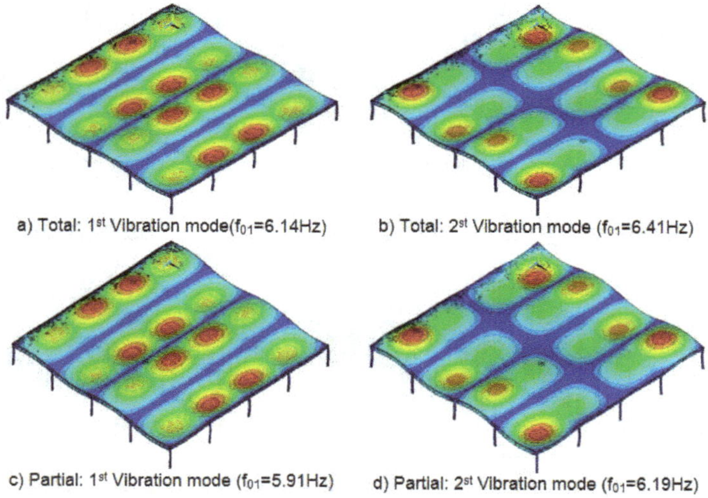

a) Total: 1st Vibration mode(f_{01}=6.14Hz)   b) Total: 2st Vibration mode (f_{01}=6.41Hz)

c) Partial: 1st Vibration mode (f_{01}=5.91Hz)   d) Partial: 2st Vibration mode (f_{01}=6.19Hz)

Figure 12. Investigated structural model vibration modes (total and partial interaction).

5.2. Maximum accelerations (peak accelerations) analysis

The present study proceeded with the evaluation of the structural model performance in terms of human comfort and vibration serviceability limit states. The peak acceleration analysis was focused in aerobics and considered a contact period carefully chosen to simulate this human rhythmic activity on the analysed composite floor.

The present work considered a contact period, simulating aerobics on the composite floor, T_c, equal to 0.34s (T_c = 0.34s) and the period without contact with the structure, T_s, of 0.10s (T_s = 0.10s). Based on the experimental results [7], the floor dynamic behaviour was evaluated keeping the impact coefficient value, K_p, equal to 2.78 (K_p = 2.78). Figures 13 and 14 illustrate the dynamic response (displacements and accelerations) related to nodes A and B (see Figure 6) when thirty two people are practising aerobics on the composite floor.

Based on the results presented in Figures 13and 14, it is possible to verify that the dynamic actions coming from aerobics, represented by the dynamic loading model (see Equation (1) and Figure 6), have generated peak accelerations higher than 0.5%g [6,9]. This trend was confirmed in several other situations [1,2], where the human comfort criterion was violated.

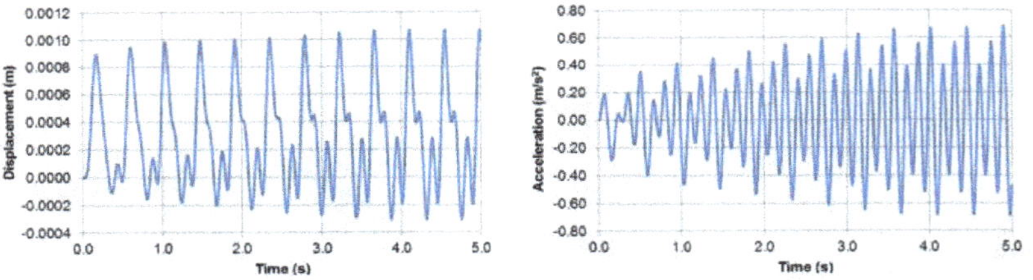

Figure 13. Composite floor dynamic response. Semi-rigid connections and partial interaction): Node A.

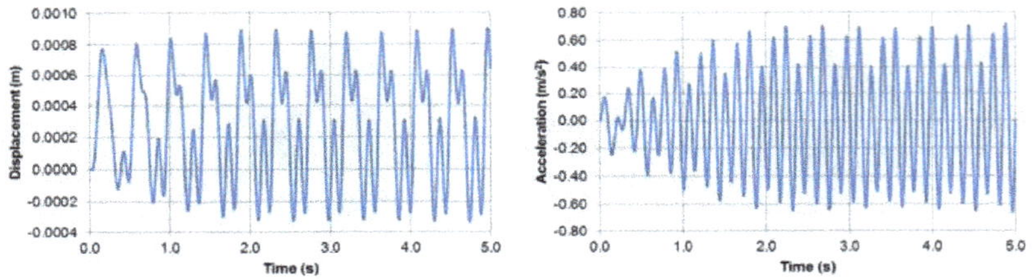

Figure 14. Composite floor dynamic response (Semi-rigid connections and partial interaction): Node B.

In sequence of the study, Tables 6 and 7 show the peak accelerations, a_p (m/s²), corresponding to nodes A to H (Figure 6), when thirty two dynamic loadings, simulating individual practising aerobics were applied on the composite floor.

Interaction	Model	a_p (m/s²) Node A	a_p (m/s²) Node B	a_p (m/s²) Node C	a_p (m/s²) Node D
Complete	Rigid	0.26	0.17	0.17	0.26
	Semi-rigid	0.28	0.20	0.20	0.28
	Flexible	0.30	0.44	0.43	0.30
Partial (50%)	Rigid	**0.53**	0.36	0.36	**0.53**
	Semi rigid	**0.62**	**0.63**	**0.63**	**0.62**
	flexible	**0.60**	**0.80**	**0.80**	**0.60**
Limiting Acceleration: a_{lim} = 0.50m/s² (5%g - g: gravity) [6,9]					

Table 6. Composite floor peak accelerations: Nodes A, B, C and D (see Figure 6).

Interaction	Model	a_p (m/s²) Node E	a_p (m/s²) Node F	a_p (m/s²) Node G	a_p (m/s²) Node H
Complete	Rigid	0.035	0.035	0.035	0.035
	Semi rigid	0.087	0.036	0.036	0.087
	flexible	0.088	0.09	0.09	0.088
Partial (50%)	Rigid	0.30	0.13	0.13	0.30
	Semi-rigid	0.40	0.14	0.14	0.40
	Flexible	0.32	0.24	0.24	0.32
Limiting Acceleration: a_{lim} = 0.50m/s² (5%g - g: gravity) [6,9]					

Table 7. Composite floor peak accelerations: Nodes E, F, G and H (see Figure 6).

The results presented in Tables 6 and 7 have indicated that when the joints flexibility (rigid to flexible) and steel-concrete interaction degree (total to partial) decreases the composite floor peak accelerations become larger. These variations (joints flexibility and steel-concrete interaction) were very relevant to the composite floor non-linear dynamic response when the human comfort analysis was considered.

It must be emphasized that individuals practising aerobics on the structural model led to peak acceleration values higher than 5%g [6,9], when the composite floor was submitted to thirty two people practising aerobics, violating the human comfort criteria (a_{max} = 0.50m/s²> a_{lim} = 0.50m/s²), see Tables 6 and 7. However, these peak acceleration values tend to decrease when the floor dynamic response obtained on the nodes E to H (see Figure 6) was compared to the response of nodes A to D (see Figure 6), see Tables 6 and 7.

6. Final Remarks

The main objective of this paper was to investigate the beam-to-beam structural connections effect (rigid, semi-rigid and flexible) and the influence of steel-concrete interaction degree (from total to various levels of partial interaction) over the non-linear dynamic behaviour of composite floors when subjected to human rhythmic activities. This way, an extensive parametric analysis was developed focusing in the determination quantitative aspects of the composite floors dynamic response.

The investigated structural model was based on a steel-concrete composite floor spanning 40m by 40m, with a total area of 1600m². The structural system consisted of a typical composite floor of a commercial building. The composite floor studied in this work is supported by steel columns and is currently submitted to human rhythmic loads. The structural system is constituted of composite girders and a 100mm thick concrete slab.

The proposed computational model adopted the usual mesh refinement techniques present in finite element method simulations, based on the ANSYS program. The numerical model enabled a complete dynamic evaluation of the investigated steel-concrete composite floor especially in terms of human comfort and its associated vibration serviceability limit states.

The influence of the investigated connectors (Stud Bolts: 13mm and 19mm) on the composite floor natural frequencies was very small, when the steel-concrete interaction degree (from total to partial) was considered in the analysis. The largest difference was approximately equal to 5% to 7%.

On the other hand, when the joints flexibility (rigid to flexible) and steel-concrete interaction degree (from total to partial) decreases the composite floor natural frequencies become smaller. This fact is very relevant because the system becomes more susceptible to excessive vibrations.

The composite floor vibration modes didn't present significant modifications when the connections flexibility and steel-concrete interaction was changed. The investigated structure presented vibration modes with predominance of flexural effects. The results have indicated that when the joints flexibility (rigid to flexible) and steel-concrete interaction degree (total to partial) decreases the composite floor peak accelerations become larger.

The maximum acceleration value found in this work was equal to 0.80m/s^2 (a_p = 0.80 m/s^2: flexible model) and 0.63m/s^2 (a_p = 0.63 m/s^2: semi-rigid model), while the maximum accepted peak acceleration value is equal to 0.50m/s^2 (a_{lim} = 0.50m/s^2) [6,9]. The structural system peak accelerations were compared to the limiting values proposed by several authors and design standard [6,9]. The current investigation indicated that human rhythmic activities could induce the steel-concrete composite floors to reach unacceptable vibration levels and, in these situations, lead to a violation of the current human comfort criteria for these specific structures.

Acknowledgements

The authors gratefully acknowledge the support for this work provided by the Brazilian Science Foundation CAPES, CNPq and FAPERJ.

Author details

José Guilherme Santos da Silva*, Sebastião Arthur Lopes de Andrade,
Pedro Colmar Gonçalves da Silva Vellasco, Luciano Rodrigues Ornelas de Lima,
Elvis Dinati Chantre Lopes and Sidclei Gomes Gonçalves

*Address all correspondence to: jgss@uerj.br

State University of Rio de Janeiro (UERJ), Rio de Janeiro-RJ, Brazil

References

[1] Lopes, EDC. (2012). Effect of the steel-concrete interaction over the composite floors non-linear dynamic response (In development). *PhD Thesis (In Portuguese)*, Pontifical Catholic University of Rio de Janeiro, PUC-Rio, Rio de Janeiro/RJ, Brazil.

[2] Gonçalves, SG. (2011). Non-linear dynamic analysis of composite floors submitted to human rhythmic activities. *MSc Dissertation (In Portuguese)*, Civil Engineering Post-graduate Programme, PGECIV, State University of Rio de Janeiro, UERJ, Rio de Janeiro/RJ, Brazil.

[3] Silva, J. G. S., da Vellasco, P. C. G. S., Andrade, S. A. L., de Lima, L. R. O., & de Almeida, R. R. (2008). Vibration analysis of long span joist floors when submitted to dynamic loads due to human activities. Athens, Greece. *Proceedings of the 9th International Conference on Computational Structures Technology, CST*, CD-ROM, 1-11.

[4] Langer, N. A. dos S., Silva, J. G. S. da, Vellasco, P. C. G. da S., Lima, L. R. O. de, & Neves, L. F. da C. (2009). Vibration analysis of composite floors induced by human rhythmic activities. Ilha da Madeira, Portugal. *Proceedings of the 12th International Conference on Civil, Structural and Environmental Engineering Computing, CC, Funchal*, CD-ROM, 1-14.

[5] Bachmann, H., & Ammann, W. (1987). Vibrations in structures induced by man and machines, IABSE Structural Engineering Document 3E. *International Association for Bridges and Structural Engineering*, 3-85748-052-X.

[6] Murray, T. M., Allen, D. E., & Ungar, E. E. (2003). Floor Vibrations due to Human Activity. *Steel Design Guide Series*, American Institute of Steel Construction, AISC, Chicago, USA.

[7] Faisca, R. G. (2003). Characterization of Dynamic Loads due to Human Activities. *PhD Thesis (In Portuguese)*, Civil Engineering Department, COPPE/UFRJ, Rio de Janeiro/RJ, Brazil.

[8] ANSYS Swanson Analysis Systems, Inc. (2007). P. O. Box 65, Johnson Road, Houston, PA, 15342-0065. Release 11.0, SP1 UP20070830, ANSYS, Inc. is a UL registered ISO 9001:2000 Company. Products ANSYS Academic Research, Using FLEXlm v10.8.0.7 build 26147, Customer 00489194

[9] International Standard Organization. (1989). Evaluation of Human Exposure to Whole-Body Vibration, Part 2: Human Exposure to Continuous and Shock-Induced Vibrations in Buildings (1 to 80Hz). *ISO 2631-2*.

[10] Oliveira, T. J. L. (2007). Steel to concrete composite floors with semi-rigid connections: non-linear static and dynamic, experimental and computational analysis. , PhD Thesis (In Portuguese) COPPE/UFRJ, Federal University of Rio de Janeiro Rio de Janeiro/RJ, Brazil

[11] Ellobody, E., & Young, B. (2005). Performance of shear connection in composite beans with profiled steel sheeting. *Journal of Constructional Steel Research,* 62.

[12] Tristão, GA. (2002). Behaviour of shear connectors in composite steel-concrete beams with numerical analysis of the response. *MSc Dissertation (In Portuguese),* School of Engineering of Sao Carlos. University of São Paulo, São Carlos/SP, Brazil.

Chapter 12

Vibration of Satellite Solar Array Paddle Caused by Thermal Shock When a Satellite Goes Through the Eclipse

Mitsushige Oda, Akihiko Honda, Satoshi Suzuki and Yusuke Hagiwara

Additional information is available at the end of the chapter

1. Introduction

Remote sensing satellites take images of the earth's surface in observing various activities by humans or nature. In order to obtain precise and high-resolution images from a satellite in Low Earth Orbit (LEO) at an attitude of 500 to 900 kilometers, the satellite's attitude must be stable when the onboard camera sensors take images of surface activities on the earth. Should the attitude stability of the satellite be disturbed while such images are being taken, poor image quality would probably result.

The fact that images taken when a satellite goes into or out from an eclipse do not provide good accuracy—due to degraded altitude stability at such timings—has been known for years. Such phenomena has long been attributed to the deformation and vibration of the satellite's solar array paddle that occurs when the satellite go into or out from an eclipse, along with the instantaneous change in solar energy received by the satellite. Several trials were conducted in the past to identify the phenomena, but all failed to achieve reasonable results.

2. Measurement of solar array paddle motion

The reason why past trials failed to observe the phenomena described above might be that the motion of the solar array paddle is too small or slow to be observed by such onboard sensors as an accelerometer. Therefore, JAXA decided to measure the phenomena by using an onboard camera that was originally mounted on the satellite to monitor solar array paddle deployment (Fig. 2). (For GOSAT, the solar array paddle consists of three solar panels, a yoke, wires that control deploying speed, and hinge/latch mechanisms that connect the panels and yoke.) A small CCD/CMOS camera (Fig. 4) is mounted on most JAXA satellites to monitor solar array paddle deployment, as failure to deploy the solar array paddle would become critical failure of the satellite itself. We thus developed a system that uses this CMOS camera to measure the distortion and vibration of the solar array paddle, with said vibration being measured as follows:

1. Attach small reflective target markers at the end of the paddle (as shown in Fig. 3).
2. Take images from this camera and transmit them to the on-ground station. Fig. 2 is an image taken by the camera.
3. Identify the locations of the target markers in the camera view with image processing.
4. Calculate displacement from the predicted target marker locations.

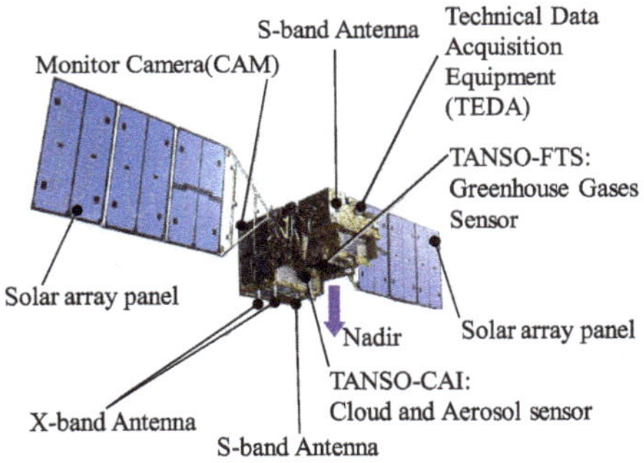

Figure 1. JAXA's earth observation satellite "GOSAT"

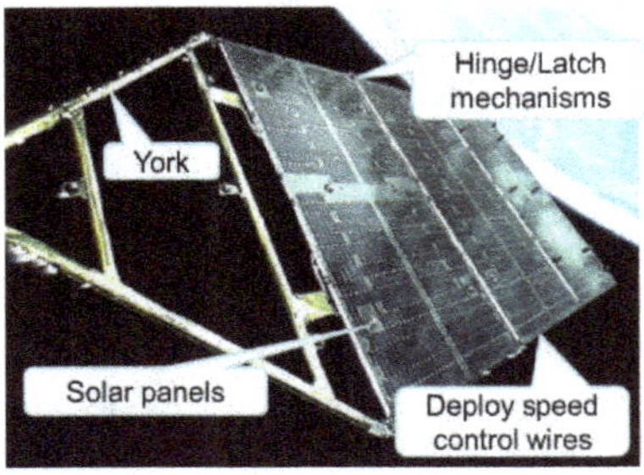

Figure 2. Image taken by camera

For GOSAT, the solar array paddle consists of three solar panels, a yoke, wires that control deploying speed, and hinge/latch mechanisms that connect the panels and yoke.

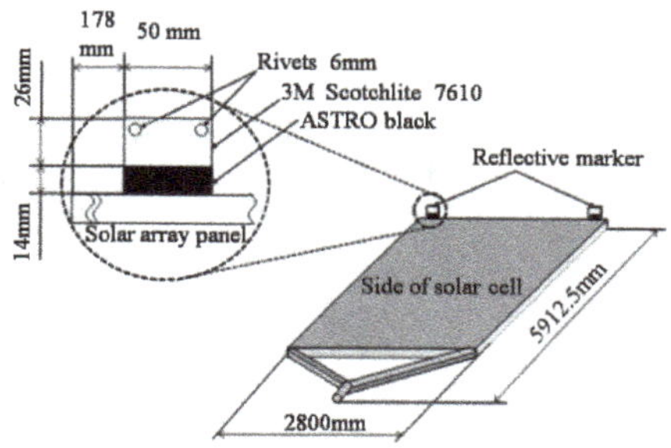

Figure 3. Target markers on the solar array paddle

Figure 4. CMOS camera and LED lights

3. Measurement algorithm

As the onboard camera used for this experiment was originally designed to monitor the deployment of a folded solar array paddle, its field of view is thus as wide as 90 [deg] (Fig. 5), while the camera's number of pixels is limited to SXGA (1280 X 1024 pixels). The size of the markers is also limited to 50 [mm] X 26 [mm], while distance from the camera to the target markers is as far as 6 [m]. These constraints mean that one pixel of the camera is equivalent to 7 [mm] at the target marker's position. Therefore, a technique for processing sub-pixel level image data is required to identify deformation of the solar array paddle.

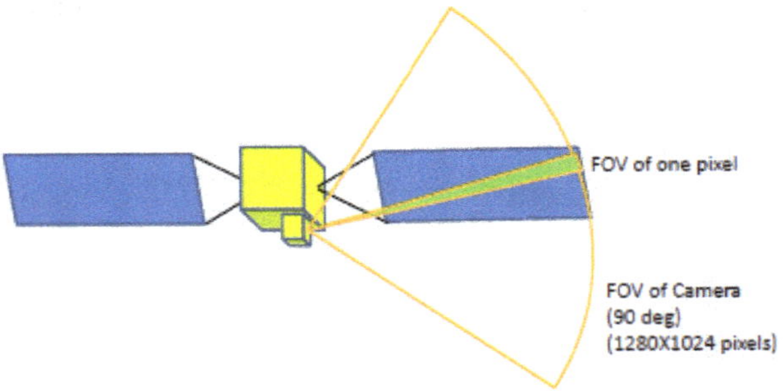

Figure 5. Constraints on image data processing

4. Previous results

The initial results of these measurements were reported in another paper in this series (Oda et al., 2011). Fig. 6 shows typical results. It shows an offset of a few millimeters appeared when solar array paddle illuminated by the Sun and when in an eclipse. However, these results pose certain difficulties in explaining the phenomena, as the tendency of the solar array paddle to bend does not agree with the observation results and conventional research.

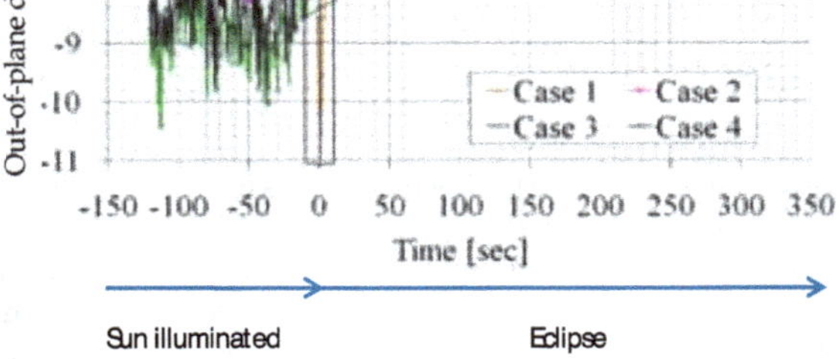

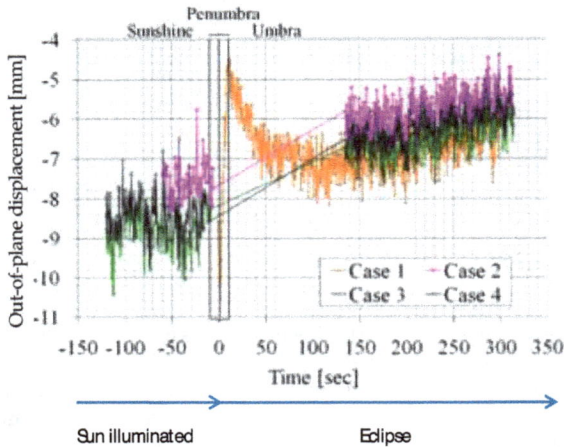

Figure 6. Motion of the solar array paddle's tip position as estimated from onboard camera images

Since solar energy from the Sun is the primary energy provided on the satellite and its solar array paddle, deformation of the solar array paddle should occur so that its structure on the Sun side is extended and causes bending of the solar panels toward the rear side of the solar array paddle. Previous results have indicated that the solar array paddle will bend toward its solar cell side, however. When the initial results of these measurements were recorded, we were unable to identify the source of the errors.

5. Revised results

Fig. 6 shows a sudden offset in the motion of the solar array paddle. When the solar array paddle is straight, the value of displacement may indicate -12 or -11 [mm] in Fig.6. Although we attempted to identify cause of this offset, we could not imagine a proper mechanism that would produce such a direction. We therefore assumed that the previous image data processing contained unidentified data processing errors, and consequently modified the algorithm that identifies the target markers. The earlier version of the algorithm used to identify locations of the target markers assumed such highly illuminated areas as those of the target markers. This algorithm works well when the markers are brightly illuminated. When the target markers are weakly illuminated, however, we found that this algorithm produces some data processing errors.

Fig. 7 and Fig. 8 illustrate the difference described above. Fig. 7 is based on the previous algorithm. The areas enclosed by a yellow line are pixels that are brighter than the threshold level and thus can be assumed to be the target markers as based on sub-pixel level image data processing. The red cross indicates the center position of the marker.

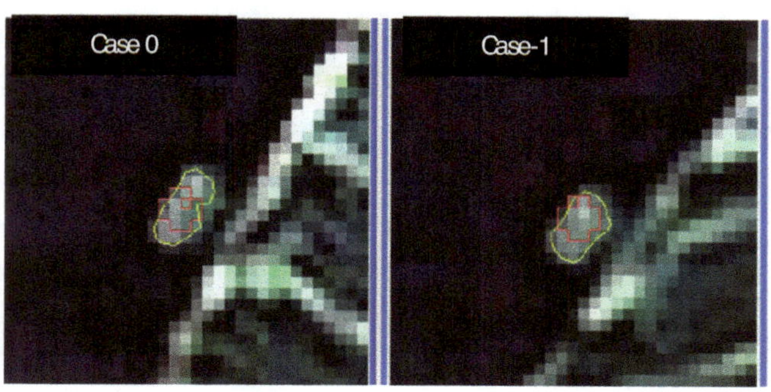

Figure 7. Target markers estimated by the previous algorithm under weak illumination. The areas of target markers assumed by the previous algorithm are smaller than the actual target marker size.

Fig. 7 shows that the sizes of the estimated target markers are smaller than the actual target markers. We therefore modified the algorithm used to estimate the area of a target marker so that the size of the predicted target marker is similar to that calculated from the actual target marker. This modification worked well to identify the target marker and its displacement. Fig. 8 is a result based on the revised algorithm. In this revised algorithm, the areas to be considered target makers are decided based on the brightness level of each pixel and also on the size of the areas considered to be target markers. When an assumed target marker is too small, then the threshold level of brightness is automatically adjusted to meet the possible size of the target markers.

Finally, Fig. 9 shows the motion of a target marker as based on the revised algorithm. We can see that the estimated motion of the solar array paddle has less dispersion.

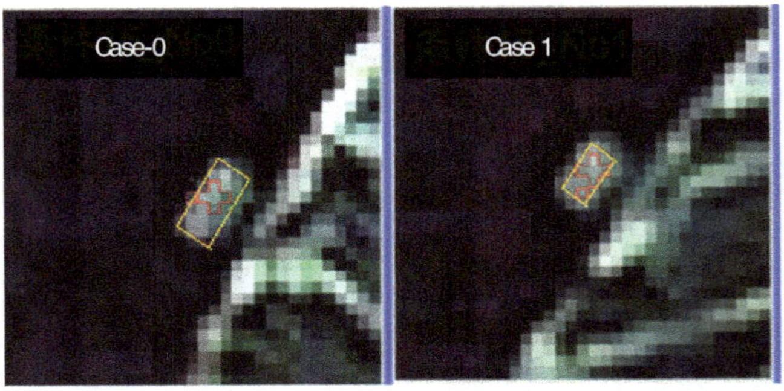

Figure 8. Target markers estimated by the revised algorithm

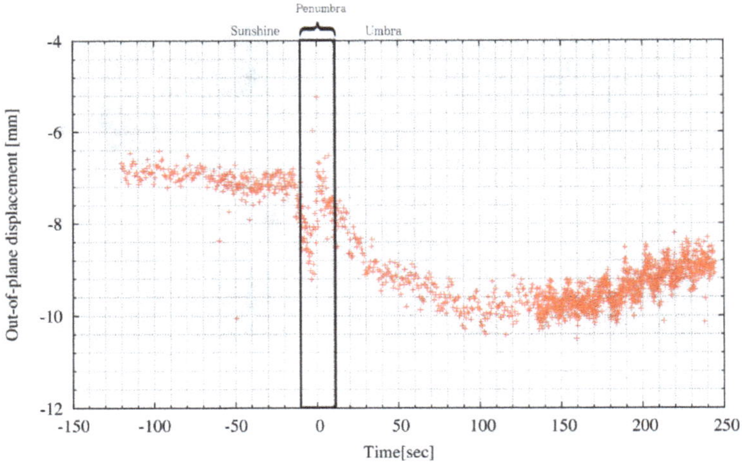

Figure 9. Displacement of GOSAT's solar array paddle when going into eclipse

From these data, we can conclude that the temperature of the solar array paddle changes steadily but not drastically, depending on its thermal capacity. And we also assume like as following.

1. When the solar array paddle is illuminated by the Sun, its temperature is governed by solar energy from the Sun (as well as that reflected from Earth).

2. When in an eclipse, the solar array paddle shows lower temperature depending on its thermal capacity. When the satellite's LST (Local Sun Time) is around noon, the solar cell side of the panels faces the Earth and receives solar radiation from Earth. In the evening and the early morning of satellite LST, however, both sides of the solar panels face toward space, resulting in a rapid drop in temperature.

6. Numerical simulation

In order to verify whether the observation results described above are correct, and to understand the features of thermal snap on the solar array paddle, we conducted numerical simulation of the solar array paddle. There are many studies analyzing the thermal snap (Thornton, 1996; Boley, 1972; Johnston, 1998; Lin, 2004; Xue2007), we develop the method using a thermal model and a structural model and revising these models by applying the observed data. This section presents the analytical system that we constructed, the result of a thermal-structural analysis and its problems. To solve the problems, we have developed a new model in considering the effects of hinge/latch mechanisms and friction. The result of using new model is also introduced.

6.1. Thermal snap analysis procedure

This section describes the thermal snap analysis procedure. Fig. 12 shows a flowchart of thermal snap analysis. The analysis can be broken down into to three parts: construction of the structural model, calculation of temperature distribution, and thermal snap analysis for the penumbra.

In the first part, a structural model of the solar array paddle is constructed for thermal snap analysis. To verify the structural model, we conducted modal analysis to obtain the natural frequency and vibrational mode of the solar array paddle model. These results will be compared with an on-orbit preliminary experiment, and if necessary, we will then revise the model (see Section 6.2 for details). In the second part, a thermal model is developed for the solar array paddle. Thermal analysis for the whole orbit is then conducted to verify the thermal model. The thermal analysis results will be applied to GOSAT's trajectory information, and also compared with data obtained by GOSAT, in order to verify accuracy. After accuracy is verified, thermal analysis will focus on GOSAT during its integration and testing. In order to improve the accuracy of thermal analysis, a profile of thermal input was determined based on the brightness of the solar paddles (see Section 6.3 for details).

In the third part, the thermal snap analysis is conducted using the structural FEM model of GOSAT.

Fig. 10 and Fig. 11 below show the structural model and the thermal model, respectively.

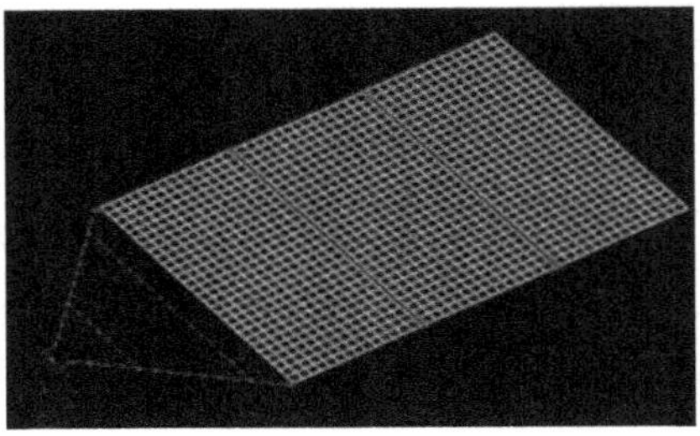

Figure 10. Structural model

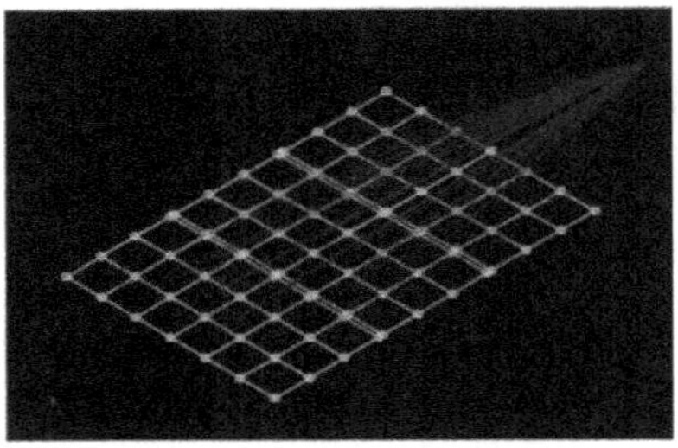

Figure 11. Thermal model

6.2. Improvement of the structural model

In constructing the structural model of GOSAT's two solar cell paddles, we were not allowed to access the detailed satellite design data. Therefore, the structural model was based on partly assumed data. Moreover, it is difficult to estimate the production errors on GOSAT. We thus compared the structural model design data and the observation data. The preliminary observation of solar cell paddle motion was made as GOSAT conducted orbit-raising maneuvers using the 20 Newton Gas Jet thrusters. This maneuvering applied relatively large force to the satellite's main body, thereby causing large bending of the solar cell paddles, followed by the induced vibration thereof. An earlier paper in this series reported the measurement results.

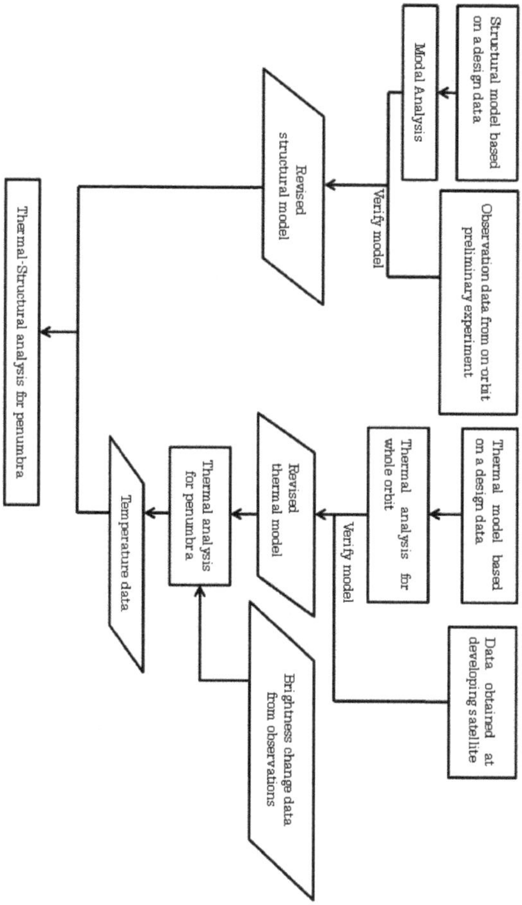

Figure 12. Flowchart of thermal snap analysis using observed data

In this experiment, we severely shook the solar paddle with the 20N thrusters, in order to observe its behavior. The FFT data obtained from this experiment on solar paddle vibration is available. Fig. 13 shows an example of the obtained FFT data. From these data, the natural frequencies of the actual solar array paddle were assumed to be 0.215 [Hz] (out-of-plane) and 0.459 [Hz] (in-plane). These natural frequencies are identical to those obtained by analysis during development.

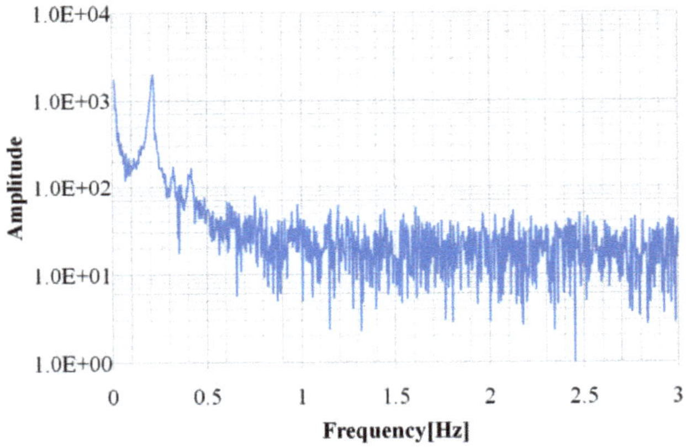

Figure 13. Example of FFT data on out-of-plane solar array paddle vibration as obtained from observation

Next, we conducted modal analysis of the structural model of the solar array paddle to calculate the natural frequency. In case of any differences between the analysis results and observation results, we adjusted the structural data around the root of the solar cell paddles.

By iterating this process, the structural model was revised to generate data similar to that obtained from the solar array paddle in orbit.

Tab. 1 below lists shows the final modal analysis results of natural frequency; Fig. 14 shows the modal shapes of the structural model. From these data, the modal analysis results of the revised structural model match the observed data. Hence, we expect that the revised model could sufficiently simulate the actual solar array paddle.

Mode number	Frequency [Hz]	Modal shapes
1	0.218	First order out-of-plane
2	0.454	First order in-plane
3	1.262	First order twist
...	...	...

Table 1. Modal analysis results

6.3. Detailed thermal input profile from observed data

During the time when passing the boundary between the sunlight area and umbra area, the thermal environment changes gradually. To conduct thermal snap simulation, a detailed thermal input profile is needed, but estimating changes from trajectory information alone is difficult due to atmospheric effects. We thus estimated the value of solar light incident (i.e. dominant factor in thermal input) from the brightness value of images taken in observation. By using gray-scale processed images, we assume that if the average gray scale value of an image is max, then the value of solar light incident will also be max. If it is minimum, then the value will be minimum. Fig. 15 shows the estimated solar light incident and total thermal input profile. From the result, we can see that GOSAT's actual optical environment changes gently three times, as long as the value is forecast based on geometric calculation of the satellite's trajectory.

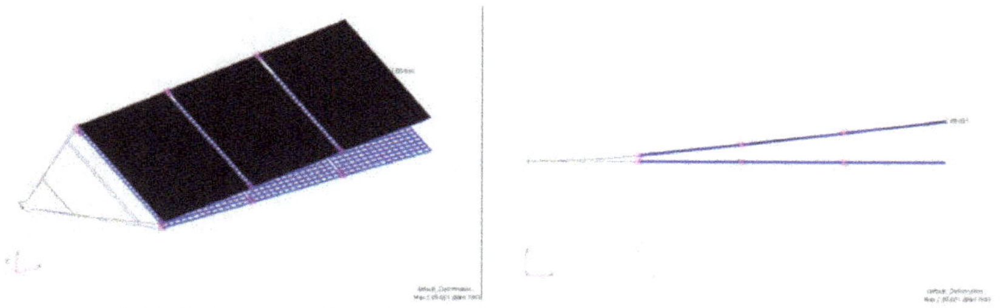

Modal shape for first order, out-of-plane vibration

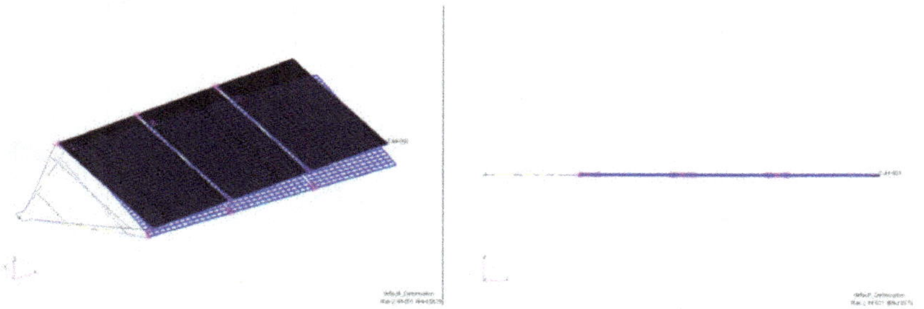

Modal shape for first order, in-plane vibration

Figure 14. Modal shapes of the structural model (Out of Plane and In-Plane Vibration)

By using the thermal input profile thus revealed, we conducted thermal analysis for the penumbra. Fig. 16 shows the simulation results of solar paddle temperature and the values of thermometers. The thermometers are attached at two points. However, no temperature sensor was attached on the front (solar cell; optical incidence) side of the solar paddle, but one was attached only on a backside plane (radiation plane; indicated as Bottom CFRP in Fig. 16). The tendency of simulated solar paddle temperature is similar, however, to the actual data. We therefore conclude that our thermal analysis can sufficiently simulate actual thermal changes.

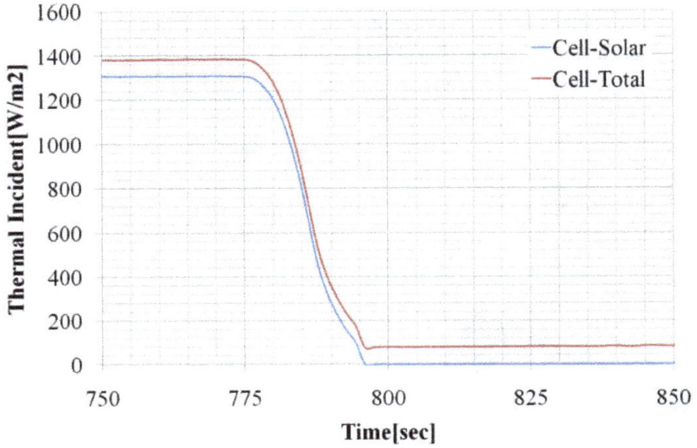

Figure 15. Estimated solar light incident and total thermal input profile from images

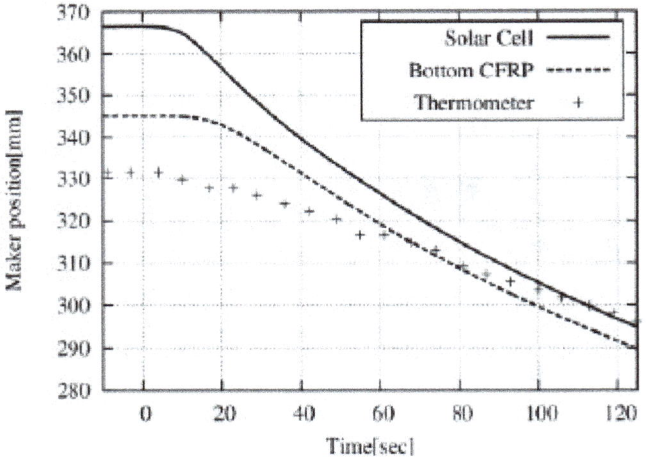

Figure 16. Estimated solar array paddle temperature and values of thermometers attached on a backside plane (Bottom CFRP)

6.4. Results of thermal snap analysis

Thermal snap analysis is conducted using the revised structural model and temperature distribution. Fig. 17 shows the results of thermal snap analysis in the penumbra. The simulation data shows that the solar array paddles will bend in quasi-static while the paddle is illuminated by the Sun. And from the deformation plot, it is assumed that the wires have a large effect on a deformation.

By using the models, we conducted thermal deformation analysis for the whole orbit—a task not possible through camera observation alone. Fig. 18 shows the results. It shows that much larger deformation occurred in the penumbra than at any other time.

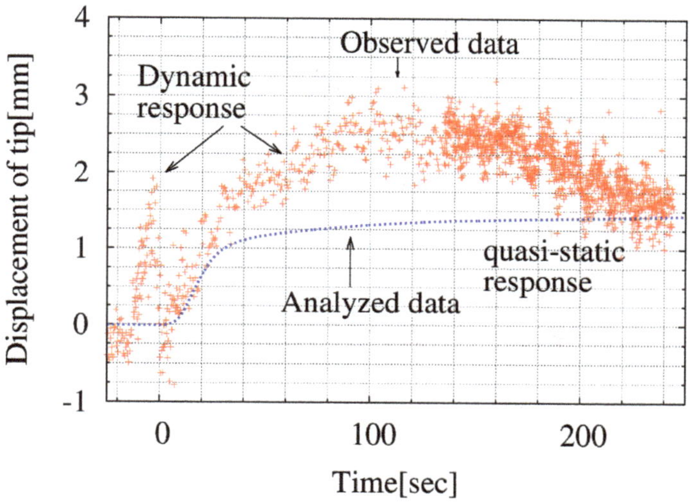

Figure 17. Analysis results of deformation in penumbra

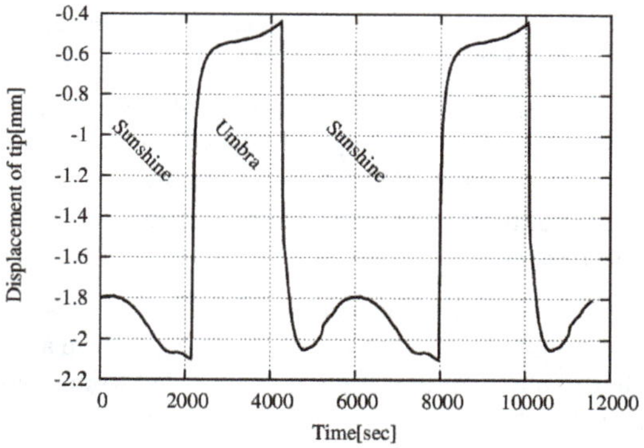

Figure 18. Solar array paddle deformation for two round orbit (The orbital period of GOSAT is 6000 [sec].)

6.5. Detailed modeling for hinges

As dynamic response offers the possibility of adding more awkward disturbances to attitude stability than quasi-static deformation, we decided to check the penumbra data in detail. The observed data obtained in the penumbra indicates that the solar array paddles have a much lower natural frequency. Fig. 19 shows the FFT data on out-of-plane deformation in the penumbra.

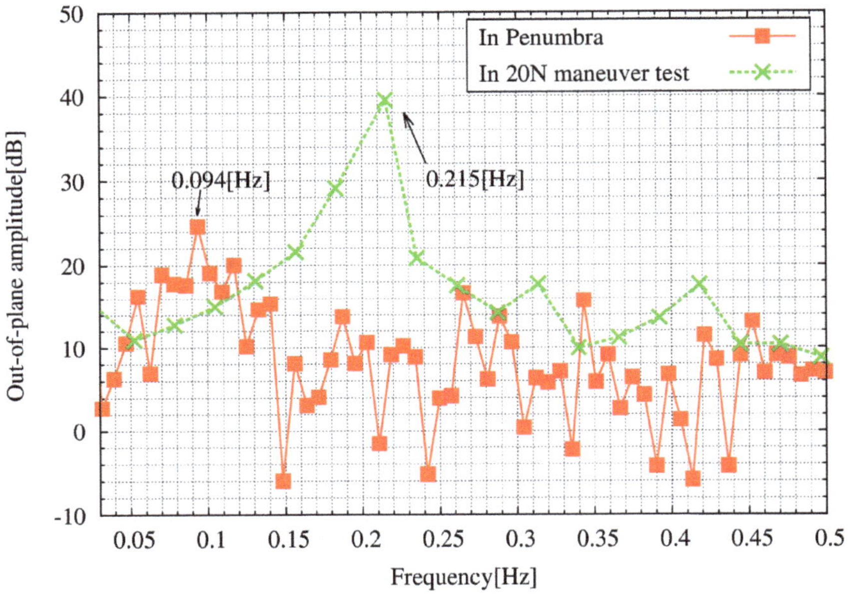

Figure 19. FFT analysis of out-of-plane deformation in the penumbra and same data obtained when maneuvering to raise orbital altitude using powerful 20N thrusters

Fig. 19 shows the data obtained when using the powerful 20N thrusters to raise GOSAT's orbital altitude, with vibration of 0.215 [Hz] being observed. When the satellite goes into eclipse, however, the data obtained did not reveal vibration of 0.215 [Hz], but instead showed a lower frequency of 0.094 [Hz]. Similar data was not reported during the satellite integration and qualification tests.

From these reasons, we assumed that the small load (e.g. solar pressure) acts on the solar panels, and that this load acts as rotation torque at each hinge. As all solar cell panels of GOSAT are linked together to maintain the angle of deployment to the deployed position, the small motion generated at each hinge will also act on all the hinges.

Vibration will occur at each hinge that connects each solar cell panel. Hinges and wires are used to interconnect the solar cell panels, in order to deploy the solar cell panels and maintain the open position of each panel. As each hinge will have some backlash and in order to pull each panel to the deployed position, wires are used to maintain the deployed position after each panel is deployed, and to connect the solar panels, resulting in only a small load under microgravity conditions. Fig. 20 and 21 shows the concept.

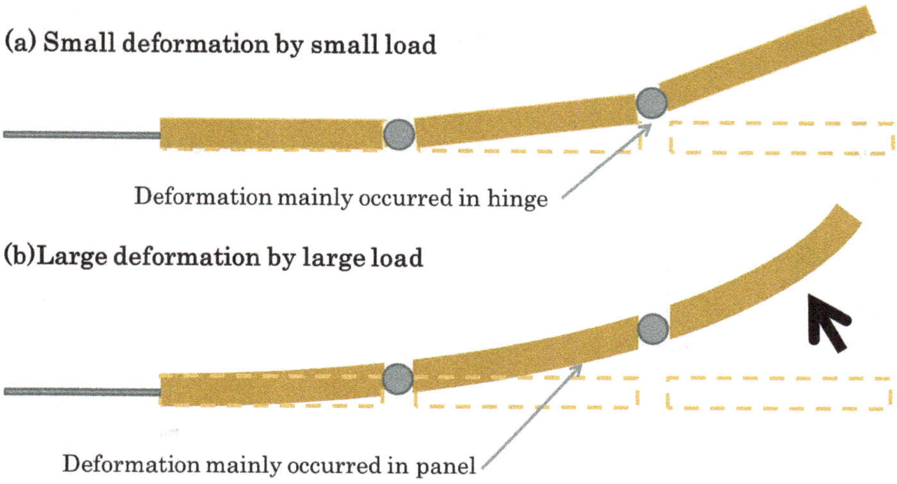

Figure 20. Concept of vibration affected by the hinge gap

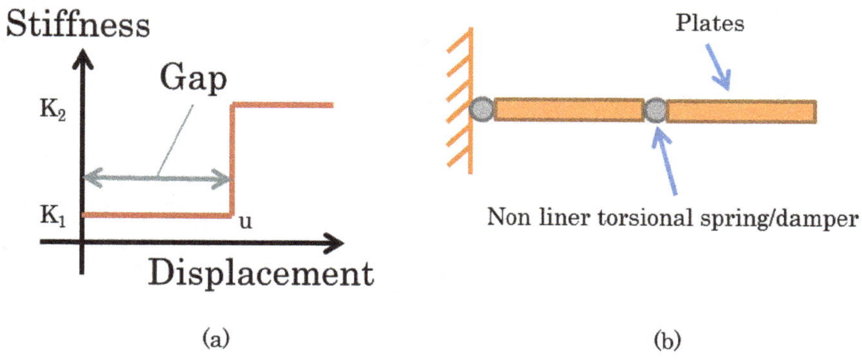

Figure 21. Concept of a hinge modeled using a nonlinear torsional spring. In Fig. 21, (a) shows how to define the stiffness of torsional springs; (b) shows the model for the solar paddle.

Should a small load (e.g. thermal stress) be applied to the solar paddle, deformation will mainly occur in the hinges that have small gaps. Conversely, in case a large load (e.g. inertia force at maneuver) is applied, deformation of the solar paddle is typically larger than the gaps; therefore, such deformation is mainly caused by the elastic deformation of the panels.

To simulate the effect caused by hinges with a gap, we modeled the hinges with nonlinear torsional springs. In Fig. 21, (a) shows how we added nonlinear characteristics to the torsional springs. The length of u is used to define the gap size. And stiffness within the gap is defined by K_1. When the whole dynamic response shown in Fig. 17 is assumed to have occurred within gap area, the gap width is easily defined. Then, the stiffness of the gap is defined as shown below.

To estimate stiffness, the modeling method for the model of spring-connected plates is used (Kojima et al., 2004). The rotational motion equation for the model of spring-connected plates as shown in Fig. 21 (b) is:

$$I\theta + K\theta = 0 \tag{1}$$

where, I denotes the inertia moment of plates and K the stiffness of torsional springs. Here, given harmonic vibration in which frequency is , the stiffness of gap area K is expressed as:

$$K = I\omega^2 \tag{2}$$

1 is 0.094 [Hz]. Therefore, we assumed the stiffness of gap area K as about 1.21 [Nm/rad] at first, and adjust the value until the vibration properties in penumbra coincide with the value of the observation results.

By using the new model with a detailed hinge, we conducted frequency response analysis. The analysis targets were both a small load condition and a large load condition shown as Tab.2. Fig. 22 shows the results. From the results, we can see that the new model could simulate the vibration occurring in the 20N maneuver test and in the penumbra at the same time.

Excitation force	Condition 1	0.4 [m/s²] inertia force
	Condition 2	7.2 [μPa] pressure on top side
Solver		General-purpose non-liner analysis software "MSC. Marc 2011"
Frequency interval		0.1[Hz]
BC		Tip of york : Fixed
Number of nodes		7903
Number of elements		8292

Table 2. Details of frequency response analysis

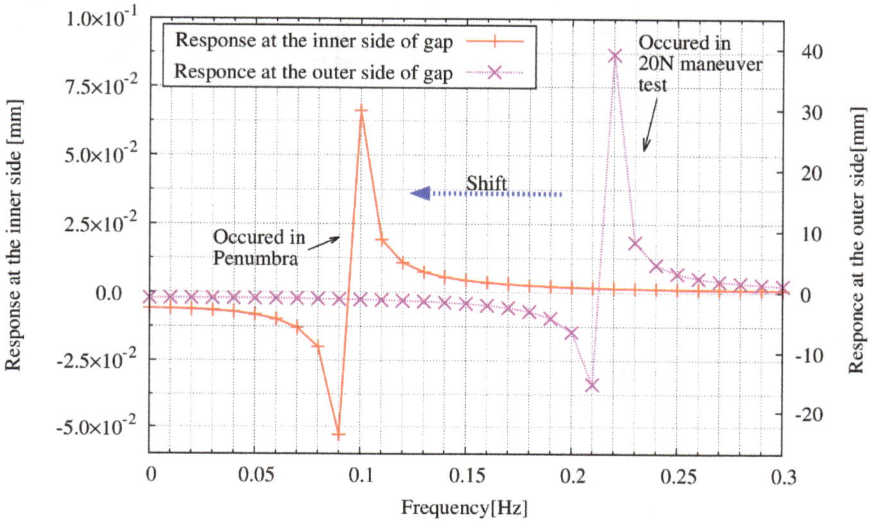

Figure 22. Frequency response analysis results of a structural model with a hinge gap. The analysis of response at the inner side of the gap simulates the dynamic regime in the penumbra. The analysis of response at the outer side of the gap simulates the dynamic regime in the preliminary experiment using the 20N thrusters of GOSAT (see Section 6.2). From the results, the new model with a nonlinear hinge demonstrated that it could simulate the peak shift in response showed in the penumbra (Fig. 19).

7. Modeling of stick-slip motion between wires and pulleys

The conventional simulator also had a problem about amplitude of the dynamic responses shown as Fig.17. The slowness of the temperature change in the paddle might be the reason for that. As shown in Fig.16, the significant decreasing of temperature began 5-10 seconds after going into penumbra. While discussing about the dynamic responses, we take notice of some particular measured data (showed in Fig.23). In the data, the solar array paddle shows the characteristic triangular wave before eclipse and the smaller dynamic responses than ordinary one. These triangular waves are found commonly in cantilever vibration affected "Stick-Slip phenomenon" (Maekawa, et.al., 2008). The Stick-Slip is a phenomenon that it arise the "stick" and "slip" behavior continuously to objects shared sliding surfaces. The phenomenon arise on the ground that the change of friction coefficient. And in slip phase, the stored strain energy is released at once. The Stick-Slip phenomenon occurred at space systems had precedents in Hubble Space Telescope (Thomton, 1993). Then, we assumed that the Stick-Slip phenomenon was occurred at deploy-speed-control wires and pulleys. That is because that the wires have small heat capacity and will be great affected by thermal environment changes. And the wires were empirically-deduced that govern the deformation of solar array paddle. The picture of them is shown in Fig. 24, and the position relation is indicated in Fig. 25. When simulating the thermal snap, sometime these specific dynamic conditions are needed to be considered. For example, when the analysis for Hubble Space Telescope was conducted, the effect from specific cross-section shape of boom was estimated. (Foster, 1995) In case of ADEOS, the tension control mechanism was evaluated in detail. (Taniwaki, 2007)

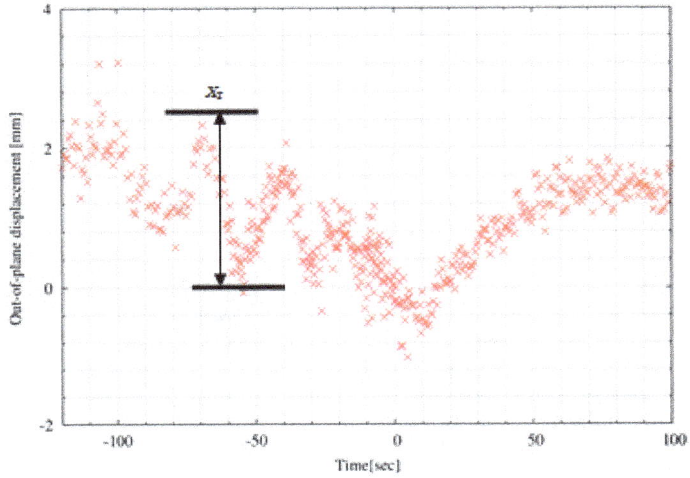

Figure 23. Particular example that solar array paddle shows unexpected behavior.

Figure 24. Similar product of GOSAT's Deploy-speed-control wires and pulleys.

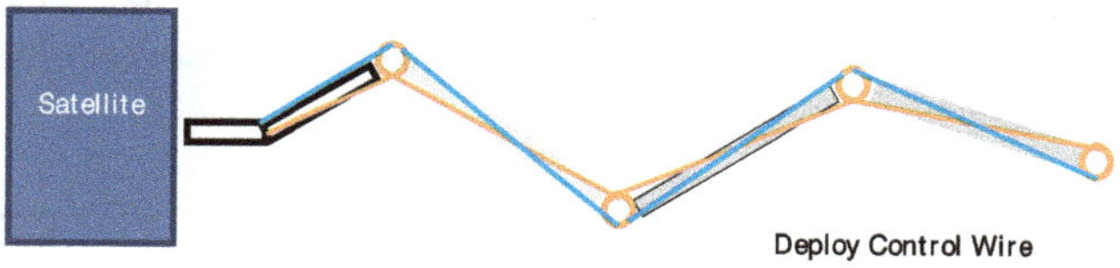

Figure 25. Position relation of GOSAT's Deploy-speed-control wires and pulleys.

To verify the effect of Stick-Slip phenomenon, the phenomenon was introduced to the structural model. In that time, the detailed information of wires and pulleys were not available. Therefore the value from measured data was used to demonstrate the effect of Stick-Slip phenomenon. Using the Bowden and Leben's equation (Bowden & Leben, 1939), the force change while the transition to the slip phase was estimated like below.

$$F_d = x_r k \tag{3}$$

Where the x_r is a displacement defined like showed in Fig.23, and the k is a stiffness of the deformation.

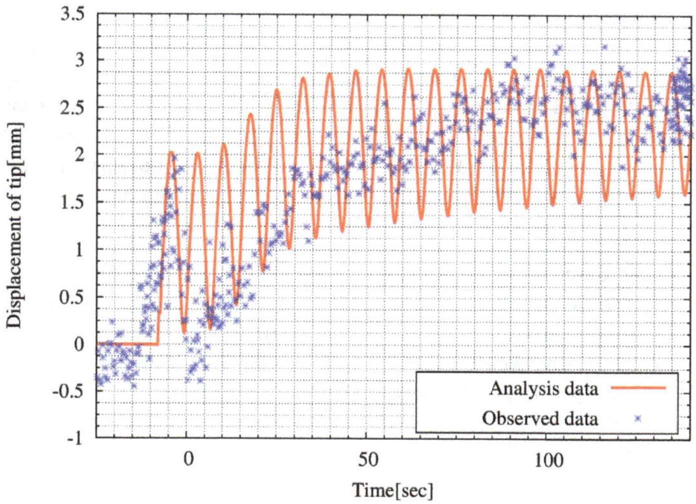

Figure 26. Time transient response simulation with detailed hinge model and stick-slip motion modeling.

From the Fig. 23, the released force was estimated like $F_d \fallingdotseq 0.029$[N]. Using this value, the new simulator was constructed, and verification analysis was conducted. In this analysis, the damping element of paddles was not concerned. The details of analysis are showed in

Tab.3. And the result of analysis is in Fig.26. As shown in Fig.26., the revised simulator could indicate roughly same value of amplitude of dynamic responses with consideration of Stick-Slip phenomenon

Analysis type	Thermal-Structure Time transient
Solver	MSC.Marc 2011
Time step	0.1[sec]
Analysis time	360[sec] (After 770[sec] of static state analysis)
BC	Tip of york : Fixed
	Temperatures of Top panel, Bottom panel, wires. Given by thermal model

Table 3. Details of thermal snap analysis

8. Conclusions

When a satellite in Low Earth Orbit (LEO) goes through an eclipse, sudden changes in the thermal environment will occur. These sudden thermal changes caused by a difference in temperature between the Sun side and the opposite side of the satellite's solar array paddle. This temperature difference causes bending and then vibration of the solar array paddle. However, the bending and vibration amplitude are too small to be observed by conventional satellite attitude sensors. To solve this problem, we used an onboard camera to directly measure solar paddle bending. The previously reported methods of observing the motion of the satellite's solar array paddle were not free from errors in image data processing, especially when the object was weakly illuminated by the Sun.

This chapter paper described the improved image data processing algorism applied to measure bending and deformation of the satellite's solar array paddle that are believed to cause degraded satellite attitude stability, which occurs when the satellite goes into an eclipse. The revised observation data shows both quasi-static deformation and rapid dynamic vibration in the penumbra. We also conducted numerical analysis to verify the observed data and understand the features of thermal snap on the solar array paddle. From this analysis, we found that sudden changes in solar array paddle temperature induce quasi-static deformation, and that the wires controlling solar array paddle deployment have a large influence on solar array paddle vibration. Additionally, we also found that a specific lower frequency vibration appearing in the penumbra can help to explain this vibration mechanism. Developing the detailed hinge model, we succeeded to simulate the vibration property change related to dynamical regime. Then, we focused on the specific results of observation and assumed that Stick-Slip phenomenon have a great influence on behavior of paddles. To simulate the influence, we revised the structural model and conducted the thermal snap analysis. The result indicated the gap of hinges and Stick-Slip phenomenon mainly effect on dynamical response of thermal snap on GOSAT.

As ongoing work, we will model the Stick-Slip phenomenon more in detail and introduce the damping element using observed data.

Author details

Mitsushige Oda[1], Akihiko Honda[2], Satoshi Suzuki[3] and Yusuke Hagiwara[4]

1 JAXA and Tokyo Institute of Technologies, Japan

2 Tokyo Institute of technology, Japan

3 Advanced Engineering Services Co.Lted, Japan

4[4]Mitsubishi Heavy Industries, Ltd, Japan

References

[1] B., A. Boley(1972), Approximate Analyses of Thermal Induced Vibrations of Beam and Plates, *Journal of Applied Mechanics*, vol. 39, no. 1, pp212-216

[2] Chijie Lin, Ramesh B. Malla(2004), Coupled Thermo-Structural Analysis of an Earth Orbiting Flexible Structure, in 45th AIAA/ASME/ASCE/AHS/ASC Structures, Structural Dynamics & Materials Conference, AIAA2004-1793

[3] Earl, A. Thomton, Yool A. Kim(1993): Thermally Induced Bending Vibrations of a Flexible Rolled-Up Solar Array, Journal of Spacecraft and Rockets, Vol.30, No.4, pp438-448

[4] E., A. Thornton (1996), Thermal Structures for Aerospace Application, AIAA, pp. 343-354

[5] F. P. Bowden & L. Leben (1939): The Nature of Sliding and the Analysis of Friction, *Proceedings of Royal Society London,* vol.A169, pp371-391

[6] Foster, C. L., Tinker, G. S., Nurre, W. & Till, W. A. (1995). NASA Technical Paper, The Solar

[7] Array-Induced Disturbance of the Hubble Space Telescope Pointing System, 3556

[8] Japan Society of Mechanical Engineering (February 2007). Mechanical Engineers' Handbook Applications 11: Space Equipment and Systems, Japan Society of Mechanical Engineering, ISBN 978-4-88898-154-5, Japan

[9] John D.Johnston, Earl A. Thornton(1998), Thermally Induced Attitude Dynamics of a Spacecraft with a Flexible Appendage, *Journal of Guidance, Control, and Dynamics*, vol. 21, no. 4, pp. 581-587

[10] Kojima, Y., Taniwaki, S., & Ohkami, Y. (2004): Attitude Vibration Caused by a Stick-Slip Motion for Flexible Solar Array of Advanced Earth Observation Satellite. *Journal of Vibration and Control*, Vol. 10, No. 10, 2004, pp.1459-1472.

[11] Ming-De Xue, Jin Duan, Zhi-Hai Xiang(2007), Thermally-Induced bending-torsion coupling vibration of large space structures, Comput Mech, vol. 40, pp.707-723

[12] Mobara, M. (1994). Introduction to aerospace engineering guidance and control of satellite and rocket, Baifukan, ISBN 978-4-56303-493-1, Japan

[13] Oda, M., Hagiwara, Y., Suzuki, S., Nakamura, T., Inaba, N., Sawada, H., Yoshii, M., & Goto, N. (2011): Measurement of Satellite Solar Array Panel Vibrations Caused by Thermal Snap and Gas Jet Thruster Firing. *Recent Advances in Vibrations Analysis*, In-Tech, ISBN 978-953-307-696-6, Croatia.

[14] S, Maekawa, K, Nakano (2008): Relationship between Schallamach waves in contact surfaces and stick-slip in sliding systems, Manuscript of JAST Tribology Conference 2008, pp81-82

[15] Taniwaki, S., Kojima, Y., Ohkami, Y. (2007), Attitude Stability Analysis for Stick-Slip-Induced Disturbances of Extended Structure with Tension Control Mechanism, Journal of System Design and Dynamics, vol. 1, No. 4, pp714-723

Chapter 13

Parametric Vibration Analysis of Transmission Mechanisms Using Numerical Methods

Nguyen Van Khang and Nguyen Phong Dien

Additional information is available at the end of the chapter

1. Introduction

Transmission mechanisms are frequently used in machines for power transmission, variation of speed and/or working direction and conversion of rotary motion into reciprocating motion. At high speeds, the vibration of mechanisms causes wear, noise and transmission errors. The vibration problem of transmission mechanisms has been investigated for a long time, both theoretically and experimentally. In dynamic modelling, a transmission mechanism is usually modelled as a multibody system. The differential equations of motion of a multibody system that undergo large displacements and rotations are fully nonlinear in n generalized coordinates in vector of variable q [1–4].

$$\boldsymbol{M}(q,\ t)\ddot{q} + \boldsymbol{k}(\dot{q},\ q,\ t) = \boldsymbol{h}(\dot{q},\ q,\ t) \tag{1}$$

It is very difficult or impossible to find the solution of Eq. (1) with the analytical way. Nevertheless, the numerical methods are efficient to solve the problem [5-9].

Besides, many technical systems work mostly on the proximity of an equilibrium position or, especially, in the neighbourhood of a desired motion which is usually called "programmed motion", "desired motion", "fundamental motion", "input–output motion" and etc. according to specific problems. In this chapter, the term "desired fundamental motion" is used for this object. The desired fundamental motion of a robotic system, for instance, is usually described through state variables determined by prescribed motions of the end-effector. For a mechanical transmission system, the desired fundamental motion can be the motion of working components of the system, in which the driver output rotates uniformly and all components are assumed to be rigid. It is very convenient to linearize the equations of motion about this configuration to take advantage of the linear analysis tools [10-18]. In other words, linearization makes it possible to use tools for studying linear systems to analyze the behavior of multibody systems in the vicinity of a desired fundamental motion. For this reason, the linearization of the equations of motion is most useful in the study of control [12-13], machinery vibrations [14-19] and the stability of motion [20-21]. Mathematically, the linearized equations of motion of a multibody system form usually a set of linear differential equations with time-varying coefficients. Considering steady-state motions of the multibody system only, one obtains a set of linear differential equations having time-periodic coefficients.

$$\boldsymbol{M}(t)\ddot{\boldsymbol{q}}(t) + \boldsymbol{C}(t)\dot{\boldsymbol{q}}(t) + \boldsymbol{K}(t)\boldsymbol{q}(t) = \boldsymbol{d}(t) \tag{2}$$

Note that Eq. (2) can be expressed in the compact form as

$$\dot{\boldsymbol{x}} = \boldsymbol{P}(t)\boldsymbol{x} + \boldsymbol{f}(t) \tag{3}$$

where we use the state variable x

$$\boldsymbol{x} = \begin{bmatrix} \boldsymbol{q} \\ \dot{\boldsymbol{q}} \end{bmatrix}, \quad \dot{\boldsymbol{x}} = \begin{bmatrix} \dot{\boldsymbol{q}} \\ \ddot{\boldsymbol{q}} \end{bmatrix} \tag{4}$$

and the matrix of coefficients $P(t)$, vector $f(t)$ are defined by

$$\boldsymbol{P}(t) = \begin{bmatrix} \boldsymbol{0} & \boldsymbol{I} \\ -\boldsymbol{M}^{-1}\boldsymbol{K} & -\boldsymbol{M}^{-1}\boldsymbol{C} \end{bmatrix}, \quad \boldsymbol{f}(t) = \begin{bmatrix} \boldsymbol{0} \\ \boldsymbol{M}^{-1}\boldsymbol{d} \end{bmatrix}, \tag{5}$$

where I denotes the $n \times n$ identity matrix.

In the steady state of a machine, the working components perform stationary motions [14-18], matrices $\boldsymbol{M}(t)$, $\boldsymbol{C}(t)$, $\boldsymbol{K}(t)$ and vector $\boldsymbol{d}(t)$ in Eq. (2) are time-periodic with the least period T. Hence, Eq. (2) represents a parametrically excited system. For calculating the steady-state periodic vibrations of systems described by differential equations (1) or (2) the harmonic balance method, the shooting method and the finite difference method are usually used [8,11,14]. In addition, the numerical integration methods as Newmark method and Runge-Kutta method can also be applied to calculate the periodic vibration of parametric vibration systems governed by Eq. (2) [5-9].

Since periodic vibrations are a commonly observed phenomenon of transmission mechanisms in the steady-state motion, a number of methods and algorithms were developed to find a T-periodic solution of the system described by Eq. (2). A common approach is by imposing an arbitrary set of initial conditions, and solving Eq. (2) in time using numerical methods until the transient term of the solution vanishes and only the periodic steady-state solution remains [14,22]. Besides, the periodic solution can be found directly by other specialized techniques such as the harmonic balance method, the method of conventional oscillator, the WKB method [14-16, 23, 24].

Following the above introduction, an overview of the numerical calculation of dynamic stability conditions of linear dynamic systems with time-periodic coefficients is presented in Section 2. Sections 3 presents numerical procedures based on Runge-Kutta method and Newmark method to find periodic solutions of linear systems with time-periodic coefficients. In Section 4, the proposed approach is demonstrated and validated by dynamic models of transmission mechanisms and measurements on real objects. The improvement in the computational efficiency of Newmark method comparing with Runge-Kutta method for linear systems is also discussed.

2. Numerical calculation of dynamic stability conditions of linear dynamic systems with time-periodic coefficients: An overview

We shall consider a system of homogeneous differential equations

$$\dot{\boldsymbol{x}} = \boldsymbol{P}(t)\boldsymbol{x} \tag{6}$$

where $\boldsymbol{P}(t)$ is a continuous T-periodic $n \times n$ matrix. According to Floquet theory [17, 18, 20, 21], the characteristic equation of Eq. (6) is independent of the chosen fundamental set of solutions. Therefore, the characteristic equation can be formulated by the following way. Firstly, we specify a set of n initial conditions $\boldsymbol{x}_i(0)$ for $i = 1, ..., n$, their elements

$$x_i^{(s)}(0) = \begin{cases} 1 \text{ when } s = i \\ 0 \text{ otherwhile} \end{cases} \tag{7}$$

and $[\boldsymbol{x}_1(0), \boldsymbol{x}_2(0), ..., \boldsymbol{x}_n(0)] = \boldsymbol{I}$. By implementing numerical integration of Eq. (6) within interval $[0, T]$ for n given initial conditions respectively, we obtain n vectors $\boldsymbol{x}_i(T)$, $i = 1, ..., n$. The matrix $\boldsymbol{\Phi}(t)$ defined by

$$\boldsymbol{\Phi}(T) = [\boldsymbol{x}_1(T), \boldsymbol{x}_2(T), ..., \boldsymbol{x}_n(T)] \tag{8}$$

is called the monodromy matrix of Eq. (6) [20]. The characteristic equation of Eq. (6) can then be written in the form

$$|\mathbf{\Phi}(T)-\rho\mathbf{I}|=\begin{vmatrix} x_1^{(1)}(T)-\rho & x_2^{(1)}(T) & \cdots & x_n^{(1)}(T) \\ x_1^{(2)}(T) & x_2^{(2)}(T)-\rho & \cdots & x_n^{(2)}(T) \\ \cdots & \cdots & \cdots & \cdots \\ x_1^{(n)}(T) & x_2^{(n)}(T) & \cdots & x_n^{(n)}(T)-\rho \end{vmatrix}=0 \tag{9}$$

Expansion of Eq. (9) yields a n-order algebraic equation

$$\rho^n + a_1\rho^{n-1} + a_2\rho^{n-2} + ... + a_{n-1}\rho + a_n = 0 \tag{10}$$

where unknowns $\rho_k(k=1, ..., n)$, called Floquet multipliers, can be determined from Eq. (10). Floquet exponents are given by

$$\lambda_k = \frac{1}{T}\ln\rho_k, \ (k=1, ..., n) \tag{11}$$

When the Floquet multipliers or Floquet exponents are known, the stability conditions of solutions of the system of linear differential equations with periodic coefficients can be easily determined according to the Floquet theorem [17–20]. The concept of stability according to Floquet multipliers can be expressed as follows.

If $|\rho_k|1$, the trivial solution $\boldsymbol{x}=0$ of Eq. (6) will be asymptotically stable. Conversely, the solution $\boldsymbol{x}=0$ of Eq. (6) becomes unstable if at least one Floquet multiplier has modulus being larger than 1.

If $|\rho_k| \le 1$ and Floquet multipliers with modulus 1 are single roots of the characteristic equation, the solution $\boldsymbol{x}=0$ of Eq. (6) is stable.

If $|\rho_k| \le 1$ and Floquet multipliers with modulus 1 are multiple roots of the characteristic equation, and the algebraic multiplicity is equal to their geometric multiplicity, then the solution $\boldsymbol{x}=0$ of Eq. (6) is also stable.

3. Numerical procedures for calculating periodic solutions of linear dynamic systems with time-periodic coefficients

3.1. Numerical procedure based on Runge-Kutta method

Now we consider only the periodic vibration of a dynamic system which is governed by a set of linear differential equations with periodic coefficients. As already mentioned in the previous section, these differential equations can be expressed in the compact matrix form

$$\dot{x} = P(t)x + f(t) \tag{12}$$

where x is the vector of state variables, matrix $P(t)$ and vector $f(t)$ are periodic in time with period T. The system of homogeneous differential equations corresponding to Eq. (12) is

$$\dot{x} = P(t)x \tag{13}$$

As well known from the theory of differential equations, if Eq. (13) has only non-periodic solutions except the trivial solution, then Eq. (12) has an unique T-periodic solution. This periodic solution can be obtained by choosing the appropriate initial condition for the vector of variables x and then implementing numerical integration of Eq. (12) within interval $[0, T]$. An algorithm is developed to find the initial value for the periodic solution [18, 19]. Firstly, the T-periodic solution must satisfy the following condition

$$x(0) = x(T) \tag{14}$$

The interval $[0, T]$ is now divided into m equal subintervals with the step-size $h = t_i - t_{i-1} = T/m$. At the discrete times t_i and t_{i+1}, $x_i = x(t_i)$ and $x_{i+1} = x(t_{i+1})$ represent the states of the system, respectively. Using the fourth-order Runge-Kutta method, we get a numerical solution [5]

$$x_i = x_{i-1} + \frac{1}{6}\left[k_1^{(i-1)} + 2k_2^{(i-1)} + 2k_3^{(i-1)} + k_4^{(i-1)}\right] \tag{15}$$

where

$$\begin{aligned}
k_1^{(i-1)} &= h\left[P(t_{i-1})x_{i-1} + f(t_{i-1})\right],\\
k_2^{(i-1)} &= h\left[P(t_{i-1} + \frac{h}{2})(x_{i-1} + \frac{1}{2}k_1^{(i-1)}) + f(t_{i-1} + \frac{h}{2})\right],\\
k_3^{(i-1)} &= h\left[P(t_{i-1} + \frac{h}{2})(x_{i-1} + \frac{1}{2}k_2^{(i-1)}) + f(t_{i-1} + \frac{h}{2})\right],\\
k_4^{(i-1)} &= h\left[P(t_i)(x_{i-1} + k_3^{(i-1)}) + f(t_i)\right].
\end{aligned} \tag{16}$$

Substituting Eq. (16) into Eq. (15), we obtain

$$x_i = A_{i-1}x_{i-1} + b_{i-1} \tag{17}$$

where matrix A_{i-1} is given by

$$\begin{aligned}A_{i-1}=I+\frac{1}{6}\Big\{&h\Big[P(t_{i-1})+4P(t_{i-1}+\frac{h}{2})+P(t_i)\Big]\\&+h^2\Big[P(t_{i-1}+\frac{h}{2})P(t_{i-1})+P^2(t_{i-1}+\frac{h}{2})+\frac{1}{2}P(t_i)P(t_{i-1}+\frac{h}{2})\Big]\\&+\frac{h^3}{2}\Big[P^2(t_{i-1}+\frac{h}{2})P(t_{i-1})+\frac{1}{2}P(t_i)P^2(t_{i-1}+\frac{h}{2})\Big]\\&+\frac{h^4}{4}P(t_i)P^2(t_{i-1}+\frac{h}{2})P(t_{i-1})\Big\}(i=1,\ ...,\ m),\end{aligned}\tag{18}$$

and vector b_{i-1} takes the form

$$\begin{aligned}b_{i-1}=\frac{1}{6}\Big\{&h\Big[f(t_{i-1})+4f(t_{i-1}+\frac{h}{2})+f(t_i)\Big]\\&+h^2\Big[P(t_{i-1}+\frac{h}{2})f(t_{i-1})+P(t_{i-1}+\frac{h}{2})f(t_{i-1}+\frac{h}{2})+\frac{1}{2}P(t_i)f(t_{i-1}+\frac{h}{2})\Big]\\&+\frac{h^3}{2}\Big[P^2(t_{i-1}+\frac{h}{2})f(t_{i-1})+P(t_i)P(t_{i-1}+\frac{h}{2})f(t_{i-1}+\frac{h}{2})\Big]\\&+\frac{h^4}{4}P(t_i)P^2(t_{i-1}+\frac{h}{2})f(t_{i-1})\Big\}.\end{aligned}\tag{19}$$

Expansion of Eq. (17) for $i=1$ *to* m yields

$$\begin{aligned}&x_1=A_0x_0+c_1\\&x_2=A_1A_0x_0+c_2\\&\dots\dots\dots\dots\dots\\&x_m=\left(\prod_{i=m-1}^{0}A_i\right)x_0+c_m\end{aligned}\tag{20}$$

where $c_0=0,\ c_1=A_0c_0+b_0,\ \ c_2=A_1c_1+b_1\ ,...,\ c_m=A_{m-1}c_{m-1}+b_{m-1}$. Using the boundary condition according to Eq. (14), the last equation of Eq. (20) yields a set of the linear algebraic equations

$$\left(I-\prod_{i=m-1}^{0}A_i\right)x_0=c_m.\tag{21}$$

The solution of Eq. (21) gives us the initial value for the periodic solution of Eq. (12). Finally, the periodic solution of Eq. (12) with the corresponding initial value can be calculated using the computational scheme according to Eq. (15).

3.2. Numerical procedure based on Newmark integration method

The procedure presented below for finding the T-periodic solution of Eq. (2) is based on the Newmark direct integration method. Firstly, the interval $[0,\ T]$ is also divided into m equal subintervals with the step-size $h=t_i-t_{i-1}=T\,/\,m$. We use notations $q_i=q(t_i)$ and $q_{i+1}=q(t_{i+1})$ to represent the solution of Eq. (2) at discrete times t_i and t_{i+1} respectively. The T-periodic solution must satisfy the following conditions

$$q(0)=q(T),\ \dot{q}(0)=\dot{q}(T),\ \ddot{q}(0)=\ddot{q}(T). \tag{22}$$

Based on the single-step integration method proposed by Newmark, we obtain the following approximation formulas [6-7]

$$q_{i+1}=q_i+h\,\dot{q}_i+h^2\left(\tfrac{1}{2}-\beta\right)\ddot{q}_i+\beta h^2\ddot{q}_{i+1}, \tag{23}$$

$$\dot{q}_{i+1}=\dot{q}_i+(1-\gamma)h\,\ddot{q}_i+\gamma h\,\ddot{q}_{i+1}, \tag{24}$$

Constants β, γ are parameters associated with the quadrature scheme. Choosing $\gamma=1/4$ and $\beta=1/6$ leads to linear interpolation of accelerations in the time interval $[\ t_i,\ t_{i+1}\ ]$. In the same way, choosing $\gamma=1/2,\ \beta=1/4$ corresponds to considering the acceleration average value over the time interval [6, 7].

From Eq. (2) we have the following iterative computational scheme at time t_{i+1}

$$M_{i+1}\ddot{q}_{i+1}+C_{i+1}\dot{q}_{i+1}+K_{i+1}q_{i+1}=d_{i+1}, \tag{25}$$

where $M_{i+1}=M(t_{i+1}),\ C_{i+1}=C(t_{i+1}),\ K_{i+1}=K(t_{i+1})$ and $d_{i+1}=d(t_{i+1})$.

In the next step, substitution of Eqs. (23) and (24) into Eq. (25) yields

$$\left(M_{i+1}+\gamma h\,C_{i+1}+\beta h^2K_{i+1}\right)\ddot{q}_{i+1}=d_{i+1}-C_{i+1}\left[\dot{q}_i+(1-\gamma)h\,\ddot{q}_i\right]-K_{i+1}\left[q_i+h\,\dot{q}_i+h^2\left(\frac{1}{2}-\beta\right)\ddot{q}_i\right]. \tag{26}$$

The use of Eqs. (23) and (24) leads to the prediction formulas for velocities and displacements at time t_{i+1}

$$\overset{*}{q}_{i+1}=q_i+h\,\dot{q}_i+h^2\left(\frac{1}{2}-\beta\right)\ddot{q}_i,\ \dot{\overset{*}{q}}_{i+1}=\dot{q}_i+(1-\gamma)h\,\ddot{q}_i. \tag{27}$$

Eq. (27) can be expressed in the matrix form as

$$\begin{bmatrix} \mathbf{q}_{i+1}^{\bullet} \\ \dot{\mathbf{q}}_{i+1}^{\bullet} \end{bmatrix} = \mathbf{D} \begin{bmatrix} \mathbf{q}_i \\ \dot{\mathbf{q}}_i \\ \ddot{\mathbf{q}}_i \end{bmatrix} \tag{28}$$

with

$$D = \begin{bmatrix} I & hI & h^2(0.5-\beta)I \\ 0 & I & (1-\gamma)hI \end{bmatrix} \tag{29}$$

where **0** represents the $n \times n$ matrix of zeros. Eq. (26) can then be rewritten in the matrix form as

$$\ddot{\mathbf{q}}_{i+1} = (\mathbf{S}_{i+1})^{-1}\mathbf{d}_{i+1} - (\mathbf{S}_{i+1})^{-1}\mathbf{H}_{i+1}\begin{bmatrix} \mathbf{q}_{i+1}^{\bullet} \\ \dot{\mathbf{q}}_{i+1}^{\bullet} \end{bmatrix}, \tag{30}$$

where matrices S_{i+1} and H_{i+1} are defined by

$$S_{i+1} = M_{i+1} + \gamma h\, C_{i+1} + h^2 \beta K_{i+1}, \tag{31}$$

$$H_{i+1} = [K_{i+1} \quad C_{i+1}]. \tag{32}$$

By substituting relationships (28) into (30) we find

$$\ddot{\mathbf{q}}_{i+1} = (\mathbf{S}_{i+1})^{-1}\mathbf{d}_{i+1} - (\mathbf{S}_{i+1})^{-1}\mathbf{H}_{i+1}\mathbf{D}\begin{bmatrix} \mathbf{q}_i \\ \dot{\mathbf{q}}_i \\ \ddot{\mathbf{q}}_i \end{bmatrix} \tag{33}$$

From Eqs. (23), (24) and (27) we get the following matrix relationship

$$\begin{bmatrix} \mathbf{q}_{i+1} \\ \dot{\mathbf{q}}_{i+1} \\ \ddot{\mathbf{q}}_{i+1} \end{bmatrix} = \mathbf{T}\begin{bmatrix} \mathbf{q}_{i+1}^{\bullet} \\ \dot{\mathbf{q}}_{i+1}^{\bullet} \\ \ddot{\mathbf{q}}_{i+1} \end{bmatrix}, \tag{34}$$

where matrix T is expressed in the block matrix form as

$$\mathbf{T} = \begin{bmatrix} \mathbf{I} & \mathbf{0} & \mathbf{I}\beta h^2 \\ \mathbf{0} & \mathbf{I} & \mathbf{I}\gamma h \\ \mathbf{0} & \mathbf{0} & \mathbf{I} \end{bmatrix} \tag{35}$$

The combination of Eqs. (28), (33) and (34) yields a new computational scheme for determining the solution of Eq. (2) at the time t_{i+1} in the form

$$\begin{bmatrix} \mathbf{q}_{i+1} \\ \dot{\mathbf{q}}_{i+1} \\ \ddot{\mathbf{q}}_{i+1} \end{bmatrix} = \mathbf{T}\begin{bmatrix} \mathbf{D} \\ -(\mathbf{S}_{i+1})^{-1}\mathbf{H}_{i+1}\mathbf{D} \end{bmatrix}\begin{bmatrix} \mathbf{q}_i \\ \dot{\mathbf{q}}_i \\ \ddot{\mathbf{q}}_i \end{bmatrix} + \mathbf{T}\begin{bmatrix} \mathbf{0} \\ \mathbf{0} \\ (\mathbf{S}_{i+1})^{-1}\mathbf{d}_{i+1} \end{bmatrix}. \tag{36}$$

In this equation, the iterative computation is eliminated by introducing the direct solution for each time step. Note that T and D are matrices of constants.

By setting

$$\mathbf{x}_i = \begin{bmatrix} \mathbf{q}_i \\ \dot{\mathbf{q}}_i \\ \mathbf{q}_i \end{bmatrix}, \quad \mathbf{A}_{i+1} = \mathbf{T}\begin{bmatrix} \mathbf{D} \\ -(\mathbf{S}_{i+1})^{-1}\mathbf{H}_{i+1}\mathbf{D} \end{bmatrix}, \quad \mathbf{b}_{i+1} = \mathbf{T}\begin{bmatrix} \mathbf{0} \\ \mathbf{0} \\ (\mathbf{S}_{i+1})^{-1}\mathbf{d}_{i+1} \end{bmatrix}, \tag{37}$$

Eq. (36) can then be rewritten in the following form

$$x_i = A_i x_{i-1} + b_i \quad (i=1,\ 2,\ \dots,\ m). \tag{38}$$

Expansion of Eq. (38) for $i=1$ *to* m yields the same form as Eq. (20)

$$\begin{aligned} x_1 &= A_1 x_0 + c_1 \\ x_2 &= A_2 A_1 x_0 + c_2 \\ &\dots\dots\dots\dots\dots\dots \\ x_m &= \left(\prod_{i=m}^{1} A_i\right) x_0 + c_m \end{aligned} \tag{39}$$

where $c_0=0,\ c_1 = A_1 c_0 + b_1,\ \ c_2 = A_2 c_1 + b_2, \dots,\ c_m = A_m c_{m-1} + b_m$.

Using the condition of periodicity according to Eq. (22), the last equation of Eq. (39) yields a set of the linear algebraic equations

$$\left(I - \prod_{i=m}^{1} A_i\right) x_0 = c_m. \tag{40}$$

The solution of Eq. (40) gives us the initial value for the periodic solution of Eq. (2). Finally, the periodic solution of Eq. (2) with the obtained initial value can be calculated without difficulties using the computational scheme in Eq. (36).

Based on the proposed numerical procedures in this section, a computer program with MATLAB to calculate periodic vibrations of transmission mechanisms has been developed and tested by the following application examples.

4. Application examples

4.1. Steady-state parametric vibration of an elastic cam mechanism

Cam mechanisms are frequently used in mechanical transmission systems to convert rotary motion into reciprocating motion (Figure 1). At high speed, the vibration of cam mechanisms causes transmission errors, cam surface fatigue, wear and noise. Because of that, the vibration problem of cam mechanisms has been investigated for a long time, both theoretically and experimentally.

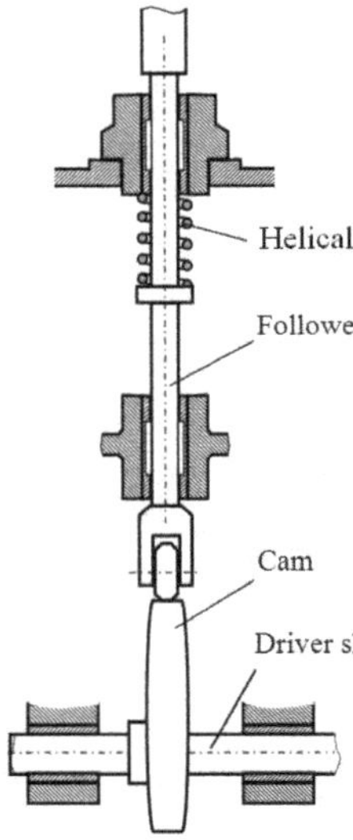

Figure 1. A cam mechanism.

The dynamic model of this system is schematically shown in Figure 2. This kind of model was also considered in a number of studies, e.g. [25-26]. The mechanical system of the elastic cam shaft, the cam with an elastic follower can be considered as rigid bodies connected by massless spring-damping elements with time-invariant stiffness k_i and constant damping coefficients c_i for $i=1, 2, 3$. Among them k_1 is the torsional stiffness of the cam shaft. Parameter k_2 is the equivalent stiffness due to the longitudinal stiffness of the follower, the contact stiffness between the cam and the roller, and the cam bearing stiffness. Parameter k_3 denotes the combined stiffness of the return spring and the support of the output link. The rotating components are modeled by two rotating disks with moments of inertia I_0 and I_1. Let us introduce into our dynamic model the nonlinear transmission function $U(\varphi_1)$ of the cam mechanism as a function of the rotating angle φ_1 of the cam shaft, the driving torque from the motor $M(t)$ and the external load $F(t)$ applied on the system.

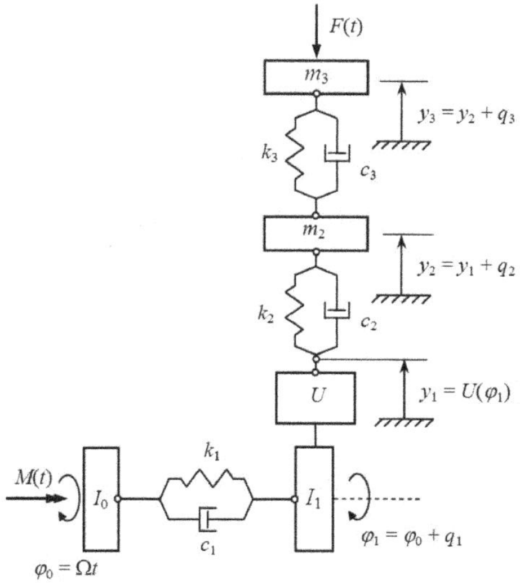

Figure 2. Dynamic model of the cam mechanism.

The kinetic energy, the potential energy and the dissipative function of the considered system can be expressed in the following form

$$T = \frac{1}{2}I_0\dot{\varphi}_0^2 + \frac{1}{2}I_1\dot{\varphi}_1^2 + \frac{1}{2}m_2\dot{y}_2^2 + \frac{1}{2}m_3\dot{y}_3^2 \tag{41}$$

$$\Pi = \frac{1}{2}k_1(\varphi_1 - \varphi_0)^2 + \frac{1}{2}k_2(y_2 - y_1)^2 + \frac{1}{2}k_3(y_3 - y_2)^2 \tag{42}$$

$$\Phi = \frac{1}{2}c_1(\dot{\varphi}_1 - \dot{\varphi}_0)^2 + \frac{1}{2}c_2(\dot{y}_2 - \dot{y}_1)^2 + \frac{1}{2}c_3(\dot{y}_3 - \dot{y}_2)^2 \tag{43}$$

The virtual work done by all non-conservative forces is

$$\sum \delta A = M(t)\delta\varphi_0 - F(t)\delta y_3 \tag{44}$$

Using the generalized coordinates $\varphi_0, \varphi_1, q_2, q_3$, we obtain the following relations

$$y_1 = U(\varphi_1),\ y_2 = y_1 + q_2,\ y_3 = y_2 + q_3 \tag{45}$$

Substitution of Eq. (45) into Eqs. (41-44) yields

$$T = \frac{1}{2} I_0 \dot{\varphi}_0^2 + \frac{1}{2} I_1 \dot{\varphi}_1^2 + \frac{1}{2} m_2 (U' \dot{\varphi}_1 + \dot{q}_2)^2 + \frac{1}{2} m_3 (U' \dot{\varphi}_1 + \dot{q}_2 + \dot{q}_3)^2, \tag{46}$$

$$\Pi = \frac{1}{2} k_1 (\varphi_1 - \varphi_0)^2 + \frac{1}{2} k_2 q_2^2 + \frac{1}{2} k_3 q_3^2, \tag{47}$$

$$\Phi = \frac{1}{2} c_1 (\dot{\varphi}_1 - \dot{\varphi}_0)^2 + \frac{1}{2} c_2 \dot{q}_2^2 + \frac{1}{2} c_3 \dot{q}_3^2, \tag{48}$$

$$\sum \delta A = M(t) \delta\varphi_0 - F(t) U' \delta\varphi_1 - F(t) \delta q_2 - F(t) \delta q_3, \tag{49}$$

where the prime represents the derivative with respect to the generalized coordinate φ_1. The generalized forces of all non-conservative forces are then derived from Eq. (49) as

$$Q^*_{\varphi_0} = M(t),\quad Q^*_{\varphi_1} = -F(t) U',\ Q^*_{q_2} = -F(t),\ Q^*_{q_3} = -F(t). \tag{50}$$

Substitution of Eqs. (46)-(48) and (50) into the Lagrange equation of the second type yields the differential equations of motion of the system in terms of the generalized coordinates $\varphi_0, \varphi_1, q_2, q_3$

$$I_0 \ddot{\varphi}_0 - c_1 (\dot{\varphi}_1 - \dot{\varphi}_0) - k_1 (\varphi_1 - \varphi_0) = M(t), \tag{51}$$

$$\begin{aligned} &\left[I_1 + (m_2 + m_3) U'^2\right] \ddot{\varphi}_1 + (m_2 + m_3) U' \ddot{q}_2 + m_3 U' \ddot{q}_3 + (m_2 + m_3) U' U'' \dot{\varphi}_1^2 \\ &+ c_1 (\dot{\varphi}_1 - \dot{\varphi}_0) + k_1 (\varphi_1 - \varphi_0) = -F(t) U', \end{aligned} \tag{52}$$

$$(m_2 + m_3) U' \ddot{\varphi}_1 + (m_2 + m_3) \ddot{q}_2 + m_3 \ddot{q}_3 + (m_2 + m_3) U'' \dot{\varphi}_1^2 + c_2 \dot{q}_2 + k_2 q_2 = -F(t), \tag{53}$$

$$m_3 U' \ddot{\varphi}_1 + m_3 \ddot{q}_2 + m_3 \ddot{q}_3 + m_3 U'' \dot{\varphi}_1^2 + c_3 \dot{q}_3 + k_3 q_3 = -F(t). \tag{54}$$

When the angular velocity Ω of the driver input is assumed to be constant in the steady state

$$\varphi_0 = \Omega t, \tag{55}$$

one leads to the following relation

$$\varphi_1 = \Omega t + q_1, \tag{56}$$

where q_1 is the difference between rotating angles φ_0 and φ_1 due to the presence of the spring element k_1 and the damping element c_1. Assuming that φ_1 varies little from its mean value during the steady-state motion, the transmission function $y_1 = U(\varphi_1)$ depends essentially on the input angle $\varphi_0 = \Omega t$. Using the Taylor series expansion around Ωt, we get

$$U(\varphi_1) = U(\Omega t + q_1) = \overline{U} + \overline{U}' q_1 + \tfrac{1}{2}\overline{U}'' q_1^2 + \ldots, \tag{57}$$

$$U'(\varphi_1) = U'(\Omega t + q_1) = \overline{U}' + \overline{U}'' q_1 + \tfrac{1}{2}\overline{U}''' q_1^2 + \ldots, \tag{58}$$

$$U''(\varphi_1) = U''(\Omega t + q_1) = \overline{U}'' + \overline{U}''' q_1 + \tfrac{1}{2}\overline{U}^{(4)} q_1^2 + \ldots. \tag{59}$$

where we used the notations

$$\overline{U} = U(\Omega t),\ \overline{U}' = U'(\Omega t),\ \overline{U}'' = U''(\Omega t),\ \overline{U}''' = U'''(\Omega t). \tag{60}$$

Since the system performs small vibrations, i.e. there are only small vibrating amplitudes q_1, q_2 and q_3, substituting Eqs. (57)-(59) into Eqs. (52)-(54) and neglecting nonlinear terms, we obtain the linear differential equations of vibration for the system

$$\begin{aligned}&\left(I_1 + (m_2+m_3)\overline{U}'^2\right)\ddot{q}_1 + (m_2+m_3)\overline{U}'\ddot{q}_2 + m_3\overline{U}'\ddot{q}_3 + \left[c_1 + 2(m_2+m_3)\Omega\overline{U}'\overline{U}''\right]\dot{q}_1\\ &+\left[k_1 + F(t)\overline{U}'' + (m_2+m_3)\Omega^2(\overline{U}'\overline{U}''' + \overline{U}''^2)\right]q_1 = -F(t)\overline{U}' - (m_2+m_3)\Omega^2\overline{U}'\overline{U}'',\end{aligned} \tag{61}$$

$$\begin{aligned}&(m_2+m_3)\overline{U}'\ddot{q}_1 + (m_2+m_3)\ddot{q}_2 + m_3\ddot{q}_3 + 2(m_2+m_3)\Omega\overline{U}''\dot{q}_1\\ &+c_2\dot{q}_2 + (m_2+m_3)\Omega^2\overline{U}'''q_1 + k_2q_2 = -F(t) - (m_2+m_3)\Omega^2\overline{U}'',\end{aligned} \tag{62}$$

$$m_3\overline{U}'\ddot{q}_1 + m_3\ddot{q}_2 + m_3\ddot{q}_3 + 2m_3\Omega\overline{U}''\dot{q}_1 + c_3\dot{q}_3 + m_3\Omega^2\overline{U}'''q_1 + k_3q_3 = -F(t) - m_3\Omega^2\overline{U}''. \tag{63}$$

In most cases, the force $F(t)$ can be approximately a periodic function of the time or a constant. Thus, Eqs. (61)-(63) form a set of linear differential equations with periodic coefficients. Finally, the linearized differential equations of vibration can be expressed in the compact matrix form as

$$\boldsymbol{M}(\Omega t)\ddot{\boldsymbol{q}} + \boldsymbol{C}(\Omega t)\dot{\boldsymbol{q}} + \boldsymbol{K}(\Omega t)\boldsymbol{q} = \boldsymbol{d}(\Omega t), \tag{64}$$

where

$$\boldsymbol{M}(\Omega t)=\begin{bmatrix} I_1+(m_2+m_3)\overline{U}'^2 & (m_2+m_3)\overline{U}' & m_3\overline{U}' \\ (m_2+m_3)\overline{U}' & (m_2+m_3) & m_3 \\ m_3\overline{U}' & m_3 & m_3 \end{bmatrix} \boldsymbol{C}(\Omega t)=\begin{bmatrix} c_1+2(m_2+m_3)\Omega\overline{U}'\overline{U}'' & 0 & 0 \\ 2(m_2+m_3)\Omega\overline{U}'' & c_2 & 0 \\ 2m_3\Omega\overline{U}'' & 0 & c_3 \end{bmatrix}$$

$$\boldsymbol{K}(\Omega t)=\begin{bmatrix} k_1+F\overline{U}''+(m_2+m_3)\Omega^2(\overline{U}'\overline{U}'''+\overline{U}''^2) & 0 & 0 \\ (m_2+m_3)\Omega^2\overline{U}''' & k_2 & 0 \\ m_3\Omega^2\overline{U}''' & 0 & k_3 \end{bmatrix}$$

$$\boldsymbol{d}(\Omega t)=\begin{bmatrix} -F\overline{U}'-(m_2+m_3)\Omega^2\overline{U}'\overline{U}'' \\ -F-(m_2+m_3)\Omega^2\overline{U}'' \\ -F-m_3\Omega^2\overline{U}'' \end{bmatrix}, \quad \boldsymbol{q}=\begin{bmatrix} q_1 \\ q_2 \\ q_3 \end{bmatrix}.$$

We consider now the function $U'(\varphi)$, called the first grade of the transmission function $U(\varphi)$, where the angle φ is the rotating angle of the cam shaft. In steady state motion of the cam mechanism, function $U'(\varphi)$ can be approximately expressed by a truncated Fourier series

$$U'(\varphi)=\sum_{k=1}^{K}(a_k\cos k\varphi+b_k\sin k\varphi). \tag{65}$$

Parameters	**Units**	**Values**
m_2	(kg)	28
m_3	(kg)	50
I_1	(kgm²)	0.12
k_1	(Nm/rad)	8×10^4
k_2	(N/m)	8.2×10^8
k_3	(N/m)	2.6×10^8
c_1	(Nms/rad)	18.5
c_2	(Ns/m)	1400
c_3	(Ns/m)	1200

Table 1. Calculation parameters.

The functions $\overline{U}'$, $\overline{U}''$, $\overline{U}'''$ in Eq. (64) can then be calculated using Eq. (65) for $\varphi=\Omega t$. Parameters used for the numerical calculation are listed in Table 1. Two set of coefficients a_k in Eq. (46) are given in Table 2 corresponding to two different cases of cam profile, coefficients $b_k=0$. Without loss of generality, the external force F is assumed to have a constant value of 100 N.

a_k (m)	Case 1	Case 2
a_1	0.22165	0.22206
a_2	0	0
a_3	0.05560	0.08539
a_4	0	0
a_5	- 0.01706	0.00518
a_6	0	0
a_7	0	- 0.00373
a_8	0	0
a_9	0	0.00345
a_{10}	0	0
a_{11}	0	- 0.00182
a_{12}	0	0

Table 2. Fourier coefficients a_k of $U'(\varphi)$.

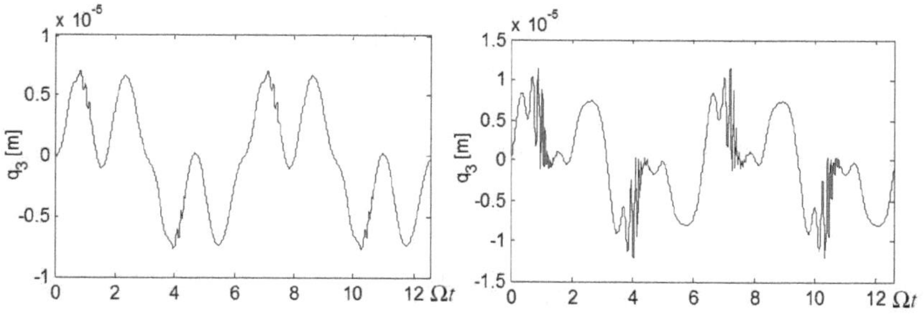

Figure 3. Dynamic transmission errors q_3 with n_{im}=100(rpm) for Case 1 (left) and Case 2 (right).

The rotating speed of the driver input n_{in} takes firstly the value of 100 (rpm) corresponding to angular velocity $\Omega \approx 10.47$ (rad/s) for the calculation. The periodic solutions of Eq. (64) are then calculated using the numerical procedures proposed in Section 3. The results of a periodic solution for coordinate q_3, which represents the dynamic transmission errors within the considered system, are shown in Figures 3 and 4. The influence of cam profile to the vibration response of the system can be recognized by a considerable difference in the vibration amplitude of both curves in Figure 3 and the frequency content of spectrums in Figure 4. In addition, the spectrums in Figure 4 shows harmonic components of the rotating frequency, such as ,Ω 3Ω, 5Ω which indicate stationary periodic vibrations.

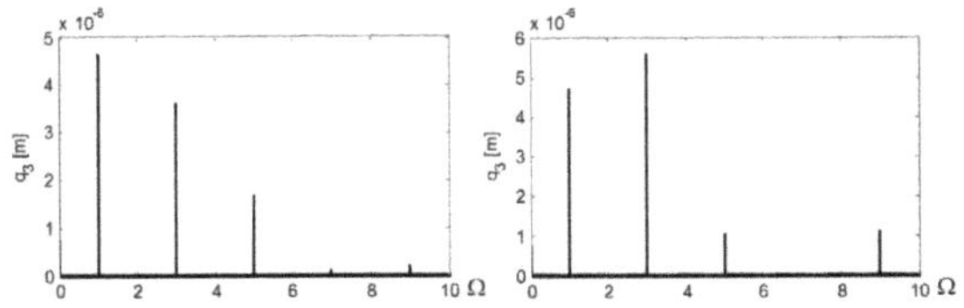

Figure 4. Frequency spectrum of q_3 with $n_{in}=100(rpm)$ for Case 1 (left) and Case 2 (right).

Figures 5 and 6 show the calculating results with rotating speed n_{in}=600 (rpm), corresponding to $\Omega \approx 62.8$ (rad/s). The mechanism has a more serious dynamic transmission error at high speeds. It can be seen clearly from the frequency spectrums that the steady state vibration at high speeds of the considered cam mechanism may include tens harmonics of the rotating frequency as mentioned in [3].

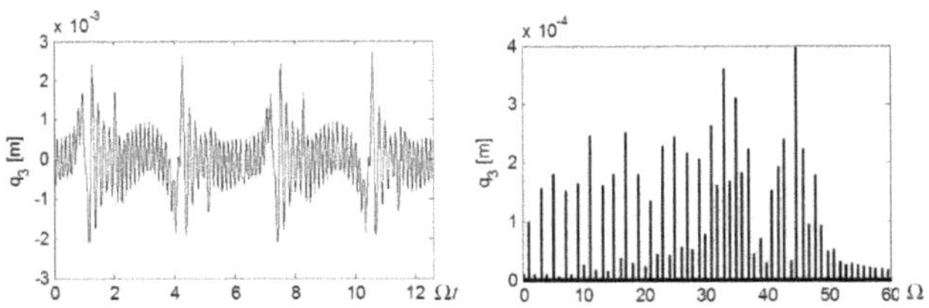

Figure 5. Dynamic transmission errors q_3 with $n_{in}=600(rpm)$ for Case 1.

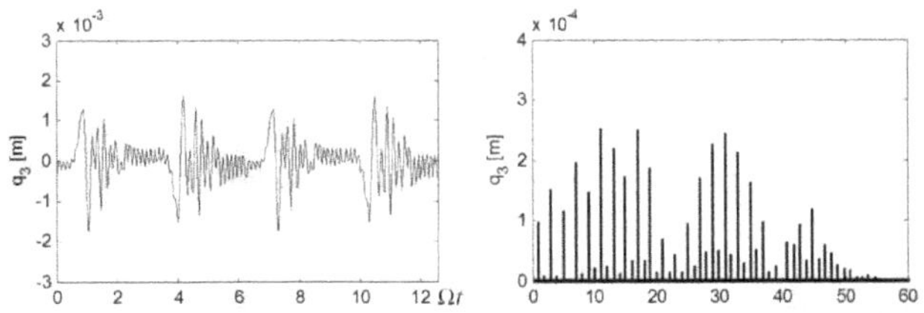

Figure 6. Dynamic transmission errors q_3 with $n_{in}=600(rpm)$ for Case 2.

The calculation of the periodic solution of Eq. (64) was implemented by a self-written computer program in MATLAB environment, and a Dell Notebook equipped with CPU Intel® Core 2 Duo T6600 at 2.2 GHz and 3 GB memory. The calculating results obtained by the numerical procedures are identical, but the computation time with Newmark method is greatly reduced in comparison with Runge-Kutta method as shown in Figure 7, especially in the cases of large number of time steps.

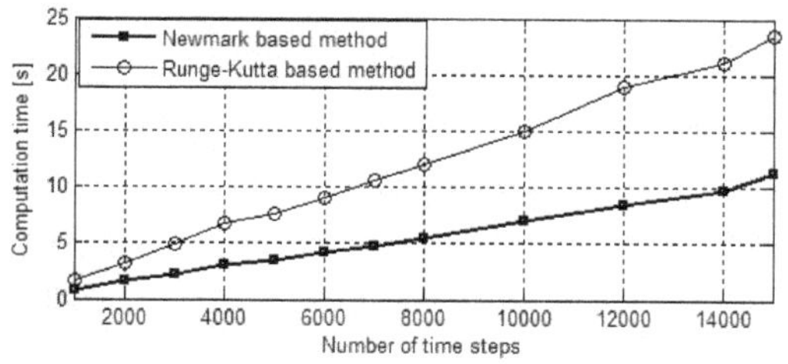

Figure 7. Comparison of the computation time for the model of cam mechanism.

4.2. Parametric vibration of a gear - pair system with faulted meshing

Dynamic modeling of gear vibrations offers a better understanding of the vibration generation mechanisms as well as the dynamic behavior of the gear transmission in the presence of gear tooth damage. Since the main source of vibration in a geared transmission system is usually the meshing action of the gears, vibration models of the gear-pair in mesh have been developed, taking into consideration the most important dynamic factors such as effects of friction forces at the meshing interface, gear backlash, the time-varying mesh stiffness and the excitation from gear transmission errors [31-33].

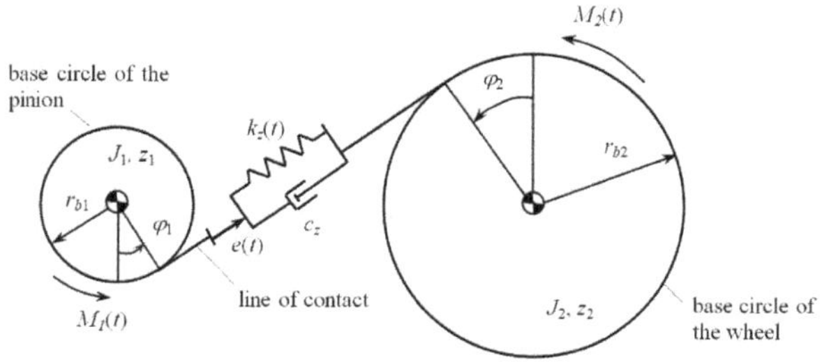

Figure 8. Dynamic model of the gear-pair system with faulted meshing.

From experimental works, it is well known that the most important components in gear vibration spectra are the tooth-meshing frequency and its harmonics, together with sideband structures due to the modulation effect. The increment in the number and amplitude of sidebands may indicate a gear fault condition, and the spacing of the sidebands is related to their source [27], [30]. However, according to our knowledge, there are in the literature only a few of theoretical studies concerning the effect of sidebands in gear vibration spectrum and the calculating results are usually not in agreement with the measurements. Therefore, the main objective of the following investigation is to unravel modulation effects which are responsible for generating such sidebands.

Figure 8 shows a relative simple dynamic model of a pair of helical gears. This kind of the model is also considered in references [24, 28, 32, 33]. The gear mesh is modeled as a pair of rigid disks connected by a spring-damper set along the line of contact.

The model takes into account influences of the static transmission error which is simulated by a displacement excitation $e(t)$ at the mesh. This transmissions error arises from several sources, such as tooth deflection under load, non-uniform tooth spacing, tooth profile errors caused by machining errors as well as pitting, scuffing of teeth flanks. The mesh stiffness $k_z(t)$ is expressed as a time-varying function. The gear-pair is assumed to operate under high torque condition with zero backlash and the effect of friction forces at the meshing interface is neglected. The viscous damping coefficient of the gear mesh c_z is assumed to be constant. The differential equations of motion for this system can be expressed in the form

$$J_1\ddot{\varphi}_1 + r_{b1}k_z(t)[r_{b1}\varphi_1 + r_{b2}\varphi_2 + e(t)] + r_{b1}c_z[r_{b1}\dot{\varphi}_1 + r_{b2}\dot{\varphi}_2 + \dot{e}(t)] = M_1(t), \tag{66}$$

$$J_2\ddot{\varphi}_2 + r_{b2}k_z(t)[r_{b1}\varphi_1 + r_{b2}\varphi_2 + e(t)] + r_{b2}c_z[r_{b1}\dot{\varphi}_1 + r_{b2}\dot{\varphi}_2 + \dot{e}(t)] = M_2(t). \tag{67}$$

where φ_i, $\dot{\varphi}_i$, $\ddot{\varphi}_i$ (i = 1,2) are rotation angle, angular velocity, angular acceleration of the input pinion and the output wheel respectively. J_1 and J_2 are the mass moments of inertia of the gears. $M_1(t)$ and $M_2(t)$ denote the external torques load applied on the system. r_{b1} and r_{b2} represent the base radii of the gears. By introducing the composite coordinate

$$q = r_{b1}\varphi_1 + r_{b2}\varphi_2. \tag{68}$$

Eqs. (66) and (67) yield a single differential equation in the following form

$$m_{red}\ddot{q} + k_z(t)q + c_z\dot{q} = F(t) - k_z(t)e(t) - c_z\dot{e}(t), \tag{69}$$

where

$$m_{red} = \frac{J_1J_2}{J_1r_{b2}^2 + J_2r_{b1}^2} \qquad F(t) = m_{red}\left(\frac{M_1(t)r_{b1}}{J_1} + \frac{M_2(t)r_{b2}}{J_2}\right). \tag{70}$$

Note that the rigid-body rotation from the original mathematical model in Eqs. (66) and (67) is eliminated by introducing the new coordinate $q(t)$ in Eq. (69). Variable $q(t)$ is called the dynamic transmission error of the gear-pair system [32]. Upon assuming that when $\dot{\varphi}_1 = \omega_1 = const$, $\dot{\varphi}_2 = \omega_2 = const$, $c_z = 0$, $k_z(t) = k_0$, the transmission error q is equal to the static tooth deflection under constant load q_0 as $q = r_{b1}\varphi_1 + r_{b2}\varphi_2 = q_0$. Eq. (69) yields the following relation

$$F(t) \approx F_0(t) = k_0q_0 + k_0e(t). \tag{71}$$

Eq. (69) can then be rewritten in the form

$$m_{red}\ddot{q} + k_z(t)q + c_z\dot{q} - f(t) = 0, \tag{72}$$

where $f(t) = k_0 q_0 - [k_z(t) - k_0]e(t) - c_z\dot{e}(t)$.

In steady state motion of the gear system, the mesh stiffness $k_z(t)$ can be approximately represented by a truncated Fourier series [33]

$$k_z(t) = k_0 + \sum_{n=1}^{N} k_n \cos(n\omega_z t + \gamma_n). \tag{73}$$

where ω_z is the gear meshing angular frequency which is equal to the number of gear teeth times the shaft angular frequency and N is the number of terms of the series.

In general, the error components are no identical for each gear tooth and will produce displacement excitation that is periodic with the gear rotation (i.e. repeated each time the tooth is in contact). The excitation function $e(t)$ can then be expressed in a Fourier series with the fundamental frequency corresponding to the rotation speed of the faulted gear. When the errors are situated at the teeth of the pinion, $e(t)$ may be taken in the form

$$e(t) = \sum_{i=1}^{I} e_i \cos(i\omega_1 t + \alpha_i). \tag{74}$$

Parameters	Pinion	Wheel
Gear type	helical, standard involute	
Material	steel	
Module (mm)	4.50	
Pressure angle (o)	20.00	
Helical angle (o)	14.56	
Number of teeth z	14	39
face width (mm)	67.00	45.00
base circle radius (mm)	30.46	84.86

Table 3. Parameters of the test gears.

Therefore, the vibration equation of gear-pair system according to Eq. (72) is a differential equation with the periodic coefficients.

According to the experimental setup which will be described later, the model parameters include J_1= 0.093 (kgm^2), J_2 = 0.272 (kgm^2) and nominal pinion speed of 1800 rpm (f_1 = 30 Hz). The mesh stiffness of the test gear pair at particular meshing position was obtained by means of a FEM software [29]. The static tooth deflection is estimated to be $q_0 = 1.2 \times 10^{-5}$ (m). The values of Fourier coefficients of the mesh stiffness with corresponding phase angles are given in Table 4. The mean value of the undamped natural frequency $\bar{\omega}_0 = \sqrt{k_0 / m_{red}} \approx 5462 s^{-1}$, corresponding to $\bar{f}_0 = \bar{\omega}_0 / 2\pi \approx 869$ (Hz). Based on the experimental work, the mean value of the Lehr damping ratio $\zeta = 0.024$ is used for the dynamic model. The damping coefficient c_z can then be determined by $c_z = 2\bar{\omega}_0 \zeta m_{red}$.

n	k_n(N/m)	γ_n(radian)
0	8.184610^{8}	
1	3.226710^{7}	2.5581
2	1.351610^{7}	-1.4421
3	8.151010^{6}	-2.2588
4	3.528010^{6}	0.9367
5	4.028010^{6}	-0.8696
6	9.710010^{5}	-2.0950
7	1.424510^{6}	0.9309
8	1.550510^{6}	0.2584
9	4.645010^{5}	-1.2510
10	1.415810^{6}	2.1636

Table 4. Fourier coefficients and phase angles of the mesh stiffness.

i	Case 1		Case 2	
	e_i(mm)	α_i(rad)	e_i(mm)	α_i(rad)
1	0.0015	-0.049	0.010	1.0470
2	0.0035	-1.7661	0.003	-1.4521
3	0.0027	-0.7286	0.0018	0.5233
4	0.0011	-0.5763	0.0011	1.4570
5	0.0005	-0.7810	0.0009	-0.8622
6	0.0013	1.8172	0.0003	1.1966

Table 5. Fourier coefficients and phase angles of excitation function $e(t)$.

Using the obtained periodic solutions of Eq. (72), the calculated dynamic transmission errors are shown in Figures 9 and 10 corresponding to different excitation functions $e(t)$ given in Table 5. The spectra in Figures 10(a) and 10(b) show clearly the meshing frequency and its harmonics with sideband structures. As expected, the sidebands are spaced by the rotational frequency f_1 of the pinion. By comparing amplitude of these sidebands in both spectra, it can be concluded that the excitation function $e(t)$ caused by tooth errors is responsible for generating sidebands.

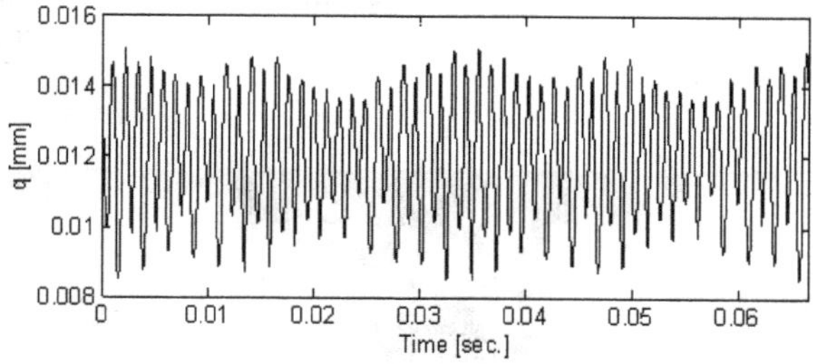

Figure 9. . Modelling result: dynamic transmission error $q(t)$.

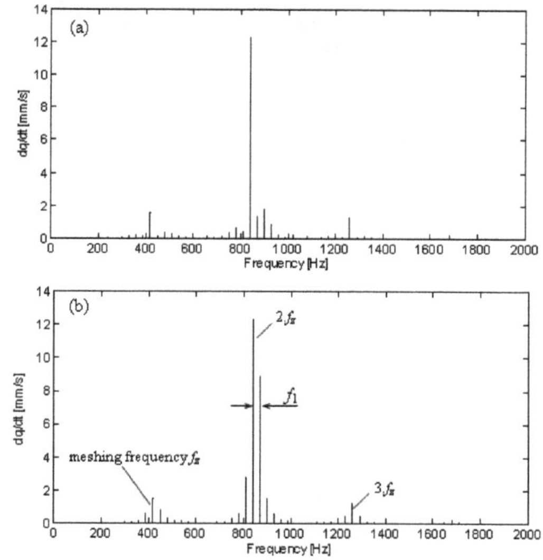

Figure 10. Modelling result: frequency spectrum of dq/dt corresponding to (a) excitation function $e(t)$ of Case 1 and b) excitation function $e(t)$ with larger coefficients (Case 2).

The experiment was done at an ordinary back-to-back test rig (Figure 11). The major parameters of the test gear-pair are given in Table 3. The load torque was provided by a hydraulic rotary torque actuator which remains the external torque constant for any motor speed. The test gearbox operates at a nominal pinion speed of 1800 rpm. (30 Hz), thus the meshing frequency f_z is 420 Hz. A Laser Doppler Vibrometer was used for measuring oscillating parts of the angular speed of the gear shafts (i.e. oscillating part of $\dot{\varphi}_1$ and $\dot{\varphi}_2$) in order to determine experimentally the dynamic transmission error. The measurement was taken with two non-contacting transducers mounted in proximity to the shafts, positioned at the closest position to the test gears. The vibration signals were sampled at 10 kHz. The signal used in this study was recorded at the end of 12-hours total test time, at that time a surface fatigue failure occurred on some teeth of the pinion.

Figure 11. Gearbox test rig.

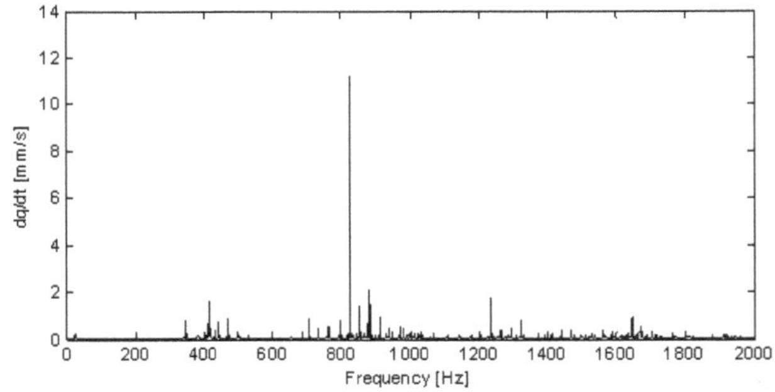

Figure 12. Experimental result: frequency spectrum of dq/dt.

Figure 12 shows a frequency spectrum of the first derivative of the dynamic transmission error $\dot{q}(t)$ determined from the experimental data. The spectrum presents sidebands at the meshing frequency and its harmonics. In particular, the dominant sidebands are spaced by the rotational frequency of the pinion and characterized by high amplitude. This gives a clear indication of the presence of the faults on the pinion. By comparing the spectra displayed in Figures 13 and 14, it can be observed that the vibration spectrum calculated by numerical methods (Figure 13) and the spectrum of the measured vibration signal (Figure 14) show the same sideband structures.

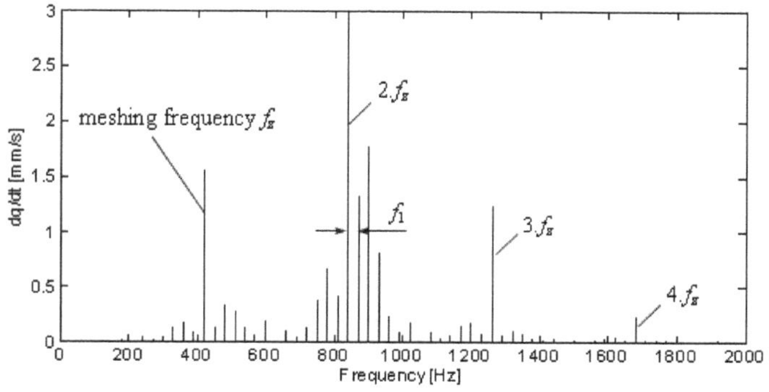

Figure 13. Calculating result: frequency spectrum of dq/dt.

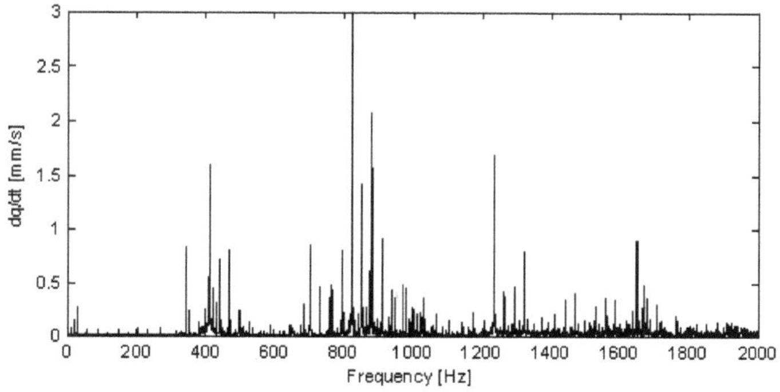

Figure 14. Experimental result: zoomed frequency spectrum of dq/dt from Figure 12.

The calculations required a large number of time steps to ensure that the frequency resolution in vibration spectra is fine enough. In comparison with the numerical procedure based on Runge-Kutta method, the computation time by the Newmark-based numerical procedure is greatly reduced for large number of time steps as shown in Figure 15, for that the same computer was used as in the previous example.

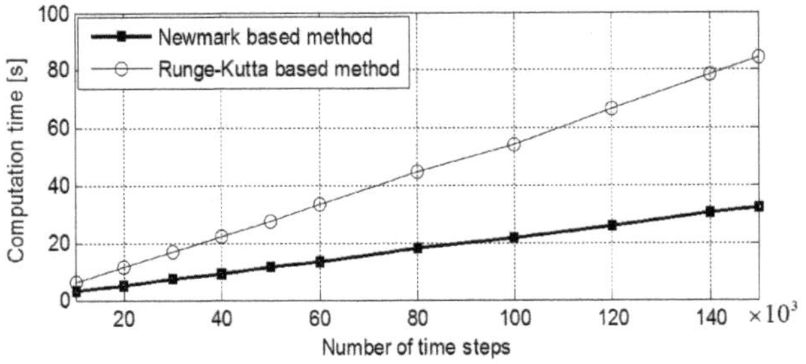

Figure 15. Comparison of the computation time for the gear-pair model.

4.3. Periodic vibration of the transport manipulator of a forging press

The most common forging equipment is the mechanical forging press. Mechanical presses function by using a transport manipulator with a cam mechanism to produce a preset at a certain location in the stroke. The kinematic schema of such mechanical adjustment unit is depicted in Figure 16.

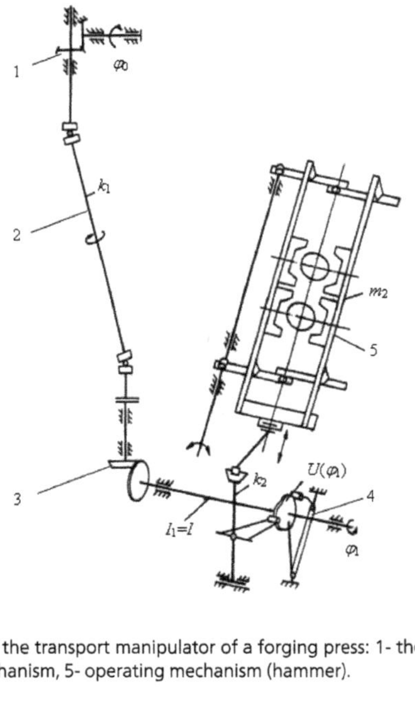

Figure 16. Kinematic schema of the transport manipulator of a forging press: 1- the first gearbox, 2- driving shaft, 3- the second gearbox, 4- cam mechanism, 5- operating mechanism (hammer).

The dynamic model of this system shown in Figure 17 is used to investigate periodic vibrations which are a commonly observed phenomenon in mechanical adjustment unit during the steady-state motion [18, 23]. The system of the driver shaft, the flexible transmission mechanism and the hammer can be considered as rigid bodies connected by spring-damping elements with time-invariant stiffness k_i and constant damping coefficients c_i, $i=1, 2$. The rotating components are modeled by two rotating disks with moments of inertia I_0 and I_1. The cam mechanism has a nonlinear transmission function $U(\varphi_1)$ as a function of the rotating angle φ_1 of the cam shaft, the driving torque from the motor $M(t)$ and the external load $F(t)$ applied on the system.

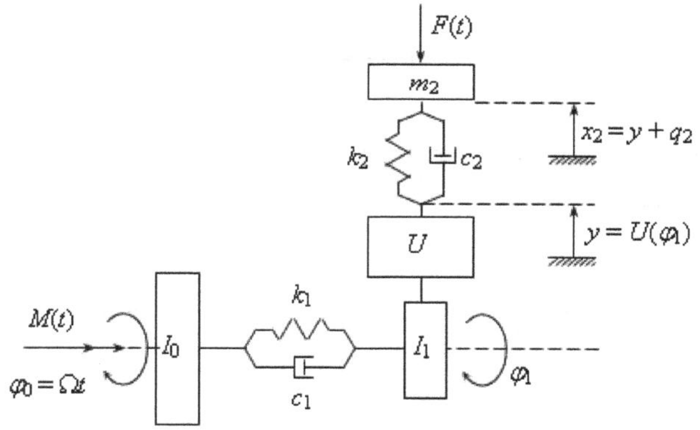

Figure 17. Dynamic model of the transport manipulator.

When the angular velocity Ω of the driver input is assumed to be constant in the steady state

$$\varphi_0=\Omega t, \tag{75}$$

one leads to the following relation

$$\varphi_1=\Omega t+q_1 \tag{76}$$

where q_1 is the difference between rotating angles φ_0 and φ_1 due to the presence of elastic element k_1 and damping element c_1, resulted from the flexible transmission mechanism.

By the analogous way as in Section 3.1, we obtain the linear differential equations of vibration for the system in the compact matrix form as

$$\boldsymbol{M}(\Omega t)\ddot{\boldsymbol{q}}+\boldsymbol{C}(\Omega t)\dot{\boldsymbol{q}}+\boldsymbol{K}(\Omega t)\boldsymbol{q}=\boldsymbol{d}(\Omega t) \tag{77}$$

where

$$\boldsymbol{M}(\Omega t)=\begin{bmatrix} I_1+m_2\overline{U}'^2 & m_2\overline{U}' \\ m_2\overline{U}' & m_2 \end{bmatrix},\ \boldsymbol{C}(\Omega t)=\begin{bmatrix} c_1+2m_2\Omega\overline{U}'\overline{U}'' & 0 \\ 2m_2\Omega\overline{U}'' & c_2 \end{bmatrix}$$

$$\boldsymbol{K}(\Omega t)=\begin{bmatrix} k_1+F\overline{U}''+m_2\Omega^2(\overline{U}'\overline{U}'''+\overline{U}''^2) & 0 \\ m_2\Omega^2\overline{U}''' & k_2 \end{bmatrix},\ \boldsymbol{d}=\begin{bmatrix} -F\overline{U}'-m_2\Omega^2\overline{U}'\overline{U}'' \\ -F-m_2\Omega^2\overline{U}'' \end{bmatrix},\ \boldsymbol{q}=\begin{bmatrix} q_1 \\ q_2 \end{bmatrix}$$

In steady state motion of the cam mechanism, function $U'(\varphi)$ takes the form [18, 23]

$$U'(\varphi)=\sum_{k=1}^{K}(a_k\cos k\varphi+b_k\sin k\varphi) \tag{78}$$

The functions $\overline{U}'$, $\overline{U}''$, $\overline{U}'''$ in Eq. (77) can then be calculated using Eq. (78) for $\varphi=\Omega t$.

The following parameters are used for numerical calculations: Rotating speed of the driver input n=50 (rpm) corresponding to $\Omega=5.236(1/s)$, stiffness $k_1=7692$ Nm; $k_2=10^6$ N/m, damping coefficients $c_1=18.5$ Nms; $c_2=2332$ Ns/m, $I_1=1.11$ kgm² and $m_2=136$ kg.

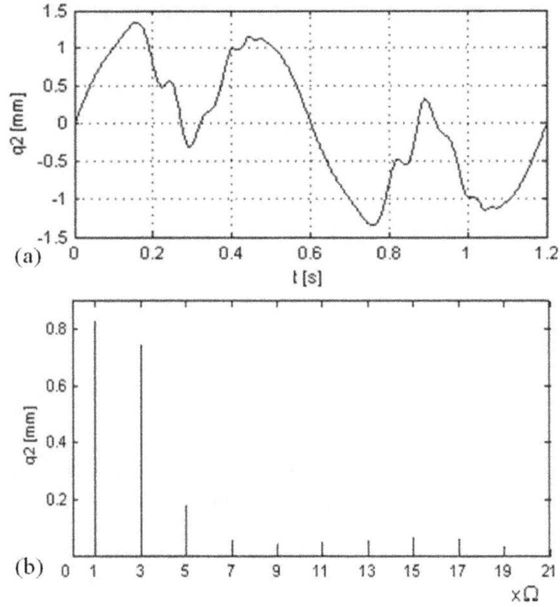

Figure 18. Calculating result of q_2 for case 1, (a) time curve, (b) frequency spectrum.

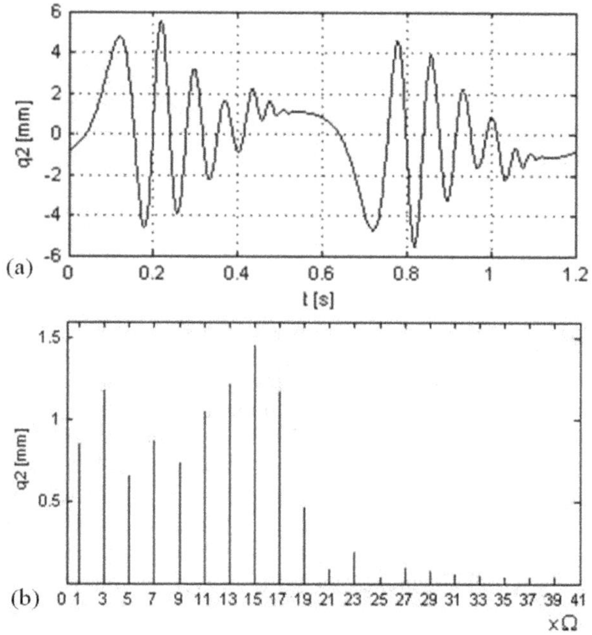

Figure 19. Calculating result of q_2 for case 2, (a) time curve, (b) frequency spectrum.

The Fourier coefficients a_k in Eq. (78) with $K = 12$ are given in Table 2 for two different cases and coefficients $b_k = 0$. We consider only periodic vibrations which are a commonly observed phenomenon in the system. The periodic solutions of Eq. (77) can be obtained by choosing appropriate initial conditions for the vector of variables q.

To verify the dynamic stable condition of the vibration system, the maximum of absolute value $|\rho|_{max}$ of the solutions of the characteristic equation, according to Eq. (10), is now calculated. The obtained values for both cases are $|\rho|_{max} = 0.001992$ (case 1) and $|\rho|_{max} = 0.001623$ (case 2). It can be concluded that the system is dynamically stable for both two cases since $|\rho|_{max} < 1$.

Calculating results of periodic vibrations of the mechanical adjustment unit, i.e. periodic solutions of Eq. (77), are shown in Figures 18-19 for two cases of the cam profile. Comparing both time curves, the influence of cam profiles on the vibration level of the hammer can be recognized. In addition, the frequency spectrums show harmonic components of the rotating frequency at Ω, 3Ω, 5Ω. These spectrums indicate that the considered system performs stationary periodic vibrations only.

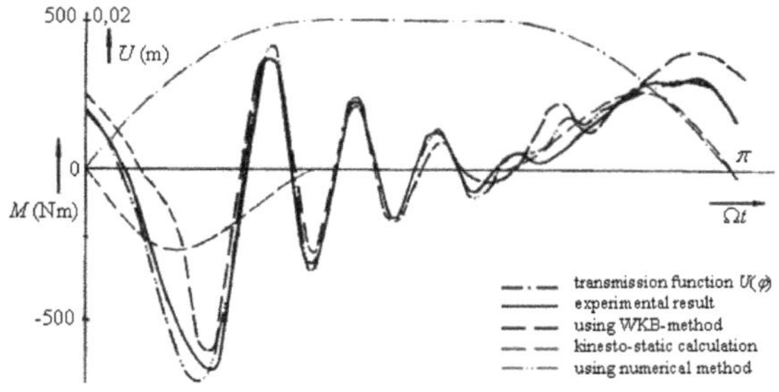

Figure 20. Dynamic moment acting on the driving shaft of the mechanical adjustment unit

To verify the calculating results using the numerical methods, the dynamic load moment of the mechanical adjustment unit was measured on the driving shaft (see also Figure 16). A typical record of the measured moment is plotted in Figure 20, together with the curves calculated from the dynamic model by using the WKB-method [18, 34], the kinesto-static calculation and the proposed numerical procedures based on Newmark method and Runge-Kutta method. Comparing the curves displayed in this figure, it can be observed that the calculating result using the numerical methods is more closely in agreement with the experimental result than the results obtained by the WKB-method and the kinesto-static calculation.

5. Concluding remarks

The calculation of dynamic stable conditions and periodic vibrations of elastic mechanisms and machines is an important problem in mechanical engineering. This chapter deals with the problem of dynamic modelling and parametric vibration of transmission mechanisms with elastic components governed by linearized differential equations having time-varying coefficients.

Numerical procedures based on Runge-Kutta method and Newmark integration method are proposed and applied to find periodic solutions of linear differential equations with time-periodic coefficients. The periodic solutions can be obtained by Newmark based procedure directly and more conveniently than the Runge-Kutta method. It is verified that the computation time with the Newmark based procedure reduced by about 60%-65% compared to the procedure using the fourth-order Runge-Kutta method (see also Figures 7 and 15). Note that this conclusion is only true for linear systems.

The numerical methods and algorithms are demonstrated and tested by three dynamic models of elastic transmission mechanisms. In the last two examples, a good agreement is obtained between the model result and the experimental result. It is believed that the proposed approaches can be successfully applied to more complicated systems. In addition, the proposed numerical procedures can be used to estimate approximate initial values for the shooting method to find the periodic solutions of nonlinear vibration equations.

Acknowledgements

This study was completed with the financial support by the Vietnam National Foundation for Science and Technology Development (NAFOSTED).

Author details

Nguyen Van Khang* and Nguyen Phong Dien

*Address all correspondence to: nvankhang@mail.hut.edu.vn

Department of Applied Mechanics, Hanoi University of Science and Technology, Vietnam

References

[1] Schiehlen, W., & Eberhard, P. (2004). Technische Dynamik (2. Auflage). Stuttgart, B.G. Teubner.

[2] Shabana, A. A. (2005). Dynamics of multibody systems. 3. Edition). Cambridge, Cambridge University Press.

[3] Josephs, H., & Huston, R. L. (2002). Dynamics of mechanical systems. Boca Raton, CRS Press.

[4] Wittenburg, J. (2008). Dynamics of Multibody Systems. 4. Edition). Berlin, Springer.

[5] Stoer, J., & Bulirsch, R. (2000). Numerische Mathematik 2 (4. Auflage). Berlin, Springer.

[6] Newmark, N. M. (1959). A method of computation for structural dynamics. *ASCE Journal of Engineering Mechanics; Division*, 85-67.

[7] Géradin, M., & Rixen, D. (1994). Mechanical Vibrations. Chichester, Wiley.

[8] Nayfeh, A. H., & Balachandran, B. (1995). Applied nonlinear dynamics. New York, John Wiley & Sons.

[9] Eich-Soellner, E., & Fuehrer, C. (1998). Numerical Methods in multibody dynamics. Stuttgart, Teubner.

[10] Cesari, L. (1959). Asymptotic behaviour and stability problems in ordinary differential equations. Berlin, Springer.

[11] Jordan, D. W., & Smith, P. (2007). Nonlinear Ordinary differential equations. (4. Edition). New York, Oxford University Press.

[12] Kortüm, W., & Lugner, P. (1994). Systemdynamik und Regelung von Fahrzeugen. Berlin, Springer.

[13] Heimann, B., Gerth, W., & Popp, K. (2007). Mechantronik (3. Auflage). München, Fachbuchverlag Leipzig in Carl Hanser Verlag.

[14] Dresig, H., & Vulfson, I. I. (1989). Dynamik der Mechanismen. Berlin, Deutscher Verlag der Wissenschaften.

[15] Vulfson, I. I. (1973). Analytical investigation of vibration of mechanisms caused by parametric impulses. *Mechanism and Machine Theory*, 10, 305-313.

[16] Vulfson, I. I. (1989). Vibroactivity of branched and ring structured mechanical drives. New York, London: Hemisphere Publishing Corporation.

[17] Müller, P. C., & Schiehlen, W. (1976). Lineare Schwingungen. Wiesbaden, Akademische Verlagsgesellschaft.

[18] Nguyen , Van Khang. (1986). Dynamische Stabilität und periodische Schwingungen in Mechanismen. *Diss. B. TH Karl-Marx-Stadt.*

[19] Nguyen , Khang Van. (1982). Numerische Bestimmung der dynamischen Stabilitätsparameter und periodischen Schwingungen ebener Mechanismen. *Rev. Roum. Sci. Tech.-Mec. Appl*, 27(4), 495-507.

[20] Malkin, J. G. (1959). Theorie der Stabilät einer Bewegung. Berlin, Akademie.

[21] Troger, H., & Steindl, A. (1991). Nonlinear stability and bifurcation theory. Wien, Springer.

[22] Cleghorn, W. L., Fenton, R. G., & Tabarrok, B. (1984). Steady-state vibrational response of high-speed flexible mechanisms. *Mechanism and Machine Theory*, 417-423.

[23] Roessler, J. (1985). Dynamik von Mechanismen-Antriebssystemen im Textil- und Verarbeitungs-maschinenbau. *Diss. B. TH Karl-Marx-Stadt.*

[24] Nguyen, Van Khang, Thai, Manh Cau, & Nguyen, Phong Dien. (2004). Modelling parametric vibration of gear-pair systems as a tool for aiding gear fault diagnosis. *Technische Mechanik*, 24(3-4), 198-205.

[25] Volmer, J. (1989). Getriebetechnik/Kurvengetriebe . (2. Auflage). Berlin Verlag der Technik

[26] Nguyen, Van Khang, Nguyen, Phong Dien, & Vu, Van Khiem. (2010). Linearization and parametric vibration analysis of transmission mechanisms with elastic links. In: Proceedings of The First IFToMM Asian Conference on Mechanism and Machine Science Taipei, Taiwan

[27] Nguyen, Phong Dien. (2003). Beitrag zur Diagnostik der Verzahnungen in Getrieben mittels Zeit- Frequenz- Analyse. *Fortschritt-Berichte VDI; Reihe*, 11(135).

[28] Padmanabhan, C., & Singh, R. (1996). Analysis of periodically forced nonlinear Hill's oscillator with application to a geared system. *Journal of the Acoustial Society of America*, 99(1), 324-334.

[29] Boerner, J. (1999). Rechenprogramm LVR: Beanspruchungsverteilung an evolventischen Verzahnungen. *Foschungsberichte TU Dresden, Institut für Maschinenelemente und Maschinen-konstruktion.*

[30] Dalpiaz, G., Rivola, A., & Rubini, R. (2000). Effectiveness and sensitivity of vibration processing techniques for local fault detection in gears. *Mechanical System and Signal Processing*, 14(3), 387-412.

[31] Howard, I., Shengxiang, Jia., & Wang, J. (2001). The dynamic modelling of a spur gear in mesh including friction and a crack. *Mechanical System and Signal Processing*, 15(5), 831-853.

[32] Parker, G. R., Vijayakar, S. M., & Imajo, T. (2000). Non-linear dynamic response of a spur gear pair: Modelling and experimental comparisons. *J. Sound and Vibration*, 237(3), 435-455.

[33] Theodossiades, S., & Natsiavas, S. (2000). Non-linear dynamics of gear-pair systems with periodic stiffness and backlash. *J. Sound and Vibration*, 229(2), 287-310.

[34] Nguyen, Van Khang , Nguyen, Phong Dien, & Hoang , Manh Cuong. (2009). Linearization and parametric vibration analysis of some applied problems in multibody systems. *Multibody System Dynamics*, 22-163.

Chapter 14

Optimal Vibrotactile Stimulation Activates the Parasympathetic Nervous System

Nelcy Hisao Hiraba, Motoharu Inoue, Takako Sato, Satoshi Nishimura, Masaru Yamaoka, Takaya Shimano, Ryuichi Sampei, Katuko Ebihara, Hisako Ishii and Koichiro Ueda

Additional information is available at the end of the chapter

1. Introduction

We currently live in an environment that has become more and more stressful. Escaping from the stress in society through various activities (e.g., acupuncture, massage, listening to classic music or natural sounds, etc.) is important for our mental health. We previously reported that optimal facial vibrotactile stimulation (i.e., 89 Hz frequency and 1.9 µm amplitude [89 Hz-S]) might activate the parasympathetic nervous system (Hiraba et al. 2008, 2011). Specifically, we showed that 89 Hz-S stimulation of the face led to increased salivation and a feeling of mental well-being through parasympathetic activity based on functional near-infrared spectroscopy (fNIRS) oxyhaemoglobin (oxyHb) activity. Namely, brain blood flow (BBF) oxyHb in the frontal cortex was near zero (Hiraba et al. 2011). We investigated adaptation to the continuous use of vibrotactile stimuli for 4 or 5 days in the same subjects to determine whether this resulted in decreased salivation (Despopoulos and Silbernagel, 2003; Principles of Neural Science, 2000a). Then, we compared resting and stimulated salivation and investigated the most effective frequency for increasing salivary secretion. Increased salivation in normal subjects was defined as a difference between resting and stimulated salivation (Hiraba et al. 2011).

Furthermore, to study the mechanism of increased salivation evoked by vibrotactile stimulation, we recorded changes in heartbeat frequency and pupillary reflex during stimulation. We reported that pulse frequency changes during vibrotactile stimulation. A decrease in pulse frequency and a contraction in pupil diameter suggest parasympathetic activity (Principles of Neural Science, 2000b). We believe these reflexes are coordinated by a highly interconnected set of structures in the brainstem and forebrain that form a central autonomic network (Principles of Neural Science, 2000b).

We found that vibrotactile stimulation increased salivation, as reported by Hiraba et al. (2008). Furthermore, Hiraba et al. (2011) reported that increased salivation due to facial vibrotactile stimulation might be due to parasympathetic stimulation based on frontal cortex BBF measurements. Particularly, vibrotactile stimulation at 89 Hz-S using a single motor was most effective in increasing salivation without adaptation following continuous daily use. We know that autonomic activity changes heart rate and pupil diameter. Thus, we believe that heart rate and pupil diameter measurements during 89 Hz-S stimulation represent the effects of the autonomic nervous system. In this study, we demonstrated that 89 Hz-S stimulation led to mental stability due to parasympathetic activity.

2. Material and Methods

2.1. Vibrotactile stimulation apparatus

The vibrotactile stimulation apparatus consisted of an oscillating body and a control unit, as shown in Hiraba et al. (2008) and Yamaoka et al. (2007). The oscillating body was composed of the following two parts: (1) a headphone headset equipped with vibrators in the positions of the bilateral microphones and (2) a vibration electric motor (VEM; Rekishin Japan Co., LE12AOG) covered in silicon rubber (polyethyl methacrylate, dental mucosa protective material; Shyofu Co.) for conglobating the stimulation parts and preventing VEM warming due to long periods of vibration (Hiraba et al. 2008).

We examined the amount of salivation during vibrotactile stimulation on the bilateral masseter muscle belly (on the parotid glands) and on bilateral parts of the submandibular angle (on the submandibular glands). We determined the amount of salivation using a dental cotton roll (1 cm width, 3 cm length) positioned at the opening of the secretory ducts (i.e., the right and left parotid glands and right and left submandibular and sublingual glands) during vibrotactile stimulation of the bilateral parotid and submandibular glands. The weights of the wet cotton rolls after 3 min of use were compared with their dry weights measured previously (Hiraba et al. 2008).

2.2. Stimulating salivation and frontal cortex BBF

We determined that a 3-min salivation measurement with a 5-min recovery time was sufficient from a previous experiment (Hiraba et al. 2008, Hiraba et al. 2011). Furthermore, salivation is most effectively induced by vibrotactile parotid gland stimulation at 89 Hz-S, which was used in this experiment. We examined adaptation to vibrotactile stimulation by monitoring changes in salivation during 4 or 5 continuous days using the same time schedule (i.e., 89 Hz-S). Frontal cortex recordings were acquired using a fNIRS OEG16 instrument (Spectratech, Inc., Shelton, CT, USA) during vibrotactile stimulation. We conducted salivation tests with 19 normal subjects (six males, 13 females; average age: 22 years) and resting-stimulation examinations for adaptation with 26 normal subjects (11 males and 15 females; average age: 25 years). We also performed fNIRS in eight normal subjects (six males, two females; average age: 22 years) to examine the effects of resting state and classical music (Mozart, *Eine kleine Nachtmusik*). This experiment was performed between 3 and 5 pm in a temperature-controlled, quiet room, as described in previous papers (Hiraba et al. 2008, 2011).

2.3. HRV analysis during vibrotactile stimulation

We recorded changes in power-spectral analysis of heart rates (HRV; Heart Rate Variability module, AD Instruments, Japan, under the following conditions: (1) resting state; (2) 89 Hz-S stimulation on the face (89 Hz-S face); (3) listening to Mozart (Mozart); (4) Mozart + 89 Hz-S on the face; (5) 89 Hz-S on the nape of the neck (89 Hz-S neck); and (6) listening to noise (Noise), as shown in Figure 2E. A power-spectral analysis of HRV module data was conducted using the period histogram analysis program based on distribution of the length of the RR interval for 3 min, and typical values during various stimuli were analysed in terms of the highest value (i.e., peak value) during the recording period. For example, Figure 1A shows RR intervals (n1, n2, n3, n4 ms, and so on,) on the electrocardiogram (ECG) during vibrotactile stimulation. Figure 1B shows a peak value example (1000 ms) during vibrotactile stimulation. Heart rates during rest and during various stimuli were recorded for 3 min, and then analyses of 3-min HRV data were performed off-line. When heart rates were compared among the rest and various stimulation conditions, we used the RR-interval peak value (i.e., 1000 ms in this example) obtained from the power-spectra analysis. We conducted these examinations with 16 normal subjects (11 males, five females; average age: 25 years). This experiment was performed at 3 and 5 pm in a quiet, temperature-controlled room.

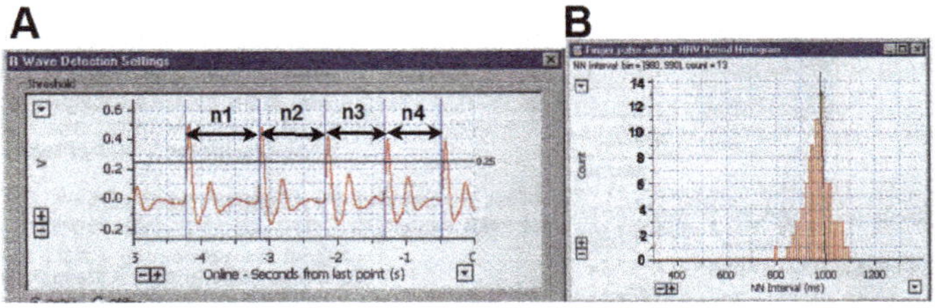

Figure 1. HRV module analysis. Method used to measure RR intervals (n1, n2, n3, n4, etc.) on ECG recordings (A) and frequency spectrum based on RR interval length over 3 min during 89 Hz-S vibrotactile stimulation (B). Horizontal line indicates RR interval (ms), and vertical line indicates number. Note that the peak frequency spectrum was 1000 ms in this experiment.

2.4. Pupillography during vibrotactile stimulation

IRIS (Iriscorder, Hamamatsu Photonics Co., [Japan]) records transverse diameter and velocity reactions and can take a picture of the eyes by illuminating visible light (infrared radiation). The resulting image can record the condition of the iris and eyeball movement on the monitor. For example, when normal subjects are exposed to continuous light stimulation for 1 sec, we can obtain a pupillogram from the IRIS apparatus; constricted pupils indicate parasympathetic activity, and pupil dilation indicates sympathetic activity. Pupil diameter in normal subjects is generally 2-5 mm, which changes under various adaptation conditions. We examined the transverse diameter and velocity of pupil constriction or dilatation after vibrotactile stimulation to explore changes in autonomic activity.

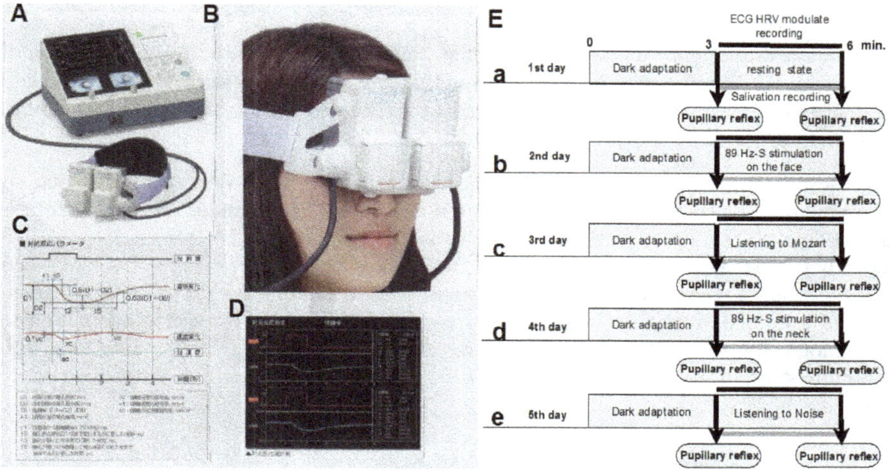

Figure 2. IRIS apparatus (A and B) and typical data acquired following light stimulation (C and D). IRIS records the bilateral pupillary reflex simultaneously. IRIS records the constricted and dilated pupil reflex for one second after light stimulation. Pupillary reflex parameters (e.g., most constricted pupil velocity) can be quantified. Experimental schedule (E).

Figure 2 shows the IRIS experimental apparatus used in Figures 2A and 2B. The pupillary light-reflex test was executed using an infrared pupillometer (Iriscorder C10641, Hamamatsu Photonics Co.), and pupil diameter in both the right and left eyes was measured after 3 min of rest or stimulation. Figure 2E shows the timeline of this experiment. Five particular experimental conditions were explored over 5 days; we recorded HRV modulation during the pupillary reflex adaptation test. However, only the right pupil was exposed to light stimulation, as shown in Figure 4. The pupillary test is non-invasive and enables real-time diagnosis. We examined the initial diameter (D1), minimum diameter (D2), constriction ratio (CR), time to total construction (T3), maximum velocity of constriction (VC), and maximum acceleration of constriction (AC) among the parameters obtained from the IRIS. Pupil diameter decreased with parasympathetic activity and increased with sympathetic activity. The IRIS records pupil parameters of the right and left eyes simultaneously. However, we adopted parameters from the right pupil because data from both sides were similar, as shown in Figure 4. We conducted these examinations with eight normal subjects (six males, two females; average age: 25 years). This experiment was performed at 3 and 5 pm in a quiet, temperature-controlled room. Furthermore, one parameter was explored each experimental day because we obtained information from adaptation to light stimulation, as shown by the experimental schedule in Figure 2E.

3. Results

3.1. Vibrotactile stimulation of the parotid and submandibular glands

We examined differences between vibrotactile stimulation of the parotid and submandibular glands. We found that the most effective frequency to induce salivation was 89 Hz-S regardless of whether vibrotactile stimulation was delivered to the parotid or submandibular glands, as shown in previous paper (Hiraba et al. 2011).

Because patients with hyposalivation often have psychiatric disorders, we conducted an experiment to realistically approximate natural conditions. We examined whether effective salivation occurred continuously when vibrotactile stimulation was performed daily. Specifically, we used the 89 Hz-S frequency with a single motor from the previous experiment. None of the glands (i.e., right and left parotid glands or right submandibular and sublingual glands) showed a reduced response. Regression curves for each gland showed non-adaptation to continuous stimulation; instead they showed parallel or increasing curves, indicating that continuous use of this apparatus should not be problematic, as shown in a previous paper (Hiraba et al. 2011).

3.2. Relationship between stimulation and fNIRS activity

The OEG16 spectroscope was used to record BBF haemoglobin concentration from areas in the frontal cortex using 16 channels. We determined the oxyHb concentration schema evoked by 89 Hz-S vibrotactile stimulation by analysing 16 channels. The results showed very weak oxyHb concentrations (i.e., near zero) during 89 Hz-S vibrotactile stimulation. Changes in oxyHb, deoxyhaemoglobin (deoxyHb), and total haemoglobin (totalHb) concentrations during salivation measurements at rest and for each vibrotactile stimulation frequency were measured. As shown in a previous paper (Hiraba et al. 2011), changes during the following six conditions were measured: (1) resting; (2) 89 Hz-S vibrotactile stimulation; (3) 89 Hz-D (89 Hz-D, 89 Hz frequency with double motors, 3.5-μm amplitude); (4) 114 Hz-S; (5) 114 Hz-D (114 Hz-D, 114 Hz frequency with double motors, 3.5-μm amplitude); and (6) "A-" phonation. Each wave was recorded for 3 min, and each 2-min vibrotactile stimulus is shown between the vertical lines (Fig. 6B and Fig. 4 in the previous paper, 2011). Although each wave measured during resting salivation, at 114 Hz-D, and during "A-" phonation showed increased activity, the 89 Hz-D and 114 Hz-S vibrotactile stimuli decreased activity. However, vibrotactile stimulation at 89 Hz-S showed a value of almost zero. Particularly, when we focused on oxyHb changes based on these results, increased oxyHb occurred during "A-" phonation, the resting condition, and at 114 Hz-D vibrotactile stimulation, whereas a decrease in oxyHb was observed during vibrotactile stimulation at 114 Hz-S and at 89 Hz-D. However, oxyHb concentration during vibrotactile stimulation at 89 Hz-D was almost zero, as were all other data (oxyHb, deoxyHb, totalHb). From these results, we computed oxyHb integral rates over 2 min, as shown by the area between the longitudinal bars (Fig. 6B and Fig. 4 in the previous paper, 2011).

Furthermore, we examined integral rates while subjects listened to classical music for 2 min. We divided the subjects into two groups: (1) one group that disliked listening to classical music and (2) one group that enjoyed listening to classical music. Although the subjects who enjoyed the music did not show a larger spread of values, the former did. Specifically, vibrotactile stimulation at 89 Hz-S led to a small, similar value spread. All integral rates during the vibrotactile stimulation at 89 Hz-S and listening to classical music showed similar averages and standard deviations (SDs), as shown in Figure 6B.

3.3. Total salivation during vibrotactile stimulation

During facial 89 Hz-S stimulation, values of total salivation and salivation in the parotid or submandibular and sublingual glands were examined in comparison with resting salivation values. Submandibular and sublingual gland total salivation values increased; however, parotid gland salivation values were similar, as shown in Figure 3. Parotid gland salivation values were 0.15 ± 0.12 m*l* on both sides during rest and 0.14 ± 0.12 m*l* under the 89 Hz-S stimulation condition. Salivation values of submandibular and sublingual glands on both sides was 0.79 ± 0.44 m*l* during rest and 1.00 ± 0.58 m*l* under the 89 Hz-S condition ($p< 0.01$; Wilcoxon singed-rank test, two-tailed). Total gland salivation values on both sides were 0.91 ± 0.55 m*l* during rest and 1.16 ± 0.60 m*l* under the 89 Hz-S condition ($p < 0.01$). Particularly, although parotid gland salivation was similar between the resting and 89 Hz-S conditions, that in the submandibular and sublingual glands showed an absolute increase between the resting and 89 Hz-S conditions, as shown in Figure 3.

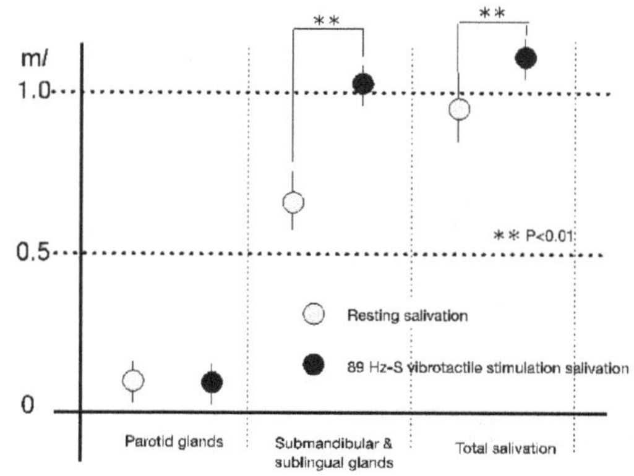

Figure 3. Salivation during the resting condition and 89 Hz-S vibrotactile stimulation over 3 min. Wilcoxon signed-rank test (two-tailed), $P< 0.01$.

3.4. Pupillary reflex after vibrotactile stimulation

Among the parameters obtained from the IRIS, we examined D1, D2, CR, T3, VC, and AC, as shown in Figures 4 A and 4B. The pupillary light reflex showed significantly decreased D1, D2, and T3 compared with the resting state. Furthermore, the pupillary light reflex showed increased AC, as shown in Figure 4B. Data from the right and left pupils were similar following light stimulation, as shown in Figure 4B. Thus, we employed data from the right pupillary reflex for data analysis.

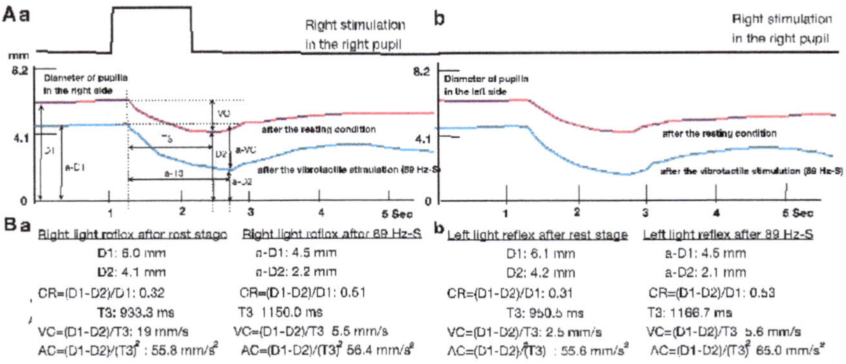

Figure 4. Typical example of data from the pupillary reflex. Right pupillary reflex data after light stimulation in the right pupil (A-a and B-a) and left pupillary reflex data after light stimulation in the left pupil (A-b and B-b).

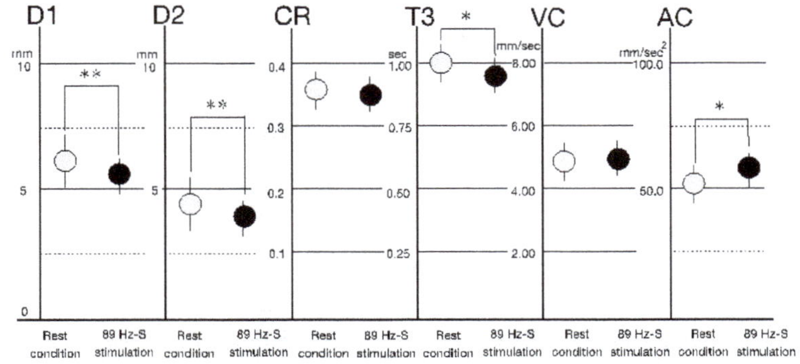

Figure 5. Effect of the pupillary reflex following right-side light stimulation between the resting condition and 89 Hz-S vibrotactile stimulation.

We analysed right pupillary reflex data from eight normal subjects, as shown in Figure 5. D1 was 6.15 ± 0.64 mm under the resting condition and 5.20 ± 1.12 mm under the 89 Hz-S condition ($p< 0.01$: Wilcoxon signed-rank test, two-tailed). D2 was 4.03 ± 0.79 mm under the resting condition and 3.47 ± 0.84 mm under the 89 Hz-S condition ($p< 0.01$). CR was 0.37 ± 0.10 ms under rest and 0.35 ± 0.09 under the 89 Hz-S condition. T3 was 1.089 ± 0.094 ms under rest and 0.973 ± 0.175 ms under the 89 Hz-S condition ($p< 0.05$). VC was 4.90 ± 0.95 mm/s under rest and 5.03 ± 0.89 mm/s under the 89 Hz-S condition. AC was 52.4 ± 16.8 mm/s^2 under rest and 56.7 ± 17.0 mm/s^2 under the 89 Hz-S condition ($p< 0.05$). Of particular note, D1, D2, and T3 showed an absolute decrease between the resting state and 89 Hz-S stimulation, and AC increased between the resting and 89 Hz-S conditions. CR and VC did not change between the resting and 89 Hz-S conditions, as shown in Figure 5.

3.5. Analysis of HRV during vibrotactile stimulation

We recorded typical heart rate changes and performed a power-spectral analysis (HRV module, AD Instruments, Japan) under the following six conditions: (1) resting state; (2) 89 Hz-S face; (3) listening to Mozart (Mozart); (4) Mozart + 89 Hz-S face; (5) 89 Hz-S neck; and (6) listening to noise (Noise), as shown in Figure 6. For example, Mozart + 89 Hz-S face "pulse (+)" indicates that participants were listening to Mozart classical music while receiving 89 Hz-S vibrotactile stimulation on the face. When comparing heart rates between the rest state and under various stimuli, we used the RR interval peak value from the power-spectral analysis (Fig. 6A and 6C). RR interval peak values (ms) from the power-spectral analysis were compared. The values were as follows: resting state, 757.5 ± 57.0 ms; 89 Hz-S face, 905.1 ± 189.5 ms, Mozart, 771.7 ± 86.7 ms; Mozart + 89 Hz-S face, 875.3 ± 188.3 ms; 89 Hz-S neck, 901.7 ± 188.4 ms; and Noise, 831.7 ± 114.6 ms (Fig. 6C). Significant differences were observed between resting state and 89 Hz-S face (paired *t*-test, $P< 0.01$) and between resting state and Mozart + 89 Hz-S face, 89 Hz-S neck, and Noise (paired *t*-test, $P< 0.05$; Fig. 6C).

The resting-state peak value had the lowest frequency. The Mozart-listening peak value was closest to the resting-state value, which might be because the majority of subjects disliked listening to classical music (three subjects favourite music was classical, and seven people reported classical music was not their favourite), as shown in Figure 6C. The 89 Hz-S stimulation led to the highest heart-beat frequency in comparison with the resting condition, as shown in Figures 6A and 6C. However, 89 Hz-S face stimulation was effective in many subjects, as 89 Hz-S face had the smallest SD, as shown in Figure 6C. On the other hand, heart rates during the resting condition and while listening to noise were similar. We generated noise with fractioned foam polystyrene. Many subjects may have felt discomfort due to the noise; however, we believe that discomfort induced by this noise was unlikely.

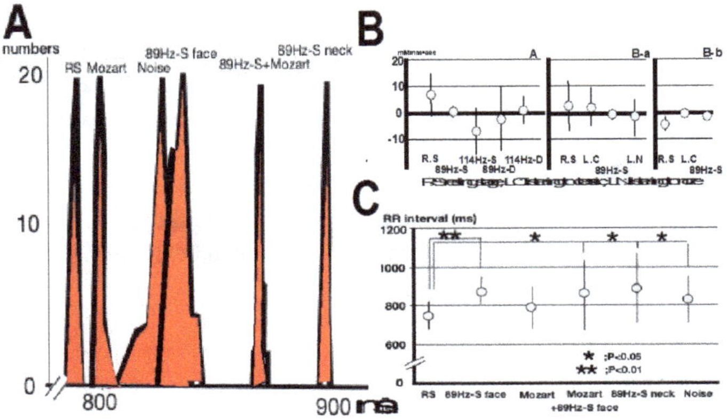

Figure 6. Changes in power spectrums (A) and HRV modulation (C) during various stimuli. B. fNIRS OxyHb concentration during various stimuli (this graph was described in a previous article, Hiraba et al. 2011). RS, 89 Hz-S face, Mozart, Mozart + 89 Hz-S face, 89 Hz-S neck, and Noise indicate 89 Hz-S on the face, listening to Mozart, both listening to Mozart and 89 Hz-S vibrotactile stimulation on the face, 89 Hz-S vibrotactile stimulation on the nape of the neck, and listening to noise, respectively. There were significant differences between RS and 89 Hz-S face (paired *t*-test, $P < 0.01$), between RS and Mozart + 89 Hz-S face, and between 89 Hz-S neck and Noise (paired *t*-test, $P < 0.05$).

4. Discussion

4.1. Relaxation produced by 89 Hz-S vibrotactile stimulation

We reported that 89 Hz-S vibrotactile stimulation evoked rest and increased salivation, as shown in previous papers (Hiraba et al. 2008, 2011). We further investigated increased salivation during 89 Hz-S. We were the first to show that increased salivation during 89 Hz-S stimulation was due to increased salivation from the submandibular and sublingual glands but not from the parotid glands, as shown in Figure 3. We knew that the amylase-rich parotid glands were principally responsible for the increased salivation. Salivation also occurs during mechanical stimulation during mastication when eating (Matuo, 2003). In addition to hunger- and mastication-induced salivation, salivation was also increased through 89 Hz-S vibrotactile stimulation of the facial and intraoral structures. This increased salivation may be different from salivation produced by hunger, as increased salivation during 89 Hz-S caused salivation in the submandibular and sublingual glands. In particular, increased salivation evoked by 89 Hz-S vibrotactile stimulation may be due to somatosensory input from the facial skin and intraoral cavity. Vibrotactile stimulation at 89 Hz-S may evoke a different perception from masticatory mechanical stimuli.

The frontal cortex is associated with cognitive function, including memory, attention, abstract reasoning, and higher cognitive processes (Principles of Neural Science, 2000a). We recorded changes in frontal cortex BBF to examine typical changes in fNIRS parameters based on increased oxyHb and totalHb and decreased deoxyHb, as reported by Sakatani et al. (2006). The effect of 89 Hz-S vibrotactile stimulation was almost zero for oxyHb, deoxyHb, and totalHb, as shown in a previous paper (Hiraba et al. 2011). The fNIRS activity focuses on excitatory behaviours that increase oxyHb. In animal experiments, changes in oxyHb and BBF are related, and fNIRS activity changes in oxyHb are used as a marker of neuronal activity (Hoshi et al. 2001). Thus, changes in oxyHb produced by 89 Hz-S vibrotactile stimulation may indicate mental stability. This may be due to the trend in oxyHb concentration between the 89 Hz-S vibrotactile stimulation in subjects who liked to listen to classical music (Fig. 6B). People relax when they listen to classical music, so we think that 89 Hz-S vibrotactile stimulation elicits excitation of the parasympathetic system. In particular, although the 89 Hz-S vibrotactile stimulation always led to parasympathetic excitation, listening to classical music caused different activity depending on music preference (Fig. 6B). Those subjects who enjoyed Mozart classical music found it relaxing, whereas those who disliked it perceived it as noise. However, 89 Hz-S vibrotactile stimulation may lead to a balanced mental condition, regardless of preference. This phenomenon suggests that the effect caused by the 89 Hz-S vibrotactile stimulation and the feeling experienced by those listening to Mozart who enjoyed it may be the same. Thus, we suggest that these feelings were produced by parasympathetic activity. We further investigated pupillary reflex and heart rate during 89 Hz-S stimulation.

4.2. Parasympathetic effect produced by 89 Hz-S vibrotactile stimulation

Our heartbeat increases when we are frightened (Principles of Neural Science, 2000b). The parasympathetic nervous system is responsible for rest, digestion, basal heart rate maintenance, respiration, and metabolism under normal resting conditions (Principles of Neural Science, 2000b). We examined parasympathetic effects by observing changes in the amount of salivation, HRV modulation of heart rate, and pupillary reflex induced by light stimulation during various stimuli.

We verified increased salivation induced by 89 Hz-S vibrotactile stimulation; the higher RR frequency was induced by the 89 Hz-S neck stimulation (Fig. 6C), and the greatest pupil contraction following light stimulation was induced by 89 Hz-S vibrotactile stimulation (Fig. 5). Furthermore, the 89 Hz-S face stimulation increased salivation the most (Fig. 3). Specifically, increased salivation was due to saliva secretion by the submandibular and sublingual glands but not the parotid glands. This was likely not due to hunger because the amount of salivation in the parotid glands did not increase. From these results, as for 89Hz-S neck, the relaxation effect might be big, however the salivated promotion effect was very weak. Conversely, the difference between pupil diameter before (D1) and after (D2) light stimulation during the resting condition was great, but the difference between pupil diameter before (a-D1) and after (a-D2) light stimulation during 89 Hz-S face was small, as shown in Figure 4. Furthermore, as shown in Figure 5, the AC (acceleration) was significantly different between the resting and the 89 Hz-S conditions ($P< 0.05$). These results suggested that the 89 Hz-S vibrotactile stimulation may elicit parasympathetic activity and greater pupil acceleration. This is likely because parasympathetic activity was stimulated by the 89 Hz-S face stimulation. Furthermore, vibrotactile stimulation at 89 Hz-S also led to parasympathetic activity.

4.3. Autonomic activity and anatomical projections in the central nervous system

We examined parasympathetic activity in three organs. Autonomic function must ultimately be coordinated for adaptation to environmental changes. The autonomic nervous system is composed of visceral sensory and motor system; the visceral reflexes are controlled by various local circuits in the brainstem and spinal cord. These reflexes are regulated by networks of central autonomic control nuclei in the brainstem, hypothalamus, and forebrain and are not under voluntary control, nor do they impinge on consciousness, with a few exceptions (Principles of Neural Science, 2000b). The pupillodilator muscle in the iris (pupil diameter), salivary glands, and heart rate are driven by sympathetic and parasympathetic nerves. Pupil diameter contracts, heart rate elongates, and salivation increases due to parasympathetic nerve activity. We also believe that changes in the frontal cortex BBF may represent autonomic activity. This coordination is carried out by a highly interconnected set of structures in the brainstem and forebrain that form a central autonomic network. The key component of this network is initiated by integrated information from the parabrachial nucleus of the solitary tract and trigeminal sensory complex in the brainstem. These nuclei receive inputs from somatosensory and visceral afferents of the trigeminal, facial, glossopharyngeal, and vagus nerves and then use the information to modulate autonomic function. The somatosensory and visceral sensory outputs from the trigeminal and solitary nuclei are relayed to the forebrain and amygdala by the parabrachial nucleus, which is important for behavioural responses to somatosensory, taste, and other visceral sensations (Principles of Neural Science, 2000b). Information arriving in the amygdala leads to sensations of pleasure and pain. In contrast, the parabrachial nucleus is a taste-sensation relay nucleus in rats (Scott and Small, 2009), and the rodent parabrachial nucleus sends integral limbic and reward system information (Yamamoto et al. 2009). Although their functions in humans are unknown, we think that these nuclei may play roles as relay nuclei of the autonomic system. On the other hand, we showed that a projection from the trigeminal sensory complex (e.g., the parabrachial nucleus) can also record the response to tactile stimuli from facial skin (Chiang et al. 1994). Furthermore, somatosensory information projects to the primary somatosensory cortex and is then relayed to the frontal cortex via the parietal association area (Handbook of Neuropsychology, 1994).

What does BBF activity in the frontal cortex indicate? We think that the information transmitted via the parabrachial nucleus dominates by way of the parietal association area. Thus, information in the frontal cortex is assumed to arrive via the parabrachial nucleus. The hypothalamus is the centre of the autonomic system. We perceive emotional experiences such as fear, pleasure, and contentment, and these perceptions reflect the interplay between higher brain centres and sub-cortical regions such as the hypothalamus and amygdala (Principles of Neural Science, 2000a). Patients in whom the prefrontal cortex or the cingulate gyrus has been removed are no longer bothered by pain but exhibit appropriate autonomic reactions; however, the sensation is not perceived as a powerfully unpleasant experience (Principles of Neural Science, 2000a). Furthermore, the anatomical connections between the amygdala and the temporal (cingulate gyrus) and frontal (prefrontal) association cortices provide the means by which visceral and somatosensory sensations trigger a rich assortment of associations or the cognitive interpretation of emotional states (Principles of Neural Science, 2000b).

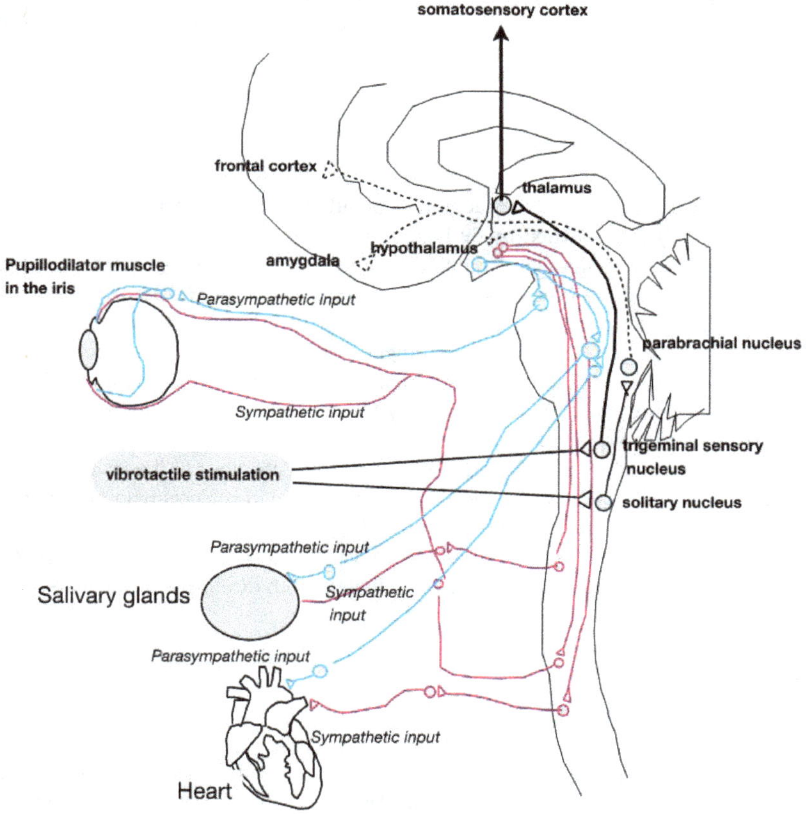

Figure 7. Pathways responsible for somatosensory information in the brain. Somatosensory information evoked by vibrotactile stimulation is relayed by the trigeminal sensory complex and solitary and parabrachial nuclei, which arrive at the hypothalamus, thalamus, amygdala, and frontal cortex, respectively. The autonomic system (particularly the parasympathetic nervous system) produces increased salivation. The lateral branch of the trigeminal sensory nucleus projects to the parabrachial nucleus. Information from the parabrachial nucleus is received by the amygdala and frontal cortex. Furthermore, somatosensory information is projected to the primary somatosensory cortex and relayed to the frontal cortex via the parietal association area. Thus, this information finally leads to a relaxed feeling, and BBF waves reflect parasympathetic activity (modified from schemas in Principles of Neural Science 2000b and Handbook of Neuropsychology 1994).

The 89 Hz-S stimulation evokes parasympathetic activity. This conclusion was based on increased salivation and heart rate and decreased pupil diameter during 89 Hz-S vibrotactile stimulation. Increased parotid gland salivation may be due to appetite, and increased salivation in the submandibular and sublingual glands may be associated with feelings of calm. On the other hand, we examined changes in the HRV module (RR intervals) between the resting state and exposure to various stimuli, and we showed increased heart rates during 89 Hz-S stimulation in comparison with the resting state (Billman, 2011). Furthermore, we think that the pupillary light-reflex test (Brashow, 1968) and heart rate HRV module (Billman, 2011) are the best tests for examining autonomic nervous system function (Gadner and Martin, 2000).

5. Conclusion

We showed that the most effective changes in salivation, pupil contraction, and HRV modulation (RR interval) were elicited by 89 Hz-S vibrotactile stimulation on the face. We thus conclude that 89 Hz-S vibrotactile stimulation affected parasympathetic activity based on changes observed in three organs. We also investigated autonomic activity by observing fNIRS waves. Because increased salivation was only observed in the submandibular and sublingual glands, it was likely not due to hunger. Furthermore, pupil constriction due to 89 Hz-S stimulation was less than that due to light following the resting condition. This likely indicated parasympathetic activity induced by 89 Hz-S stimulation. Changes in heart rate (RR intervals) during various stimuli were as effective as changes due to various stimuli combined with the 89 Hz-S stimulation. BBF oxyHb concentrations in the frontal cortex during 89 Hz-S vibrotactile stimulation were the same as those in subjects who preferred listening to classical music. Thus, 89 Hz-S vibrotactile stimulation may produce relaxation; salivation increases, pupil diameter constricts, and the heart rate (RR interval) is prolonged due to parasympathetic excitation. Thus, we believe that fNIRS in the frontal cortex reflects autonomic activity.

Acknowledgements

This work was supported by a Sogoshigaku research grant and the Sato Fund, as well grants from the Ministry of Education and a Grant-in-Aid for Scientific Research (21592539).

Author details

Nelcy Hisao Hiraba[1*], Motoharu Inoue[1], Takako Sato[2], Satoshi Nishimura[2], Masaru Yamaoka[3], Takaya Shimano[1], Ryuichi Sampei[1], Katuko Ebihara[1], Hisako Ishii[1] and Koichiro Ueda[1]

*Address all correspondence to: hiraba@dent.nihon-u.ac.jp

1 Departments of Dysphasia Rehabilitation, Japan

2 Oral and Maxillofacial Surgery, Japan

3 Nihon University, School of Dentistry, 1-8-13 Kanda-surugadai, Chiyoda-ku, Japan

References

[1] Brashow, J. I. (1968). Pupillary changes and reaction time with varied stimulus uncertainty. *Psychonomic Science*, 13, 69-70.

[2] Burdette, B. H., & Gale, E. N. (1988). The effects of treatment on masticatory muscle activity and mandibular posture in myofascial pain-dysfunction patients. *J Dent Res*, 67, 1126-1130.

[3] Chiang, C. Y., Hu, W., & Sessle, B. J. (1994). Parabrachial area and nucleus raphe magnus-induced modulation of nociceptive and non-nociceptive trigeminal subnucleus caudalis neurons activated by cutaneous or deep inputs. *J. Neurophysiol.*, 71, 2430-2445.

[4] Despopoulos, A., & Silbernagl, S. (2003). Nutrition and digestion. In: Color Atlas of Physiology. 5 th ed. New York Medical Science International,Ltd. , 226-265.

[5] Gardner, E. P., & Martin, J. H. (2000c). Coding of sensory information. In: Kandel ER., Schwartz JH., Jessell TM., editors, Principles of neuronal science, 4th ed, New York: McGraw-Hill , 411-429.

[6] Billman , George E. (2011). Heart rate variability-a historical perspective. *Frontiers in Physiology*, 1-13.

[7] Hiraba, H., Yamaoka, M., Fukano, M., Ueda, K., & Fujiwara, T. (2008). Increased secretion of salivary glands produced by facial vibrotactile stimulation. *Somatosensory and Motor Research*, 25, 222-229.

[8] Hiraba, Hisao, Sato, Takako, Nishimura, Satoshi, Yamaoka, Masaru, Inoue, Motoharu, Sato, Mitsuyasu, Iida, Takatoshi, Wada, Satoko, Fujiwara, Tadao, & Ueda, Koichiro. (2011). Changes in brain blood flow on frontal cortex depending on facial vibrotactile stimuli. In Vibration Analysis and Control-New Trends and Developments. Francisco Beltran-Carbajal ed. In Tech Croatia http://www.intechweb.org. , 337-352.

[9] Hoshi, Y., Kobayashi, N., & Tamura, M. (2001). Interpretation of near-infrared spectroscopy signals: a study with a newly developed perfused rat brain model. *J Appl Physiol.*, 90, 1657-1662.

[10] Ivarsen, S., Ivarsen, L., & Saper, C. B. (2000b). The Autonomic nervous system and the hypothalamus. In: Kandel ER., Schwartz JH., Jessell TM., eds, Principles of Neuronal Science 4 th ed, New York McGraw-Hill. , 960-981.

[11] Ivarsen, S., Kupfermann, F., & Kandel, E. R. (2000a). Emotional states and feelings. In: Kandel ER., Schwartz JH., Jessell TM., editors, Principles of Neuronal Science, 4 th ed, New York McGraw-Hill , 982-997.

[12] Matuo, R. (2003). Daeki, Daekisen. In Nakamura Y., Morimoto T. And Yamada Y, editors, Basic Physiology for Dental Students, 4 th ed, Tokyo Ishiyaku Co. in Japanese), 381-398.

[13] Petrides, M. (1994). Frontal lobes and working memory: evidence from investigations of the effects of cortical excisions in nonhuman primates. In: Handbook of Neuropsychology, F. Boller and J. Grafman (eds.), Elsevier Science B.V., Amsterdam, , 9, 50-82.

[14] Sakatani, K., Lichty, W., & Xie, Y. (1999). Effects of aging on language-activated cerebral blood oxygenation changes of the left prefrontal cortex. Near infrared spectroscopy study. *J Stroke Cerebrovascular Dis.*, 8, 398-403.

[15] Sakatani, K., Yamashita, D., & Yamanaka, T. (2006). Changes of cerebral blood oxygenation and optical path length during activation and deactivation in the prefrontal cortex measured by time-resolved near-infrared spectroscopy. *Life Sciences*, 78, 2734-2741.

[16] Scott, T. R., & Small, D. M. (2009). The role of the parabrachial nucleus in taste processing and feeding. *Ann NY Acad Sci.*, 1170, 372-377.

[17] Ueda, K. (2005). Sessyoku enge rehabilitation. In: Uematsu H, Inaba S, Watanabe M, editors. Koureishya Shika guidebook. Tokyo Ishiyaku in Japanese)., 248-275.

[18] Yamamoto, T., Takemura, M., Inui, T., Torii, K., Maeda, N., Ohmoto, M., Matumoto, I., & Abe, K. (2009). Functional organization of the rodent parabrachial nucleus. *Ann N Y Acad Sci.*, 1170, 378-382.

[19] Yamaoka, M., Hiraba, H., Ueda, K., & Fujiwara, T. (2007). Development of a vibrotactile stimulation apparatus for orofacial rehabilitation. Nihondaigaku Shigakubu Kiyou , in Japanese), 35, 13-18.

Permissions

The contributors of this book come from diverse backgrounds, making this book a truly international effort. This book will bring forth new frontiers with its revolutionizing research information and detailed analysis of the nascent developments around the world.

We would like to thank Francisco Beltrán Carbajal, for lending his expertise to make the book truly unique. He has played a crucial role in the development of this book. Without his invaluable contribution this book wouldn't have been possible. He has made vital efforts to compile up to date information on the varied aspects of this subject to make this book a valuable addition to the collection of many professionals and students.

This book was conceptualized with the vision of imparting up-to-date information and advanced data in this field. To ensure the same, a matchless editorial board was set up. Every individual on the board went through rigorous rounds of assessment to prove their worth. After which they invested a large part of their time researching and compiling the most relevant data for our readers. Conferences and sessions were held from time to time between the editorial board and the contributing authors to present the data in the most comprehensible form. The editorial team has worked tirelessly to provide valuable and valid information to help people across the globe.

Every chapter published in this book has been scrutinized by our experts. Their significance has been extensively debated. The topics covered herein carry significant findings which will fuel the growth of the discipline. They may even be implemented as practical applications or may be referred to as a beginning point for another development. Chapters in this book were first published by InTech; hereby published with permission under the Creative Commons Attribution License or equivalent.

The editorial board has been involved in producing this book since its inception. They have spent rigorous hours researching and exploring the diverse topics which have resulted in the successful publishing of this book. They have passed on their knowledge of decades through this book. To expedite this challenging task, the publisher supported the team at every step. A small team of assistant editors was also appointed to further simplify the editing procedure and attain best results for the readers.

Our editorial team has been hand-picked from every corner of the world. Their multi-ethnicity adds dynamic inputs to the discussions which result in innovative outcomes. These outcomes are then further discussed with the researchers and contributors who give their valuable feedback and opinion regarding the same. The feedback is then collaborated with the researches and they are edited in a comprehensive manner to aid the understanding of the subject.

Apart from the editorial board, the designing team has also invested a significant amount of their time in understanding the subject and creating the most relevant covers. They scrutinized every image to scout for the most suitable representation of the subject and create an appropriate cover for the book.

The publishing team has been involved in this book since its early stages. They were actively engaged in every process, be it collecting the data, connecting with the contributors or procuring relevant information. The team has been an ardent support to the editorial, designing and production team. Their endless efforts to recruit the best for this project, has resulted in the accomplishment of this book. They are a veteran in the field of academics and their pool of knowledge is as vast as their experience in printing. Their expertise and guidance has proved useful at every step. Their uncompromising quality standards have made this book an exceptional effort. Their encouragement from time to time has been an inspiration for everyone.

The publisher and the editorial board hope that this book will prove to be a valuable piece of knowledge for researchers, students, practitioners and scholars across the globe.

List of Contributors

J. G. Detoni, F. Impinna, N. Amati and A. Tonoli
Department of Mechanical and Aerospace Engineering, Politecnico di Torino, Turin, Italy

Andrea Tonoli, Angelo Bonfitto, Mario Silvagni and Lester D. Suarez
Mechanics Department, Mechatronics Laboratory - Politecnico di Torino, Italy

Francisco Beltran-Carbajal
Universidad Autonoma Metropolitana, Unidad Azcapotzalco, Departamento de Energia, Mexico, D.F., Mexico

Gerardo Silva-Navarro
Centro de Investigacion y de Estudios Avanzados del I.P.N., Departamento de Ingenieria Electrica, Seccion de Mecatronica, Mexico, D.F., Mexico

Manuel Arias-Montiel
Universidad Tecnologica de la Mixteca, Instituto de Electronica y Mecatronica, Huajuapan de Leon, Oaxaca, Mexico

Seung-Bok Choi, Sung Hoon Ha and Juncheol Jeon
Smart Structures and Systems Laboratory, Department of Mechanical Engineering, Inha University, Incheon 402-751, Korea

Seyed M. Hashemi and Omar Gaber
Department of Aerospace Eng., Ryerson University, Toronto (ON), Canada

N. M. M. Maia, Y. E. Lage and M. M. Neves
IDMEC-IST, Technical University of Lisbon, Department Mechanical Engineering, Lisboa, Portugal

Zissimos P. Mourelatos
Mechanical Engineering Department, Oakland University, USA

Dimitris Angelis
Beta CAE Systems S. A., Greece

John Skarakis
Beta CAE Systems, U.S.A., Inc.

Tiejun Yang and Lu Dai
College of Power and Energy Engineering, Harbin Engineering University, Harbin, PR China

Wen L. Li
Department of Mechanical Engineering, Wayne State University, Detroit, USA

A. S. Bouboulas, S. K. Georgantzinos and N. K. Anifantis
Machine Design Laboratory, Mechanical and Aeronautics Engineering Department, University of Patras, Greece

Hideo Takabatake
Department of Architecture, Kanazawa Institute of Technology, Institute of Disaster and Environmental Science, Japan

Mitsushige Oda
JAXA and Tokyo Institute of Technologies, Japan

Akihiko Honda
Tokyo Institute of technology, Japan

Satoshi Suzuki
Advanced Engineering Services Co. Ltd., Japan

Yusuke Hagiwara
Mitsubishi Heavy Industries Ltd., Japan

Nguyen Van Khang and Nguyen Phong Dien
Department of Applied Mechanics, Hanoi University of Science and Technology, Vietnam

Nelcy Hisao Hiraba, Motoharu Inoue, Takaya Shimano, Ryuichi Sampei, Katuko Ebihara, Hisako Ishii and Koichiro Ueda
Departments of Dysphasia Rehabilitation, Japan

Takako Sato and Satoshi Nishimura
Oral and Maxillofacial Surgery, Japan

Masaru Yamaoka
Nihon University, School of Dentistry, 1-8-13 Kanda-surugadai, Chiyoda-ku, Japan

www.ingramcontent.com/pod-product-compliance
Lightning Source LLC
Chambersburg PA
CBHW070716190326
41458CB00004B/998